Seventh International Conference on Dissociative Recombination: Theory, Experiments and Applications 2007

(DR 2007)

Journal of Physics: Conference Series Volume 192

Ameland, The Netherlands
18-23 July 2007

ISBN: 978-1-61738-284-0

Printed from e-media with permission by:

Curran Associates, Inc.
57 Morehouse Lane
Red Hook, NY 12571

Some format issues inherent in the e-media version may also appear in this print version.

Printed by Curran Associates, Inc. (2010)

Institute of Physics
Dirac House, Temple Back
Bristol BS1 6BE UK

Phone: 44 1 17 929 7481
Fax: 44 1 17 920 0979

techtracking@iop.org

Additional copies of this publication are available from:

Curran Associates, Inc.
57 Morehouse Lane
Red Hook, NY 12571 USA
Phone: 845-758-0400
Fax: 845-758-2634
Email: curran@proceedings.com
Web: www.proceedings.com

TABLE OF CONTENTS

Dissociative Recombination is one of the outcomes of collisions between electrons and molecular ions. This reaction dominates many plasmas, dense as well as dilute. Therefore, knowledge about this reaction is highly relevant for astrophysics, atmospheric sciences, (non-)reactive plasmas, and fusion plasmas. A theoretical description of this process requires intensive *ab initio* quantum chemistry calculations as well as a detailed description of the electron collision process. The high density of states near the ionization energy of molecules renders calculations extremely difficult and complex. In experiments, both discharge and flowing afterglow experiments and fast beam and ion storage ring experiments, a steady development can be seen towards more detailed studies often at the quantum state resolved level including product internal state identification and very recently detailed vector properties such as fragment angular distributions. Thanks to an efficient interplay between experiment and theory dissociative recombination has progressed significantly over the last few years.

The Seventh International Conference on Dissociative Recombination: Theory, Experiments, and Applications (DR2007) was organized by the Institute of Molecules and Materials at the Radboud University Nijmegen, The Netherlands and was held on the Wadden Island of Ameland. Earlier conferences in this series were held at Lake Louise, Alberta, Canada (1988), Saint Jacut, Bretagne, France (1992), Ein Gedi, Israel (1995), Nässlingen, Stockholm Archipelago, Sweden (1999), Chicago, USA, as part of the American Chemical Society meeting (2001), and in Mosbach, Germany (2004). Ameland and the Resort d'Amelander Kaap was chosen in the spirit of previous dissociative recombination conferences. It turned out to be a venue where discussions about dissociative recombination could take place around the clock both inside and outside while walking through dunes and on the beach.

The conference was made possible by generous sponsors, whom we thank wholeheartedly: The Radboud University Nijmegen, The Institute for Molecules and Materials of the Radboud University Nijmegen, The Foundation for Fundamental Research on Matter (Stichting FOM), The Foundation PHYSICA (Stichting Physica), and The Netherlands Royal Academy of Sciences (KNAW). The organisational support by Erna Gouwens van Oss before and during the conference was essential for its success. The help of Thanja Lambrechts and Vitali Zhaunerchyk during the preparation of the proceedings is greatly appreciated. The delay in the publication of these proceedings is entirely caused by the editor. The authors of the contributions are thanked for the quality of their contributions,

Wim J van der Zande, Editor
Institute for Molecules and Materials, Radboud University Nijmegen, PO Box 9010, NL-6500 GL Nijmegen, The Netherlands
Email: w.vanderzande@science.ru.nl

Participants of the 7th International Conference on Dissociative Recombination: Theory, Experiments and Applications, taken in front of d'Amelander Kaap, the conference venue in Ameland, one of the Wadden Islands in the North of the Netherlands.

1. Dror Shafir
2. Ioan Scheider
3. Nigel Adams
21. Annemieke Petrignani
22. Johanna Roos
23. Erna Gouwens van Oss
41. Oumanou Motopan
42. Max Berg
43. Henrik Buhr

4. Hajime Tanuma
5. Jonathan Tennyson
6. Vitali Zhaunerchyk
7. Robert Continetti
8. Stefan Rosén
9. Erik Vigren
10. Magdalena Kaminska
11. Chris Greene
12. Steffen Novotny
13. Amy Schumak
14. Gerard van Rooij
15. Wim van der Zande
16. Daan Schram
17. Dahbia Talbi
18. Steven Guberman
19. Mats Larsson
20. Åsa Larson
24. Natalie de Ruette
25. Francois Wameu Tamo
26. Rainer Johnsen
27. Viatcheslav Kokoouline
28. Hidekazu Takagi
29. Hans-Jakob Wörner
30. Oskar Asvany
31. Lutz Lammich
32. Brandon Jordon-Thaden
33. Wolf Diettrich Geppert
34. Alexander Faure
35. Mathias Hamberg
36. Oldrich Novotny
37. Matthias Danielson
38. Radek Plasil
39. Richard Thomas
40. Andreas Wolf
44. Ilya Fabrikant
45. Claude Krantz
46. Michael Stenrup
47. Xavier Urbain
48. Evelyne Roueff
49. Dirk Schwalm
50. Valery Ngassam
51. Julien Lecointre
52. Ann Orel
53. Ihor Korolov
54. Romain Guerot
55. Peet Hickman
56. Christiaan Jungen
57. Jozef Varju
58. Juraj Glosik

Absent from the picture: Detlev Reiter, Brian Mitchell, Jens Hoffmann and Dave Parker

All papers published in this volume of *Journal of Physics: Conference Series* have been peer reviewed through processes administered by the proceedings Editors. Reviews were conducted by expert referees to the professional and scientific standards expected of a proceedings journal published by IOP Publishing.

Excited core states in the dissociative recombination of N_2^+

Steven L Guberman

Institute for Scientific Research, 22 Bonad Road, Winchester, MA 01890, USA

slg@sci.org

Abstract. Theoretical studies of dissociative recombination along the $b'^1\Sigma_u^+$ dissociative route of N_2 from the v=0 level of the $X^2\Sigma_g^+$ ion are described. Rydberg states having both the ground $X^2\Sigma_g^+$ and $A^2\Pi_u$ cores are included in the calculation of cross sections and rate constants. The electronic width connecting these Rydberg series is zero. Nevertheless, the two Rydberg series are connected through the dissociative state. The large electronic width for the $A^2\Pi_u$ Rydberg states leads to a factor of 20 decrease in the rate constant at 1000 K electron temperature.

1. Introduction

Dissociative recombination (DR) is driven by variants of the direct [1] and indirect [2] mechanisms. The interference between these two mechanisms determines the shapes of cross sections and rate coefficients. For many years, the indirect DR mechanism was described using only Rydberg states having an ionic core that is identical to the state of the recombining ion. However, the energetics are often such that Rydberg states with an excited ion core are also accessible by low energy electrons undergoing capture by the ground ion state. The vibrational levels of both the ground and excited core Rydberg (ECR) states which lie above the lowest vibrational level of the ground core are resonances and they are the source of considerable structure in DR cross sections.

DR through ECR states has been discussed previously for N_2^+ [3-5], CD^+ [6-9], CH^+ [7-9], OH^+ [10] and H_2^+ [11]. For H_2^+, ECR states dominate DR since all the dissociative states are of this character. In this paper, I examine the role of excited core states in the DR of N_2^+ by focusing upon DR within $^1\Sigma_u^+$ symmetry. Prior theoretical studies have shown that the $^3\Pi_u$ states of N_2^+ are the dominant routes for DR [3-5, 12]. The additional dissociative routes for the lowest 5 ion vibrational levels are states of $^1\Pi_u$, $^1\Sigma_g^+$, $^1\Sigma_u^+$, $^3\Pi_g$, $^3\Sigma_u^+$ and $^3\Sigma_g^+$ symmetries. Prior studies have shown that for $^3\Pi_u$, the addition of ECR $^3\Pi_u$ states leads to a 10% increase in the total room temperature DR rate constant [4] from v=0. For $2^1\Sigma_g^+$, the ECR states increase the room temperature DR rate constant along $2^1\Sigma_g^+$ from v=0 by a factor of 4 but because the rate constant is small [5], there is only a small change in the total rate constant due to all routes.

In the next section, the potential curve calculations for $^1\Sigma_u^+$ are described. A description of the calculation of the widths and quantum defects is in section 3. The cross sections are analyzed in section 4 and the conclusions follow in section 5.

2. Potential curves

The potential curve for the $b'^1\Sigma_u^+$ state has been calculated using the same approach described previously [5] for the $2^1\Sigma_g^+$ route with one exception. With the cc-pVQZ basis of Dunning [13], Rydberg character mixes into the wave function at energies just above the ion potential curve. However, in the model used here, all states are diabatic, i.e. they have purely valence or Rydberg character. If the dissociative state were fully optimized, i.e. if we allowed Rydberg character to mix in, the dissociative state would not cross the ion and a different model of DR would be required. The widths described in the next section are between diabatic valence and Rydberg states. In order for the widths to be appropriate for the valence dissociative state, any Rydberg character must be projected out. The projection is done in an approximate manner by removing the most diffuse contracted basis function in each of the s and p Gaussians. Each of these basis functions is a single primitive. The resulting contracted basis consisted of 4s, 3p, 3d, 2f, 1g contracted functions. The ion curve used previously [5] was shifted in energy by placing it at the experimental ionization energy [14] above the ground state of N_2, also calculated in the above basis. The curves are shown in figure 1. The dissociative curve crosses the ion ground state at the small R turning point of the v=0 level of the ion ground state. The large R asymptote of the b′ state lies at an experimental energy of 0.137 eV [14, 15] above the v=0 ion level. DR does not take place along this route until the electron energy exceeds 0.137 eV and the DR rate constant is small for room temperature electrons. Also shown in figure 1a is the $3\pi_g\ {}^1\Sigma_u^+$ Rydberg state of the $A^2\Pi_u$ core. The first accessible vibrational level for DR is v=2 since the lowest two vibrational levels lie below the v=0 ion level. For the neighboring state with the $4\pi_g$ orbital (figure 1b), the lowest accessible vibrational level is v=0. We will return to these potential curves in the discussion of the cross sections in section 4.

3. Widths and Quantum Defects

The width for electron capture into the dissociative state is given by

$$\Gamma^{\varepsilon}_{ion,d}(R)=2\pi <\{\Phi_{ion}(x, R)\ \varphi_\varepsilon(x, R)\}\ |H\ |\ \Phi_d(x, R) >^2 \quad (1)$$

where the integration is over the electron coordinates, x. H is the electronic Hamiltonian, R is the internuclear distance, φ_ε is the orbital of the "free" electron with energy ε and Φ_{ion} and Φ_d are the multiconfiguration wave functions of the ion and dissociative states. The widths are calculated as described previously [5] by substituting a high principal quantum number Rydberg orbital for the "free electron" and multiplying the matrix element in (1) by a density of states. An R dependent width of the form of (1) needs to be calculated for each of the ion states. The width connecting the dissociative state to the A core states is needed to describe the predissociation of ECR states. In terms of real orbitals, the dominant configurations in the b′ state at R=2.0 a_o are

$$\ldots 3\sigma_g^2\ 1\pi_{ux}^2\ 1\pi_{uy}\ 1\pi_{gy} + \ldots 3\sigma_g^2\ 1\pi_{ux}\ 1\pi_{uy}^2\ 1\pi_{gx} \quad (2)$$

where ... denotes the inner orbitals, $1\sigma_g^2\ 1\sigma_u^2\ 2\sigma_g^2\ 2\sigma_u^2$. The dominant configurations of the ion states plus a "free" electron, φ_ε, in $^1\Sigma_u^+$ symmetry are

$$\ldots 3\sigma_g\ 1\pi_{ux}^2\ 1\pi_{uy}^2\ \varphi_\varepsilon(\sigma_u) \quad (3)$$

for $X^2\Sigma_g^+$ and

$$\ldots 3\sigma_g^2\ 1\pi_{ux}\ 1\pi_{uy}^2\ \varphi_\varepsilon(\pi_{gx})\ +\ \ldots 3\sigma_g^2\ 1\pi_{ux}^2\ 1\pi_{uy}\ \varphi_\varepsilon(\pi_{gy}) \quad (4)$$

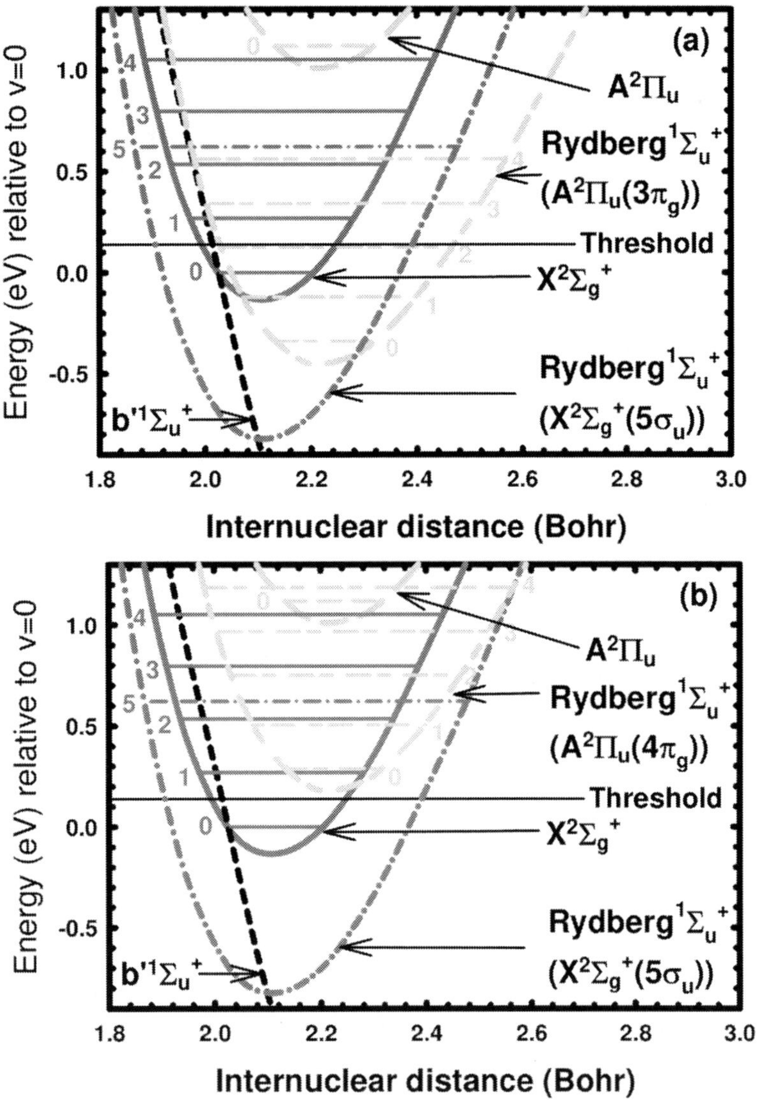

Figure 1. (a) Potential curves are shown for the ground state of the ion (solid, red online), the n=5 ground core Rydberg state (dot-dash, red online), the A state of the ion (long dash, green online) and the dissociative $b'^1\Sigma_u^+$ (short dash, black online). Numbered vibrational levels for the ion and Rydberg states are also shown. The large R asymptote of the b′ state is shown by the horizontal line labeled "threshold." Also shown is the potential curve for $^1\Sigma_u^+$ (($A^2\Pi_u$) $3\pi_g$) (long short dash, green online). (b) Same as (a) except that the $^1\Sigma_u^+$ (($A^2\Pi_u$) $4\pi_g$) (long short dash, green online) potential curve is shown in place of the $3\pi_g$ state.

for $A^2\Pi_u$. Since the dissociative state has $^1\Sigma_u^+$ symmetry and the Hamiltonian in equation (1) transforms as $^1\Sigma_g^+$, a nonzero width results only if the ion plus a free electron (or a Rydberg state) wave function also has $^1\Sigma_u^+$ symmetry and thus the σ_u "free orbital" is coupled to the X state and the $1\pi_g$ "free orbital" is coupled to the A state.

The dissociative state differs from the X state plus a continuum electron by a double excitation, i.e. two electrons must be moved from the orbitals in (2) to obtain the configuration in (3). However, only a single excitation ($\varphi_\varepsilon(\pi_g) \rightarrow (1\pi_g)$) is required to change the configurations in (2) to that for (4). Valence and electron-ion states that differ by a single excitation are connected by large widths. This is borne out by our calculations which show that the electronic width for the A core state plus a continuum electron is 4.0 eV near R=2.1 a_0. The width for the X core state plus a continuum electron is 0.36 eV near R=2.1 a_0.

An additional width used in the cross section calculations is that connecting the X and A core continuum states, i.e.,

$$\Gamma(R, \varepsilon, \varepsilon') = 2\pi < \{\Phi_X(x, R)\, \varphi_\varepsilon(x, R)\}\, |H|\, \{\Phi_A(x, R)\;\; \varphi'_{\varepsilon'}(x, R)\} >^2. \qquad (5)$$

However, using a similar approach to that described previously [5], the calculated value of the width in (5) is zero to two decimal places at R=2.3 a_0. Therefore, a width of zero at all internuclear distances has been used in the cross section calculation. The excited core states can still mix with the ground core states by a second order interaction in which the dissociative state acts as an intermediate. In this mechanism, the dissociative state, which predissociates the Rydberg states of both cores, in turn allows the Rydberg states to mix and allows the ECR levels to autoionize to the ground core continuum. This interaction is included in the Multichannel Quantum Defect Theory (MQDT) approach described below for calculating the cross sections and rate constants. The second order K matrix element is given by [18]:

$$K_{vv'}^{E_T} = \mathbf{P}\int dE(\langle\chi_v|V|F_d(E)\rangle\langle F_d(E)|V|\chi_{v'}\rangle/(E_T-(E_d+E)) \qquad (6)$$

where v and v′ run over the bound vibrational levels of the X and A core ion states or of the Rydberg levels. **P** is the principal value operator, the χ_v are the bound vibrational wave functions, F_d is the continuum vibrational wave function of the dissociative state, V is the electronic part of the matrix element given in equation (1), E_T is the total energy, E_d is the energy of the asymptote of the dissociative curve, E is the kinetic energy of the dissociating atoms so that $E_T = E_d + E$. The matrix element in equation (6) describes the mixing of vibrational levels among a single state of the ion or the mixing of levels between the X and A states. The mixing is due to the dissociative state. Further comment on equation (6) is given below.

As described previously [5], the quantum defect is determined from the energy difference, ΔE, between the Rydberg state used in the width matrix elements and the ion. The quantum defect is defined by $\mu(R) = n - 1/(2\Delta E(R))^{1/2}$ where n is the principal quantum number and ΔE is the binding energy of the Rydberg orbital. Quantum defects for the $^1\Sigma_u^+$ Rydberg series converging to the X and A ion states are both included in the calculations.

4. Cross Sections

The DR cross sections have been calculated with the MQDT approach [16-18] using the second order of the K matrix [18]. The calculations include 18 vibrational levels in the ground ion potential, 18 in the A state potential and 18 in each of the Rydberg states for both the ground and excited cores. All vibrational wave functions were calculated on a 0.001 a_0 grid from an internuclear distance of 1.0 to 8.0 a_0. Cross sections were calculated on an energy grid of 0.0001 eV from threshold to 1.0 eV.

If only the ground state ion core is included in the calculation, the cross section appears as shown in figure 2a. Note that the cross section starts at an electron energy of 0.137 eV corresponding to the position of the dissociative state asymptote. The direct cross section (dashed, green on line) has steep drops as new autoionization channels open at 0.27 eV (v=1), 0.54 eV (v=2) and 0.80 eV (v=3). The

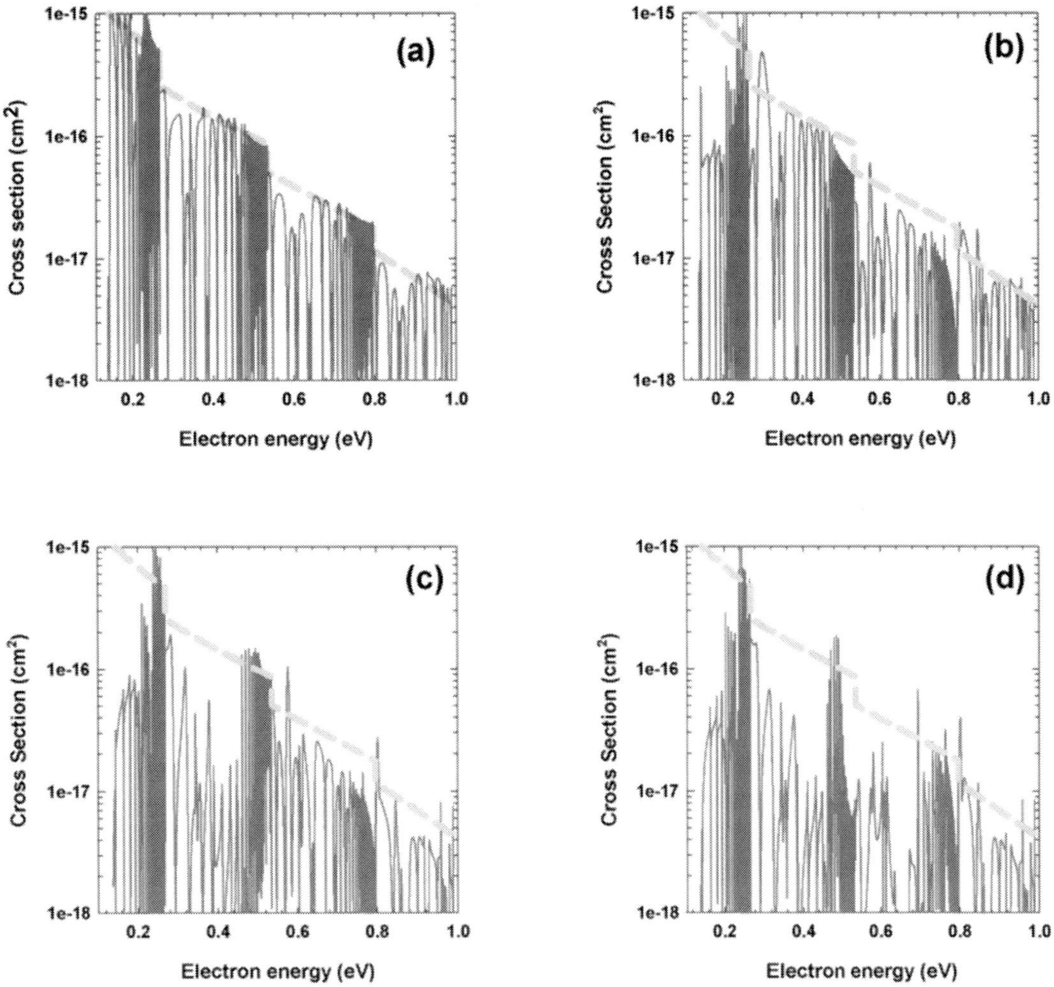

Figure 2. DR cross section for the $^1\Sigma_u^+$ dissociative route. (a) Only the ground ion core states are included. The direct cross section is the dashed line (green online) and the full cross section is the solid line (red online). (b), (c) and (d) are the same as (a) except that the v=0-2, v=0-3 and v=0-4 levels respectively of the A core Rydberg states are included.

full cross section (solid line, red online) includes structure due to the Rydberg resonances.

Below the energy of each ion vibrational threshold there is a pile up of Rydberg states having that level as the series limit. The resonances appear mostly as dips falling below the direct cross section. This is a characteristic cross section for a dissociative curve that crosses the ion within or near the turning points of v=0.

The large width for the A core states indicates that an MQDT treatment that includes K matrix terms beyond second order may be needed. However, it is important to note that the vibrational overlaps for the excited core are low and the factor of about 10 by which the excited core electronic width exceeds the ground core width is much larger than the factor separating the K matrix elements. The largest K matrix elements (unitless) are about 2.0 and correspond to mixing among excited core vibrational levels due to the dissociative state (see previous section). The largest K matrix element among ground core levels is 0.6 and corresponds to the direct interaction of the v=1 level with the dissociative state. Nevertheless, a calculation using a K matrix beyond second order would be worthwhile.

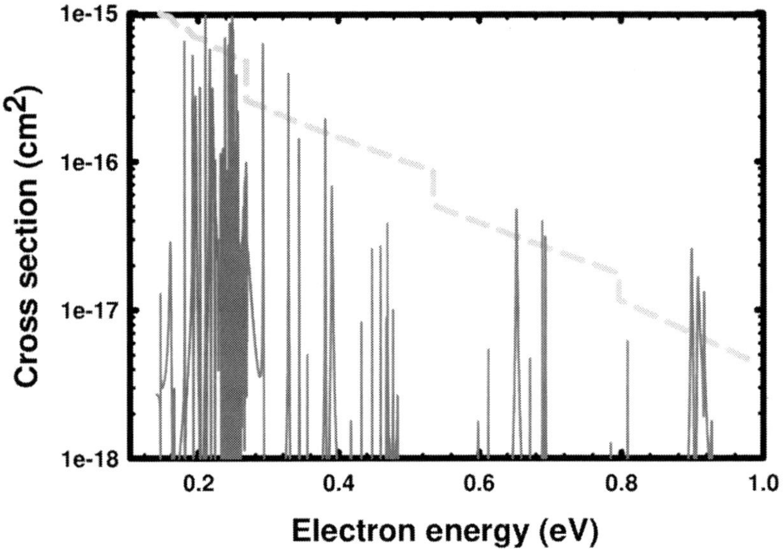

Figure 3. Same as figure 2a except that 18 Rydberg vibrational levels for both the ground core and the A state are now included.

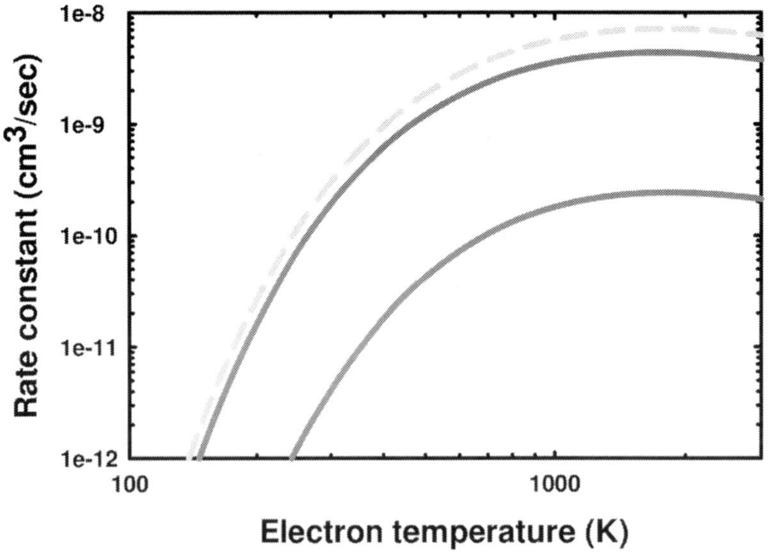

Figure 4. Rate constants for DR. The dashed curve is the direct rate constant. The upper and lower solid curves include X core resonances and X + A core resonances respectively.

In order to understand the important role of the excited core states, we explore the change induced in the total cross section due to the addition of individual ECR vibrational levels. From figure 1b it can be seen that the first v=0 Rydberg level with the A core lies at about 0.3 eV above the v=0 level of the ground ion and has n=4. However, figure 1b shows that the v=0 ECR level has little overlap with the dissociative vibrational wave function. For v=1, the first ECR level occurs at about 0.5 eV electron energy (see figure 1b). As for v=0, the low overlap of the v=1 resonances with the dissociative vibrational wave function relegates these resonances to a negligible role. The situation is however different for the ECR n=3, v=2 resonances. Figure 1a shows that this resonance is found at an electron energy near 0.137 eV, i.e. near the threshold for DR. Because the vibrational overlap with the dissociative vibrational wave function is higher than that with v=0, 1, this resonance will notably affect DR at the threshold. Figure 2b shows the cross section including 18 vibrational levels for ground ion states and the lowest three vibrational levels of the ECR states. A pronounced drop is seen in the cross section just above threshold due to the n=3, v=2 A core state. The v=2 A core states have a large K matrix element (0.28) with the v=0 ground core continuum and with the v=1 ground core levels (-0.38). These matrix elements are all of the form of the second order term given in equation (6) and involve the dissociative state acting as an intermediate. The net effect of these interactions is to lead to an increase in autoionization and a large drop in the DR cross section just above threshold.

From figure 1b, the next v=2 state occurs for n=4 at about 0.75 eV, just below the v=3 level of the ground core ion. The ground core v=3 levels have a large K matrix element with excited core v=2 levels (0.39). With the large K matrix element connecting the v=2 ECR states with the ground core continuum, the result is the destructive interference shown in figure 2b at 0.75 eV. Note that the cross section drop near 0.52 eV is due to the second order interaction (-0.38) of the A core, n=4, v=1 level with the ground core v=2 levels. The cross section drops because the A core v=1 levels have a large second order interaction (-0.45) with the v=0 ground core continuum leading to an increase in the probability for autoionization.

Addition of the v=3 ECR resonances leads to the cross section shown in figure 2c. The lowest resonance, n=3 (see figure 1a), is near 0.34 eV. Comparing figures 2b and 2c, the n=3 resonance annihilates the cross section structure near 0.34 eV by nearly an order of magnitude. The next v=3 resonance, n=4, falls at about 0.96 eV (see figure 1b), and also affects the structure in this region.

Figure 2d shows the resulting cross section that includes the v=0-4 ECR resonances. Once again, a large degree of destructive interference occurs near 0.56-0.66 eV due to the n=3, v=4 ECR resonance which has a large K matrix element (-0.57) with the v=3 ground core states. The final cross section including 18 vibrational levels in the A core states is in figure 3. The A core Rydberg states have annihilated most of the cross section due to the ground core states.

The rate constants for the two cases are shown in figure 4. As might be expected, the rate constant for the case in which the excited ion core is included is much smaller than that which results from only the ground core states. Because of the threshold of 0.137 eV for DR from v=0, the rate constant at room temperature is very small, 4.1 x 10^{-12} cm^3/sec. At 1000K electron temperature, the full rate constant is 1.8 x 10^{-10} cm^3/sec compared to 3.6 x 10^{-9} cm^3/sec if only ground core resonances are included.

5. Conclusions

DR along the $b'^1\Sigma_u^+$ dissociative state from the v=0 ion vibrational level has a low rate constant at room temperature because DR can only occur for electron energies above 0.137 eV. The addition of the A core Rydberg states decreases the rate coefficient by a factor of 20 at 1000 K electron temperature. The decrease is due to the large electronic width of the excited core states which increases the probability of autoionization. The autoionization probability is high because the ECR levels are connected to the ionization continuum of the v=0 ground ion by large second order K matrix terms. In these terms the dissociative state provides the coupling between the ECR levels and the ground ion continuum. It is clear that ECR states cannot be neglected in this case if an accurate description of DR is desired. A further study of the role of the ECR states in this case with a higher order K matrix is planned.

Acknowledgements
This material is based upon work supported by the National Science Foundation under Grant No. ATM-0225256. Any opinions, findings, and conclusions or recommendations expressed in this material are those of the author and do not necessarily reflect the views of the National Science Foundation. This research is also supported by NASA grants NNG05GO93G and NNG05GI34G.

References

[1] Bates D R 1950 *Phys. Rev.* **78** 492
[2] Bardsley J N 1968 *J. Phys. B* **1** 365
[3] Guberman S L Ab initio studies of Dissociative Recombination, in *Dissociative Recombination: Theory, Experiment and Applications*; Mitchell, J. B. A., Guberman, S. L. Eds.; World Scientific: Singapore, 1989; p 45
[4] Guberman S L New Mechanisms for Dissociative Recombination, in *The Physics of Electronic and Atomic Collisions, Invited Papers for the XIX International Conference*; Dube, L. J., Mitchell, J. B. A., McConkey, J. W., Brion, C. E., Eds.; American Institute of Physics Press: New York, 1995; p 307
[5] Guberman S L 2007 *J. Phys. Chem. A* **111** 11254
[6] Forck P et al. 1994 *Phys. Rev. Lett.* **72** 2002
[7] Amitay Z et al. 1996 *Phys. Rev. A* **54** 4032
[8] Carata L, Orel A E, Schneider I F and Suzor-Weiner A Core Excited Resonances in the Dissociative Recombination of CH^+ and CD^+, in *Dissociative Recombination: Theory, Experiment and Applications*; Larssen M, Mitchell J B A and Schneider I F, Eds.; World Scientific: Singapore, 2000; p 291.
[9] Carata L, Orel A. E., Raoult M., Schneider I F, Suzor-Weiner A 2000 Phys. Rev. A **62** 052711
[10] Amitay Z et al. 1996 *Phys. Rev. A* **53** R644
[11] Guberman S L 1983 *J. Chem. Phys.* **78** 1404
[12] Guberman S L 1991 *Geophys. Res. Lett.* **18** 1051
[13] Dunning, Jr. T H 1989 *J. Chem. Phys.* **90** 1007
[14] Lofthus A and Krupenie P H 1977 *J. Phys. Chem. Ref. Data* **6** 113
[15] Moore, C E Atomic Energy Levels, Vol. I (US Dept. of Commerce, Washington, DC, 1971)
[16] Giusti, A 1980 *J. Phys. B* **13** 3867
[17] Lee, C M 1977 *Phys. Rev. A* **16** 109
[18] Guberman S L and Giusti-Suzor A 1991 *J. Chem. Phys.* **95** 2602

Journal of Physics: Conference Series **192** (2009) 012002 doi:10.1088/1742-6596/192/1/012002

Dissociative electron attachment: threshold phenomena and multimode effects

Ilya I. Fabrikant

Department of Physics and Astronomy, University of Nebraska-Lincoln, Lincoln, NE 68588, USA

E-mail: `ifabrikant1@unl.edu`

Abstract. We present a summary of recent progress on studies of threshold phenomena in low-energy dissociative electron attachment to polyatomic molecules: vibrational Feshbach resonances and threshold cusps. The theoretical approach is based on the resonance R-matrix theory. It allows us to analyze threshold effects in other channels, particularly vibrational excitation. If the molecular long-range field is not strong enough, virtual-state cusps are observed at vibrational excitation thresholds. The second part of the paper illustrates the multimode effects in dissociative attachment using attachment to the CF_3Cl molecule as an example. Symmetric stretch C-Cl vibrations ν_3 and symmetric deformation (or so-called "umbrella") vibrations ν_2 are included. Calculations allow us to study the final-state vibrational distribution in the $CF_3(\nu_2)$ fragment.

1. Introduction

Recent progress in studies of dissociative electron attachment (DEA) processes was stimulated by development of new experimental techniques involving electron beams with high energy resolution [1]. This allowed detailed investigation of threshold phenomena in low-energy dissociative attachment to diatomic and polyatomic molecules: vibrational Feshbach resonances (VFRs) and threshold cusps. VFRs are similar to Rydberg resonances in the dissociative recombination processes and are due to capture of the incident electron into the long-range field of a vibrationally excited molecule. If the molecular long-range field is not strong enough, virtual-state cusps are observed at vibrational excitation thresholds.

Our approach to describing these phenomena is based on the resonance R-matrix theory [2] which is equivalent to the nonlocal complex potential theory [3]. In the present paper we review recent theoretical results on DEA and vibrational excitation for several halogen-containing molecules.

The theoretical analysis of threshold phenomena is carried out in the approximation including only one symmetric-stretch vibrational mode corresponding to the reaction coordinate. In the second part of the paper we discuss a generalization of the local complex potential theory involving two vibrational modes. This theory is applied to dissociative attachment to CF_3Cl molecules in the energy region of the lowest shape resonance of A_1 symmetry.

2. Resonance and threshold phenomena

The one-pole R-matrix theory [4] is very well suited for studying resonances and threshold effects in low-energy electron-molecule scattering. A challenging part in this approach is the calculation of the R-matrix surface amplitude generating the resonance width. In most calculations performed so far (see, e.g. Refs. [5, 6, 7, 8, 9, 10] the surface amplitude was adjusted empirically in order to reproduce observed vibrational excitation cross sections or thermal attachment rate coefficients. After this parameter is fixed, structures in DEA and vibrational excitation can be calculated with high accuracies using the quasiclassical version [2, 11] of the R-matrix theory. The threshold structures are determined by the long-range electron-molecule interaction which in R-matrix theory is included in the logarithmic derivative of the electron wavefunction in different vibrational channels.

We will illustrate resonance and threshold effects in electron collisions with three polar molecules which were studied recently experimentally [9, 10, 12]: CF_3I, CF_3Br, and CH_3Br. The major goal here is to show how the relevant potential energy curves and long-range interaction influence the DEA and vibrational excitation cross sections, particularly at vibrational excitation thresholds. A detailed comparison of theoretical results with experimental data was given in Refs. [9, 10, 12].

The molecules chosen for the present investigation have substantially different dipole moments at the equilibrium internuclear separation: 0.413, 0.256 and 0.717 a.u. for CF_3I, CF_3Br and CH_3Br respectively. Their potential energy curves as functions of the C-X (X=I or Br) distance exhibit the behavior shown schematically in Fig. 1. The C-X distance is the only vibrational coordinate involved in the present calculations, and it corresponds to the symmetric stretch vibration ν_3. We show only one curve representing the neutral molecule, and three anion curves demonstrating major features distinguishing the three cases. All three curves correspond to exothermic processes. The lowest curve, representing CF_3I, crosses the neutral very close to the equilibrium internuclear separation, therefore the DEA cross section in this case is expected to be very large in the low-energy region and very weakly dependent on vibrational temperature, in agreement with earlier observations [13]. No VFRs are expected in this case. The intermediate case corresponds to CF_3Br. The actual anion curve crosses the neutral at the right turning point for the $\nu_3 = 3$ state [10]. This is a typical case for a strong temperature dependence of the DEA cross section. One would also expect VFRs in this case if the long-range interaction were significant like in the case of the CH_3I molecule [14]. However, the dipole moment of CF_3Br is too small for VFR formation. In the third case the curve crossing occurs at even higher internuclear separation corresponding to the right turning point for the $\nu_3 = 5$ state. Here we expect very low cross sections for attachment to the ground state (or for attachment at low target temperatures), and the temperature effect should be even stronger than for CF_3Br. Because of the substantial dipole moment of CH_3Br, VFRs appear in this case, as was predicted theoretically [8] and confirmed recently by measurements [12].

Fig. 2 shows the DEA cross sections for these three cases and illustrates the temperature effect in DEA to CH_3Br. For CF_3I and CF_3Br cusps are observed at the vibrational excitation thresholds, whereas VFRs are clearly seen in the CH_3Br case. Note that due to the thermal distribution, several vibrational states contribute to the cross section at a given temperature, therefore one resonance, associated with the $\nu_3 = 4$ threshold, produces several peaks in the thermally-averaged cross section. The first peak (at 69 meV) is due to attachment from the $\nu_3 = 3$ state, the second (at 143 meV) due to the $\nu_3 = 2$ state, and the third (at 218 meV) due to the $\nu_3 = 1$ state. The contribution due to attachment from the $\nu_3 = 0$ state is noticeable at room temperature, but is suppressed at $T = 600$ K.

Resonances associated with lower thresholds become weaker when going from $\nu_3 = 4$ to $\nu_3 = 1$ [8, 12]. This can be also seen from the vibrational excitation cross sections. In Figs. 3 and 4 we show vibrational excitation from the ground state. The CH_3Br cross section is not enhanced compared to the CF_3Br case, although there are VFRs at higher thresholds in the former case and none in the latter. However, a huge enhancement was predicted [12, 15] for excitation from excited states of CH_3Br. In particular, the cross section for the $\nu_3 = 2 \rightarrow 3$ transition reaches

Journal of Physics: Conference Series **192** (2009) 012002 doi:10.1088/1742-6596/192/1/012002

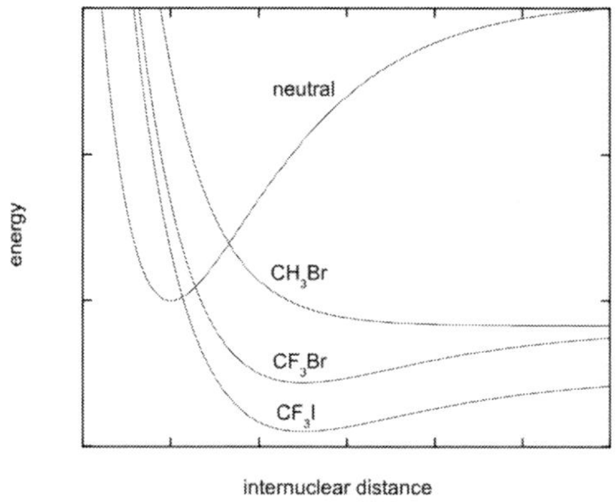

Figure 1. Schematic potential energy curves.

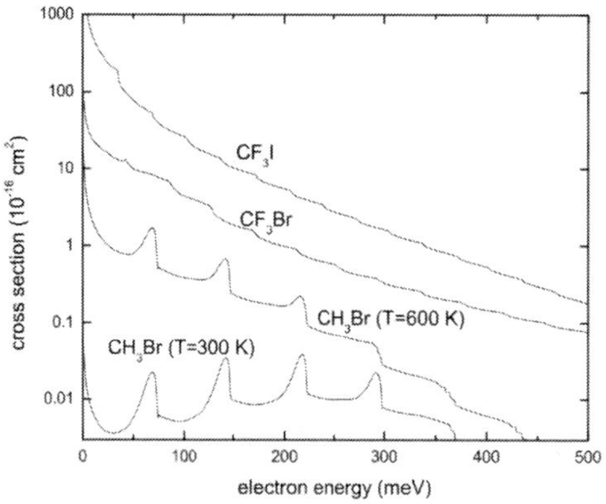

Figure 2. DEA cross sections.

90×10^{-16} cm^2 near the threshold [12]. Note that the ν_3 mode in all discussed molecules is infrared active, therefore electron-impact vibrational excitation can also occur directly due to the transition dipole moment. However, in CF_3Br and CF_3I the transition dipole moment is relatively weak. Only in CH_3Br the contribution of the direct excitation is noticeable and equals about 20% of the total cross section near the threshold [12, 15].

Another substantial difference between CF_3Br and CH_3Br is the existence of the broad peak associated with the A_1 resonance in the former case, and its absence in the latter. The vertical attachment energy for CH_3Br, 2.4 eV, is much higher than for CF_3Br, 1.0 eV, and the resonance width grows with the electron energy. In addition, the resonance width increases with the increase of the long-range interaction, as was shown for the hydrogen halides [16]. Therefore in the CH_3Br case the shape resonance is much broader and does not appear in the vibrational excitation cross section.

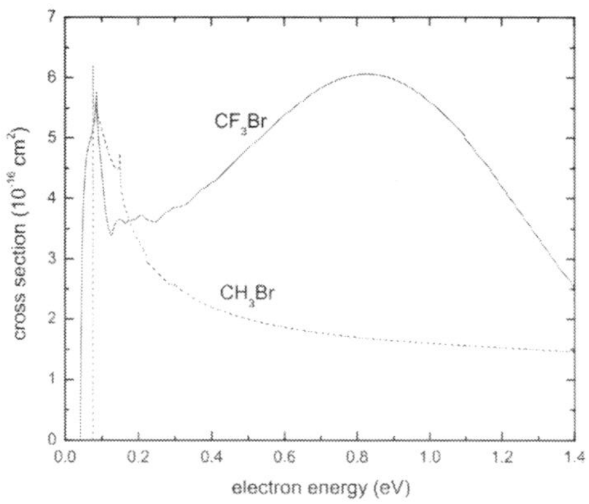

Figure 3. Vibrational excitation cross sections for CH_3Br and CF_3Br.

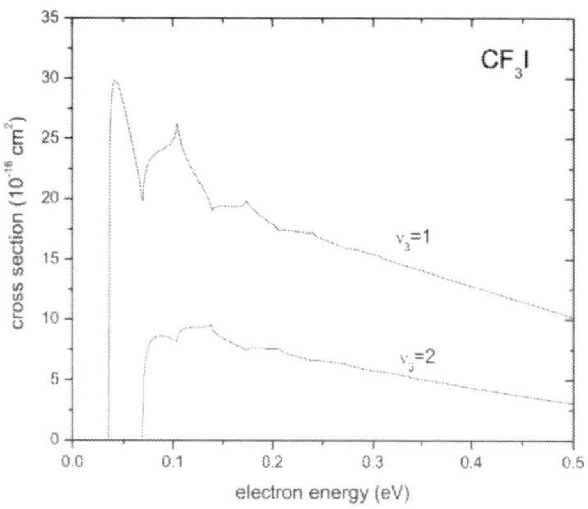

Figure 4. Vibrational excitation cross sections for CF_3I.

A large cross section for excitation from the ground vibrational state is observed for CF_3I (Fig.

4). This is because the anion curve crosses the neutral very close to the equilibrium separation, as illustrated in Fig. 1. This relative curve position makes the Franck-Condon factor favorable and enhances the cross section due to the bound/virtual state of the CF_3I^- system [17].

Resonances and cusps in the elastic cross sections were discussed in Refs. [12, 15]. An important point worth noting here is that for polar molecules resonances and threshold effects are strongly suppressed in angular-integrated cross sections due to a large contribution of the direct (nonresonant) process. However, the threshold features are well pronounced in differential cross sections at large scattering angles.

3. Multimode effects in DEA to CF_3Cl

The calculations presented in the previous section were carried out in the approximation including one vibrational mode corresponding to the reaction coordinate. This is a natural simplification in DEA treatment for polyatomic molecules in view of the tremendous amount of computational work necessary to obtain multidimensional complex potential energy surfaces and the solution of the multidimensional Schrödinger equation for the nuclear motion. Some examples of recent calculations going beyond the 1-dimensional approach are model studies of DEA to CO_2 [18] and *ab initio* calculations of DEA to water [19]. Both calculations were performed within the framework of the local complex potential theory.

Multimode effects in DEA to molecules of the type CX_3Y, where Y is a halogen atom, have not been investigated so far. In this section we present the first attempt to study these effects in the process

$$e + CF_3Cl(\nu_2, \nu_3) \rightarrow CF_3Cl^- \rightarrow CF_3(\nu_2) + Cl^-. \quad (1)$$

Here ν_2 stands for symmetric deformation vibrations (or, so-called "umbrella" mode) The inclusion of only one mode, in addition to the reaction mode ν_3, can be justified by the same reasoning as that given by Shapiro and Bersohn [20] in their studies of photodissociation of CH_3I. First, the intermediate negative-ion state of CF_3Cl responsible for DEA at low energies has A_1 symmetry, therefore degenerate vibrations ν_4, ν_5 and ν_6, having e-type, can be excited resonantly only in pairs. Second, the C-F bond length, is very close to that for the anion [21] and the radical CF_3. Therefore we can assume that the C-F symmetric stretch mode ν_1 is unlikely to be excited as well. These assumptions are confirmed by measurements of Mann and Linder [22] who observed strong electron-impact vibrational excitation of the ν_3 and ν_2 modes in the A_1 resonance region, whereas other modes were not significantly excited.

Inclusion of additional vibrational modes in the description of the attachment dynamics for CF_3X is of interest for two reasons. First, it allows us to study the distribution of the internal energy of the fragment, in our case the energy of the umbrella motion in the CF_3 radical. This energy can be significant due to, first, a significant Franck-Condon factor for transition between the initial state of CF_3X and the excited vibrational states of the intermediate negative ion CF_3X^-, and, second, due to the final-state interaction between CF_3 and X^- redistributing the internal energy of CF_3. Studies of Rydberg electron attachment to CH_3I, CF_3I and CF_3Br molecules [23] showed that at high principal quantum number n (or at low electron energies) the major portion of energy released by electron capture appears in translation, indicating insignificance of the internal energy redistribution in the final state. However, at lower n (or higher electron energies) the final-state interaction becomes important [24, 25], although there is no quantitative data, either experimental or theoretical, on the internal energy distribution in the CF_3 or CH_3 fragment.

The second motivation for DEA studies with inclusion of the umbrella vibrations is the strong sensitivity of the DEA cross section to the initial vibrational state of the target, or to the initial vibrational temperature. The one-mode approximation explains quite well the observed temperature dependence of DEA in electron collisions with methyl halides [8, 26]. However, it fails [7, 27] to give the correct quantitative description of the observed temperature dependence

[28] of low-energy peak in DEA to CF_3Cl. A recent joint experimental and theoretical paper on DEA to CF_3Br [10] demonstrates that the DEA rate calculated in the one-mode approximation gives a slower growth at high temperatures than that observed. This indicates that the umbrella mode in perfluoromethyl halides is more important than in methyl halides. This is consistent with the fact that corresponding vibrational quanta are lower in perfluoromethyl halides (e.g., compare 168 meV in CH_3Cl and 97 meV in CF_3Cl), therefore the excited umbrella mode is more populated there than in methyl halides.

3.1. Coordinates and Hamiltonian

The potential energy surfaces U^0 and U for the neutral molecule CF_3Cl and the anion CF_3Cl^- are represented using two internal coordinates: C-Cl separation R, and the distance between the C atom and the plane formed by the F atoms, $r = -R_{\mathrm{CF}} \cos\theta$, where θ is the FCCl angle. At $R \to \infty$ both surfaces are represented in the harmonic form $k(r - r_{ef})^2/2$ for the neutral and $k(r - r_{ef})^2/2 - EA$ for the anion, where EA is the electron affinity of the Cl atom, and r_{ef} is the equilibrium value of of r for the radical CF_3.

The classical kinetic energy for the nuclear motion in the approximation of two symmetric coordinates can be worked out from the general Hamiltonian for XY_3Z molecules [29]

$$T = \frac{1}{2}\mu\dot{R}^2 + \frac{1}{2}m_{\mathrm{F}_3}\left(\cot^2\theta_0 + \frac{m_{\mathrm{C}} + m_{\mathrm{Cl}}}{M}\right)\dot{r}^2 + \frac{m_{\mathrm{Cl}} m_{\mathrm{F}_3}}{M}\dot{R}\dot{r}, \tag{2}$$

where $M = m_{\mathrm{Cl}} + m_{\mathrm{C}} + 3m_{\mathrm{F}}$, $\mu = m_{\mathrm{Cl}} m_{\mathrm{CF}_3}/M$, and θ_0 is the equilibrium value of θ. In order to uncouple two motions in this expression, we introduce now the reaction coordinates ρ and r [20], where

$$\rho = R + \eta r, \quad \eta = \frac{3m_{\mathrm{F}}}{m_{\mathrm{C}} + 3m_{\mathrm{F}}}, \tag{3}$$

and r remains unchanged. The reaction coordinate ρ is simply the distance between Cl and the center of mass of CF_3. With the reaction coordinates ρ and r the kinetic energy becomes decoupled [30].

If the initial state of the neutral is not highly excited, we can use the harmonic approximation for $U^0(\rho, r)$ and diagonalize the Hamiltonian $T + U^0$. This gives us the normal frequencies and normal mode coordinates expressed as linear combinations of $\rho - (R_e + \eta r_e^0)$ and $r - r_e^0$.

3.2. Local theory of dissociative attachment

We will employ now the *local* version [3, 31] of the complex potential theory for DEA. First, the nonlocal calculations are quite complicated even in the one-mode approximation, therefore before we attempt them, it is desirable to get some results using a simpler approach. Second, the A_1 shape resonance in CF_3Cl occurs at a relatively high energy, 1.83 eV [32], and its width is rather narrow, about 0.6 eV in the Franck-Condon region. This justifies the use of the local theory for CF_3Cl.

In Fig. 5 we present comparison of two versions of the theory in the one-mode approximation, nonlocal calculations based on the resonance R-matrix approach [7] and the local version. Calculations were done for different initial vibrational quantum numbers ν_3. The comparison confirms our assumption. Both calculations employ the width function $\Gamma(E, R)$ obtained from the semiempirical R-matrix theory as described in Ref. [7].

In the one-mode calculations the local (adiabatic) width is represented as $\Gamma(E_r(R), R)$, where $E_r(R) = U(R) - U^0(R)$ is the resonance energy for a given internuclear distance R. Its extension to the two-mode case requires additional information on the dependence of the width on the second coordinate r. As a first try, we present the two-mode width $\Gamma(E_r(R, r), R, r)$ as a function of $E_r(R, r)$ only.

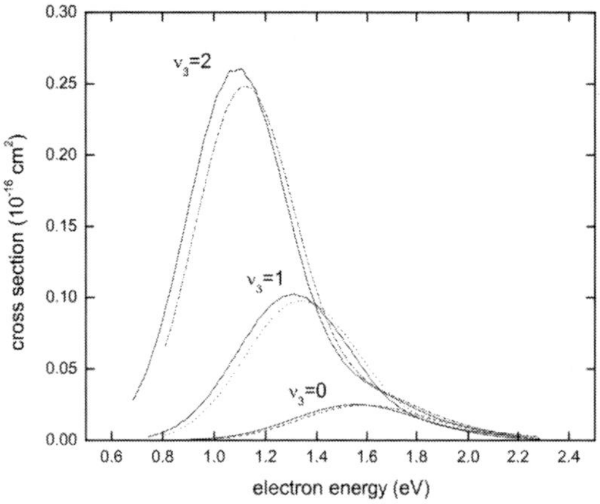

Figure 5. Dissociative attachment to CF_3Cl: comparison of the non-local (solid lines) and local (dashed lines) calculations. Numbers near the curves indicate the vibrational quantum number for the initial state.

The basic equation for the nuclear wavefunction $\chi_E(\rho, r)$ in the local approximation has the form [31]

$$[T_\rho + T_r + U(\rho, r) - i\Gamma(\rho, r)/2 - E]\chi(\rho, r) = V(\rho, r)\zeta_i(\rho, r), \tag{4}$$

where $T_\rho + T_r$ is the kinetic energy operator in the reaction-coordinate representation, $V(\rho, r) = [\Gamma(\rho, r)/2\pi]^{1/2}$ is the amplitude for electron capture into the resonance state, and $\zeta_i(\rho, r)$ is the vibrational wave function of the neutral in the initial state.

To solve Eq. (4), we expand χ in a basis depending on r only [20, 33]. Specifically, we choose eigenfunctions of the Hamiltonian corresponding to the infinite separation of CF_3 and Cl (note that $\Gamma(\infty, r) = 0$):

$$(T_r + U^0(\infty, r) - \epsilon_s)\phi_s(r) = 0, \tag{5}$$

where ϵ_s are corresponding eigenenergies, $\epsilon_s = D + \omega_2^f(s + 1/2)$, where ω_2^f is the frequency of the umbrella vibration in free CF_3, and D is the C-Cl bond dissociation energy.

Thus, the expansion has the form

$$\chi(\rho, r) = \sum_s \psi_s(\rho)\phi_s(r), \tag{6}$$

where functions $\psi_s(\rho)$ are solutions to the coupled equations

$$(T_\rho + \epsilon_s - E)\psi_s(\rho) + \sum_{s'} U_{ss'}(\rho)\psi_{s'}(\rho) = \lambda_{is}(\rho), \tag{7}$$

where

$$U_{ss'}(\rho) = \int \phi_s(r)[U(\rho, r) - i\Gamma(\rho, r)/2 - U(\infty, r)]\phi_{s'}(r)dr, \tag{8}$$

$$\lambda_{is}(\rho) = \int V(\rho, r)\zeta_i(\rho, r)\phi_s(r)dr. \tag{9}$$

Coupled equations (7) should be solved with the outgoing-wave boundary condition at $\rho \to \infty$ (or exponentially decaying condition for energetically closed channels):

$$\frac{d\psi_s(\rho)}{d\rho} \sim iK_s\psi_s(\rho), \quad \rho \to \infty, \tag{10}$$

where

$$K_s^2 = 2\mu(E - \epsilon_s). \tag{11}$$

Then the cross section for DA with the formation of the CF_3 fragment in the state s is

$$\sigma_s = \frac{2\pi^2}{k^2}\frac{K_s}{\mu}\lim_{\rho\to\infty}|\psi_s(\rho)|^2, \tag{12}$$

where $k^2/2$ is the initial electron energy.

3.3. Results of two-mode calculations

The two-dimensional potential surfaces for CF_3Cl and corresponding anion were calculated using the second-order Møller-Plesset perturbation theory employing the Dunning's aug-cc-pVTZ basis set [21, 34]. The anion surface was adjusted in the Franck-Condon region to incorporate the experimental vertical attachment energy for CF_3Cl [32].

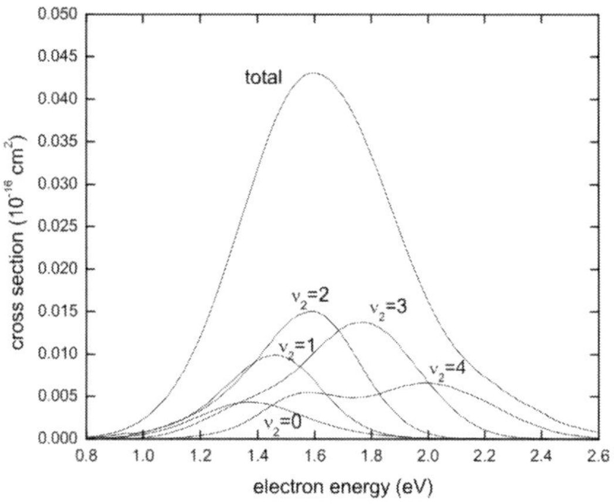

Figure 6. Dissociative attachment to CF_3Cl from the ground vibrational state. Numbers near the curves show the vibrational quantum numbers for umbrella vibrations in the final state.

In Fig. 6 we present dissociative attachment cross sections for CF_3Cl. The *total* cross section is higher than in the 1-d calculations. Our analysis of the capture amplitudes indicates that the possible reason for this effect is the lower effective resonance width in the 2-d case which leads to a higher survival probability. Of course, this effect strongly depends on the width as a function of the nuclear coordinates, therefore to find out if this effect is real we need *ab initio* calculations of the 2-d width. Both 1-d and 2-d calculations agree qualitatively with the results of the swarm [35] and beam [36] measurements for the total DEA cross sections. However, these measurements disagree in the absolute value of the peak cross sections by more than a factor of 2 with the beam measurements being closer to our 1-d results and swarm measurements closer to our 2-d results.

The shape of the cross section as a function of electron energy obtained in relative beam measurements of Hahndorf *et al* [28] agrees approximately with both 1-d and 2-d calculations. However, both theoretical curves decay faster towards lower energies than the experimental curve [28]. On the other hand, the 2-d curve decays slower at higher energies than the 1-d curve which leads to a better agreement with the experiments [35, 36].

The final-state vibrational energy distribution indicates a large contribution of the states with ν_2 = 1, 2, 3 and 4. This phenomenon is qualitatively independent of the width, and it suggests that the DEA process is an efficient way of exciting vibrational modes in CF_3.

Acknowledgments
The author is thankful to G. A. Gallup, H. Hotop, S. Roszak and P. Wielgus for fruitful collaboration. This work was supported by the US National Science Foundation, Grant Nos. PHY-0354688 and PHY-0652866, and by DFG for visits to TU Kaiserslautern.

References
[1] Hotop H, Ruf M-W, Allan M and Fabrikant I I 2003 *Adv. At. Mol. Phys.* **49** 85
[2] Fabrikant I I 1991 *Phys. Rev. A* **43** 3478
[3] Domcke W 1991 *Phys. Rep.* **208** 97
[4] Lane A M and Thomas R G 1958 *Rev. Mod. Phys.* **30** 257
[5] Fabrikant I I, Kalin S A and Kazansky A K 1991 *J. Chem. Phys.* **95** 4966
[6] Fabrikant I I, Kalin S A and Kazansky A K 1992 *J. Phys. B* **25** 2885
[7] Wilde R S, Gallup G A and Fabrikant I I 1999 *J. Phys. B* **32** 663
[8] Wilde R S, Gallup G A and Fabrikant I I 2000 *J. Phys. B* **33** 5479
[9] Marienfeld S, Fabrikant I I, Braun M, Ruf M-W and Hotop H 2006 *J. Phys. B* **39** 105
[10] Marienfeld S, Sunagawa T, Fabrikant I I, Braun M, Ruf M-W and Hotop H 2006 *J. Chem. Phys.* **124** 154316
[11] Kalin S A and Kazansky A K 1990 *J. Phys. B* **23** 4377
[12] Braun M, Fabrikant I I, Ruf M-W and Hotop H 2007 *J. Phys. B* **40** 659
[13] Hahndorf I and Illenberger E 1997 *Int. J. Mass Spectrom. Ion. Process.* **167/168** 87
[14] Schramm A, Fabrikant I I,Weber J M, Leber E, Ruf M-W and Hotop H 1999 *J. Phys. B* **32** 2153
[15] Gallup G A and Fabrikant I I 2007 *Phys. Rev. A* **75** 032719
[16] Gallup G A, Xu Y and Fabrikant I I 1998 *Phys. Rev. A* **57** 2596
[17] Fabrikant I I and Hotop H 2001 *Phys. Rev. A* **63** 022706
[18] Kazansky A K 1995 *J. Phys. B* **28** 3987
[19] Haxton D J, Zhang Z, Meyer H-D, Rescigno T N and McCurdy C W 2004 *Phys. Rev. A* **69** 062714
[20] Shapiro M and Bersohn R 1980 *J. Chem. Phys.* **73** 3810
[21] Roszak S, Koski W S, Kaufman J J and Balasubramanian K 1997 *J. Chem. Phys.* **106** 7709
[22] Mann A and Linder F 1992 *J. Phys. B* **25** 1621
[23] Walter C W, Lindsay B G, Smith K A and Dunning F B 1989 *Chem. Phys. Lett.* **154** 409
[24] Kalamarides A, Walter C W, Lindsay B G, Smith K A and Dunning F B 1989 *J. Chem. Phys.* **91** 4411
[25] Parthasarathy R, Finch C D, Wolfgang J, Nordlander P and Dunning F B 1998 *J. Chem. Phys.* **109** 8829
[26] Pearl D M, Burrow P D, Fabrikant I I and Gallup G A 1995 *J. Chem. Phys.* **102** 2737
[27] Beyer T, Nestmann B M and Peyerimhoff S D 2001 *J. Phys. B* **34** 3703
[28] Hahndorf I, Illenberger E, Lehr L and Manz J 1994 *Chem. Phys. Lett.* **231** 460
[29] Shaffer W H 1942 *J. Chem. Phys.* **10** 1
[30] van Veen G N A, Baller T, de Vries A E and Shapiro M 1985 *Chem. Phys.* **93** 277
[31] Bardsley J N, Herzenberg A and Mandl F 1966 Proc. Phys. Soc. London **89** 305; **89** 321
[32] Aflatooni K and Burrow P D 2001 *Int. J. Mass Spectrom.* **205** 149
[33] Levine R D 1969 *Quantum Mechanics of Molecular Rate Processes* (Oxford: Clarendon) chapter 36
[34] Tarana M, Wielgus P, Roszak S and Fabrikant I I 2009 *Phys. Rev. A* **79** 052712
[35] Spyrou S M and Christophrou L G 1985 *J. Chem. Phys.* **82** 2620
[36] Underwood-Lemons T, Gergel T J and Moore J H 1995 *J. Chem. Phys.* **102** 119

Dissociative recombination of H_2^+ with slow electrons for a rigorous examination of theory

H Takagi[1], **S Hara**[2] **and H Sato**[3]

[1] Center for Natural Science, Kitasato University, 1-15-1 Kitasato, Sagamihara, Kanagawa, 228-8555 Japan
[2] Tsukuba University of Technology, 4-12-7 Kasuga, Tsukuba-shi, Ibaraki, 305-8521, Japan
[3] Faculty of Science, Ochanomizu University, 2-1-1 Ootuka, Bunkyou, Tokyo, 112-8610, Japan

E-mail: [1]`takagi@kitasato-u.ac.jp`

Abstract. The Lippmann-Schwinger (LS) equation for the K matrix is numerically solved for the collision of H_2^+ + e interacting by the configuration interaction (CI), of which magnitude is carefully deduced from ab-initio calculation of electron scattering. The LS equation is extended to the negative collision energies in the context of multichannel quantum defect theory (MQDT) for the dissociative recombination (DR). Using the MQDT method for the DR, The DR cross section of H_2^+ is obtained. It has turned out that the DR at low energy is induced by the indirect process with rotational excitation and the off-the-energy-shell contribution is indispensable for the accurate description. The present calculation represents the recent state specific experiment [1], and has assured the effectiveness of the MQDT method for DR.

1. Introduction

The major mechanism of dissociative recombination (DR) of H_2^+ is considered as the incident electrons are captured into the two-electron excited resonance state of which potential curve is dissociative[2]. This mechanism, known as the 'direct process', is induced by the configuration interaction (CI) between the one- and two-electron excited states. Another mechanism named the 'indirect process' is the DR, where the incident electron is temporarily recombined into the rotationally and/or vibrationally excited Rydberg states[3]. These recombined Rydberg states finally dissociate into the two-electron excited state. The non-adiabatic interaction (NAI) is thought to be the main mechanism of the indirect process. The NAI in the Rydberg and electronic continuum states is known to be well represented by the multichannel quantum defect theory (MQDT)[4]. So called 'two-step' method[5] enabled to take account both those interactions uniformly. In the first step of this method, a scattering problem of electrons by the CI needs to be solved accurately. Next, the problem of NAI is solved by the MQDT using the electronic state obtained by the first step.

In this paper, we carefully investigate the CI strength at all over the energy. Using the resultant CI strength, we calculate the DR cross section in the framework of 'two-step method' as accurate as possible. On the recent experimental study, the DR cross section specified the vibrational and rotational quantum number of the initial state has been measured[1] for H_2^+. By comparing the calculation with the experiment, we aim at a rigorous examination of the DR theory besides theoretical analysis of the experiment. The reliable theory can make clear the mechanism of the DR at the low collision energies.

2. electron-molecule scattering by configuration interaction

A brief report of this section is seen in the progress report by one of the authors (HT)[6, 7]. We here show a minimum required formulation for the calculation.

We adopt the Born-Oppenheimer expansion in order to represent a system interacting by the CI. The CI considered now is the interaction between a two-electron excited state labeled by suffix d and one-electron excited one, which is specified by the energy of excited electron ϵ. Putting the radial coordinate of electron r, we present the wave functions of those states by $\phi_d(r)$ and $\phi_\epsilon(r)$. The radial nuclear wave functions associated with those electronic functions are respectively represented by $F_\varepsilon^J(R)$ and $\chi_v^N(R)$, where R is internuclear distance. The former is the dissociative wave function of the dissociating energy ε and the total angular momentum J. The latter is the wave function of vibrational state v with rotational state N. The strength of the CI is represented by the following matrix elements of electronic Hamiltonian of fixed nuclei at R; $H^{ele}(r;R)$.

$$V_{v\epsilon,d\varepsilon}^{NJ} = \langle \chi_v^N(R) | V_{\epsilon,d}(R) | F_\varepsilon^J(R) \rangle_R, \tag{1}$$

where

$$V_{\epsilon,d}(R) = \langle \phi_\epsilon(r;R) | H^{ele}(r;R) | \phi_d(r;R) \rangle_r. \tag{2}$$

Since the molecular orientation is fixed during the electron collision for the inner region of the MQDT, we assume the rotational transition is frozen for the CI: $N = J$. The rotational transition is taken into account at the second step of the two-step method separated from the first step. Thus, we omit the fixed parameter N and J in this section. The total energy is sum of the energies of electronic and nuclear motion; $E^T = E_v + \epsilon = E_d(R = \infty) + \varepsilon$. When we write the total energy as E_ϵ^T for the channel v, and E_ε^T for the channel d, the CI strength is represented by

$$V_{v\epsilon,d\varepsilon} = V_{v,d}(E_\epsilon^T, E_\varepsilon^T), \tag{3}$$

which is the off-the-energy-shell element for $E_\epsilon^T \neq E_\varepsilon^T$.

We use Greek character, for example α, for identifying the channel and E_α for the total energy of the channel α: $\{\alpha\} = \{v\} \oplus \{d\}$, $\{E_\alpha\} = \{E_\epsilon^T\} \oplus \{E_\varepsilon^T\}$. The energy resolved Lipmann-Schwinger (LS) equation for the K matrix becomes

$$K_{\beta\alpha}(E_\beta, E_\alpha) = -\pi V_{\beta,\alpha}(E_\beta, E_\alpha) + \sum_\gamma \wp_v \int_{-\infty}^{\infty} dE_\gamma \frac{V_{\beta\gamma}(E_\beta, E_\gamma)}{E_\alpha - E_\gamma} K_{\gamma\alpha}(E_\gamma, E_\alpha). \tag{4}$$

The above equation is the same form as equation (7.62) given by Newton[8] except for the integral region being extended to the negative energy region. In the MQDT formalism, the CI is represented at the inner region, where the boundary condition on the energy is not imposed. As the Coulombic basis functions in the inner region of the MQDT, the scattering channel functions should be free from the control of the asymptotic energy boundary condition. The Rydberg states are treated as open channels for the CI interaction.

We can solve this equation by replacing the principal value integral of E_γ to appropriate numerical integration. Using the Chebyshev polynomial T_n and n-th order Chebyshev polynomial of the second kind U_n, equation (4) is rewritten as

$$\sum_\gamma \sum_{i=1}^{M} \left[m(y_\beta, y_{\gamma_i}, y_\alpha; \beta, \gamma, \alpha) - \delta_{\beta,\gamma} \delta_{E_\beta, E_\gamma} \right] K_{\gamma\alpha}(E_{\gamma_i}, E_\beta) = \pi V_{\beta,\alpha}(E_\beta, E_\alpha). \tag{5}$$

with

$$m(y_\beta, y_{\gamma_i}, y_\alpha; \beta, \gamma, \alpha) = \frac{2\pi}{M}(1 + y_\alpha)\sqrt{\frac{1 - y_{\gamma_i}}{1 + y_{\gamma_i}}} V_{\beta\gamma}(E_\beta, E_{\gamma_i}) \sum_{n=1}^{N} T_n(y_{\gamma_i}) U_{n-1}(y_\alpha). \tag{6}$$

In the above equation, we adopted following transformation from the total energy E to y.

$$y = \frac{2C}{E - a + C} - 1, \tag{7}$$

where $a(\leq 0)$ and C is arbitrary parameter. The parameter a means the lowest energy of the integration and C controls the domain of integration.

3. Electron and H_2^+ collisions by the CI

3.1. Adiabatic CI strength

We here take into account only the lowest two-electron excited state $^1\Sigma_g(2p\sigma_u)^2$ as the dissociative states ϕ_d. The Rayleigh-Ritz variational method is employed to calculate the two-electron excited state ϕ_d. The ionizing state ϕ_ϵ is calculated in the static exchange approximation with the adiabatic polarization potential. Using those wave functions (ϕ_d, ϕ_ϵ) into equation (2), we obtain the CI strength $V_{\epsilon,d}(R)$[9, 10]. The adiabatic resonance width $\Gamma(\epsilon, R)$ is given by

$$\Gamma(\epsilon, R) = 2\pi|V_{\epsilon,d}(R)|^2. \tag{8}$$

These approximations adopted to the present calculation are appropriate to estimate the dynamical interaction between the resonance and continuum electronic stats. The boundary condition of asymptotic region is satisfied by the direct product of two-electron excited states and the dissociative nuclear motion, or the direct product of ionizing continuum states and vibrational nuclear motion. The neglected higher order coupling between the two-electron excited states and ionizing continuum states in fixed nuclei is to be recovered by taking account the CI in the dynamical process.

The crucial important quantity is the strength of interaction $V_{\epsilon,d}(R)$ or $\Gamma(\epsilon, R)$ as a function of two variables ϵ and R. The calculation was done for the internuclear distance$R = 1.0 \sim 2.6$ au with the interval of 0.2 au, and the collision energy $\epsilon =\sim 0.001 \sim 5$ au. In the present method, the calculation of adiabatic resonance width Γ is confined to positive collision energies but we need it in all energy region including the negative collision energies. This means we must extrapolate the calculated values to the all R and ϵ^T region. The asymptotic behavior of the Γ is important in the extrapolation.

(i) $\Gamma = 0$ for $\epsilon < -1/2$.
(ii) $\Gamma = 0$ for $\epsilon \longrightarrow \infty$.
(iii) $\Gamma = 0$ for $R \longrightarrow 0$.
(iv) $\Gamma = 0$ for$R > R_m$,

where for $R \geq R_m$, the two-electron excited state merges to the Rydberg state: $R_m \sim 4.5$ au[11]. The conditions for the asymptotic behavior (i)~(iv) confine the Γ in the full region of (ϵ^T, R) space, where ϵ^T is the total electronic energy: ϵ+potential energy of the ion.

We employed the bicubic B-spline interpolation [12] using the calculated (ϵ, R) points. In the outside of the calculated R region, we need extrapolate the calculated values. The B-spline works well for the extrapolation to $R < 1.0$ au because the strength monotonically decreases as R becomes zero. For the R larger than 2.6 au, the extrapolation using the calculated values diverges. Because the Γ forms an isolated island on the sea lebel as is shown in Fig. 1, we extrapolate the contour lines of some Γ values at the boundary of calculated region ($R = 2.6$ au) assuming that the contour shape could be represented by quadratic form of ϵ^T and R. We chose tree contour lines of Γ =0.005, 0.03, and 0.07 au.

Adding those values on the contour, we could interpolate and extrapolate the calculated values to the larger R by two dimensional B-spline method. The adiabatic resonance width $\Gamma(\epsilon^T, R)$ on the whole ϵ^T-R space is shown in Fig. 1. The shape of Γ is not simple, especially there is bump at the low energies of $1.7 < R < 2.4$ au, of which effect will be investigated on the low energy dissociative recombination.

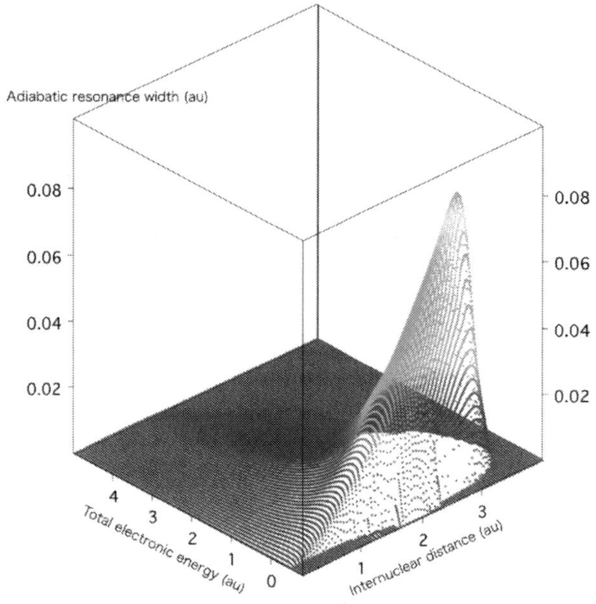

Figure 1. Adiabatic resonance width $\Gamma(\epsilon^T, R)$ as a function of the total electronic energy and the internuclear distance.

3.2. Transition probability by the CI

The K matrix elements were calculated by solving equation (5). The values of parameters C and a in equation (5) ware adopted to $C = 0.3$ and $a = -0.625au$ (energy of dissociation limit). The number of energy points M in equation (5) is adopted to $M = 100$ after confirming the obtained K matrix to be converged[7]. twelve vibrational functions ($v = 0 \sim 11$) are employed in the calculation. The energy region of the present interest is the incident electron energy lower than 1 eV for $v = 0 \sim 2$ initial vibrational states, which correspond to the total energies between -0.6 and -0.5 au. The dissociatiion probability is defined by square absolute value of the S matrix element $|S_{d,v}(E)|^2$, where S matrix element is deduced from the calculated K matrix: $S = (1 + iK)(1 - iK)^{-1}$.

The calculated dissociation probability was compared with the first order calcualtion in Fig. 2. The higher order effect drastically enhances the dissociation from the initial vibrational state $v = 0$ and reduces that from $v = 2$ around the zero collision energy. As the result, the probabilities from $v^+ = 0$ and $v = 2$ become similar magnitude. The $v = 2$ initial state is energetically more suitable for the dissociation than the $v = 0$ since the energy level of $v = 2$ exceeds to the energy of the potential-curve crossing between the ionic and dissociative states whereas the $v = 0$ energy level does not. The vibrational excitation by the CI compensates for this energetic advantage. The difference between the first order and the higher order calculation is outstanding at the low energies. The off-the-energy-shell contribution has turned out to be crucial for precise description of the dynamics.

4. Dissociative recombination

The calculation of DR cross section by the MQDT was given in the previous studies[13, 14]. We employed the same electronic parameters as the previous calculation[15, 16]: potential curves and the adiabatic quantum defect. The contribution from the two-electron excited state was carefully excluded from the adiabatic quantum defect. The minor component $s\sigma$ contributes about 30% at most to the DR [17]. We consider that the contribution of s wave is negligible at low energies because the s wave does not induce the rotational excitation, which turns out to be the main mechanism of the DR.

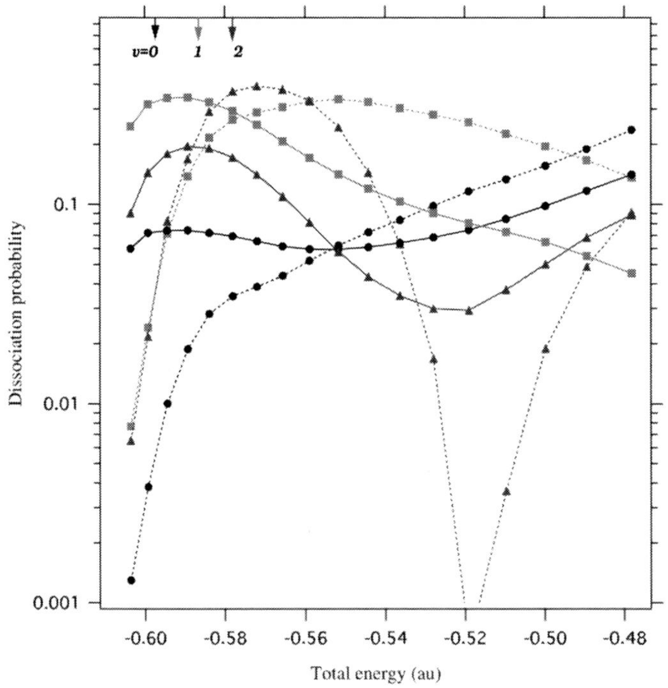

Figure 2. Dissociation probability by the CI: effect of off-the-energy-shell contribution. The full curve shows the result with off-the-energy-shell contribution, the dotted curve does the result by the first order perturbation. The initial vibrational states are indicated: circle: $v^+ = 0$, square: $v^+ = 1$, triangle: $v^+ = 2$. The arrow below the top axis indicates the position of zero energy for each v^+.

In Fig. 3, we show an example of the present DR cross section of H_2^+ of which initial vibrational state v^+ is $v^+ = 0$, and the rotational state N^+ is $N^+ = 0$. The vertical axis shows the cross section multiplied by the collision energy, which is sometimes called reduced cross section (cm^2 eV), in order to see the cross section accurately over wide range of collision energy. The energy of rotationally and/or vibrationally excited Rydberg state are shown by the bar and cross mark in the figure. Almost all peak structures appear at the energies of rotationally excited and vibrationally unexcited Rydberg states. This indicates that the indirect process through the rotational excitation is the dominant mechanism of the DR at low energies.

To see the overall feature of the DR and compare to experiments, we convolute the calculated cross section with a thermal distribution [18]. For rotational distribution, we assume the temperature of 300K. For the energy fluctuation of electron beam, we assume the temperature of 1 meV for transversal to the incident beam direction and 0.05 meV for the longitudinal.

The dependence of DR rate coefficient on the initial vibrational state v^+ is shown in Fig. 4. The rate coefficient of $v^+ = 1$ is about four times larger than that of $v^+ = 0$ and 2 over all collision energies. The magnitude of the rate agrees with those of previous study[15], where the rate is 2.31×10^{-8} for $v^+ = 0$, 1.81×10^{-7} for $v^+ = 1$, and 1.34×10^{-8} for $v^+ = 2$ at the collision energy of 4.35meV (100K). This agreement is, however, accidental coincidence because the previous study was footing on the first order perturbation on the CI and the important rotational motion was neglected. The off-the-energy-shell contribution remarkably affects the dependence on the initial vibrational state in the DR as is seen in Fig 2.

The DR cross section has been measured for H_2^+ of which the initial rotational and vibrational state is specified[1]. In Fig 5, we compare this experiment for para-H_2^+ with the result of present calculation for $v^+ = 0$ and for each N^+. We compare the DR rate coefficient multiplied by the square root of the collision energy for precise comparison. The calculated cross section is convoluted on the collision energy assuming the fluctuation temperature of 2.5 meV for the transversal direction and 0.05 meV for the longitudinal one. Except for the lowest peak around 1 meV, we can find corresponding peak structures between the calculation for $N^+ = 0$ and

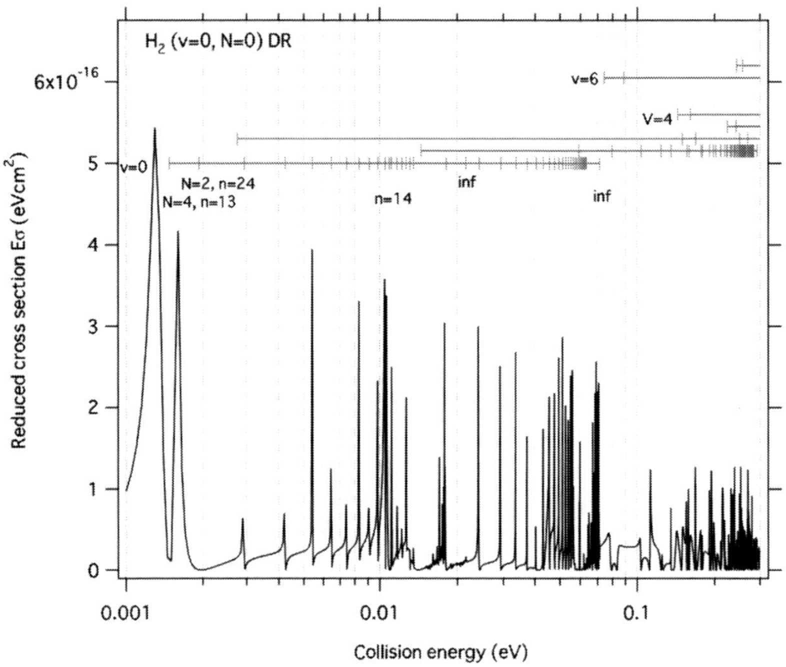

Figure 3. Reduced DR cross section of $H_2^+(v^+ = 0, N^+ = 0)$
The bars with cross marks present the energies of rotationally(N) and/or vibrationally(v) excited Rydberg states, of which principal quantum numbers are less than 58. The series limit of Rydberg states is indicated by `inf`. The lowest bar is for $v = 0$, and the higher bar presents the higher v up to $v = 7$.

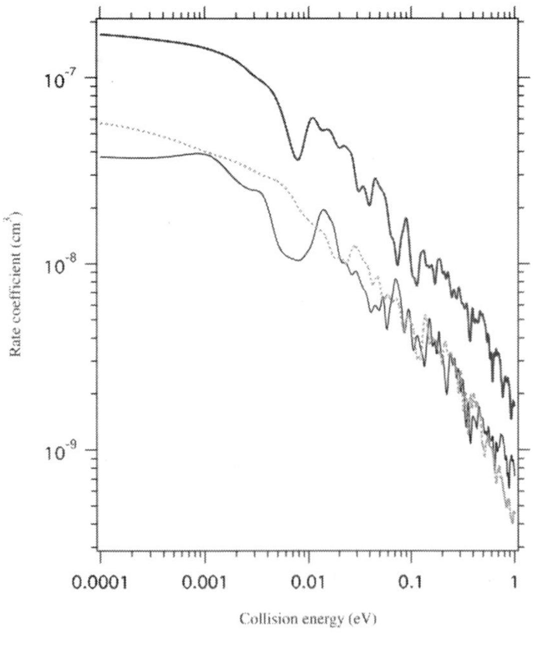

Figure 4. Dependence of DR rate coefficient for H_2^+ on the initial vibrational quantum number v^+. The kind of curves indicates the initial vibrational state, thin solid: $v^+ = 0$, bold solid: $v^+ = 1$, dotted: $v^+ = 2$. *See the text* on the electron fluctuation and the rotational state distribution.

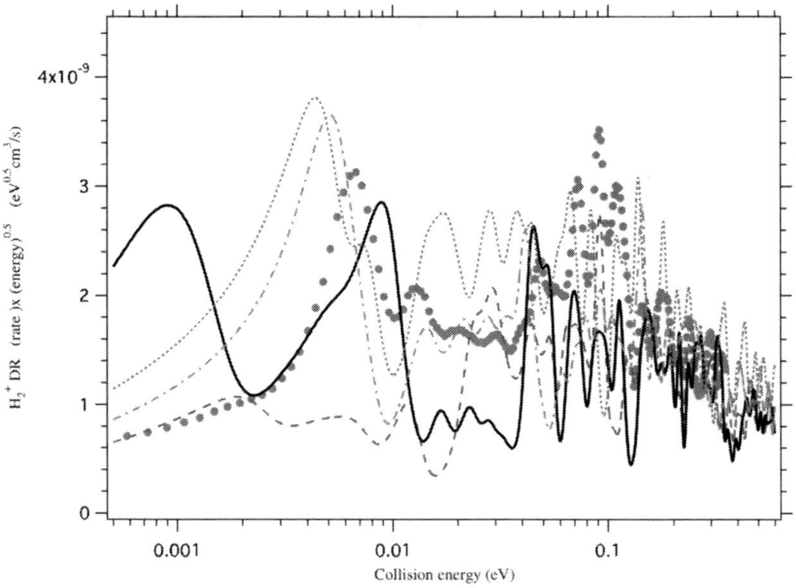

Figure 5. The rate coefficient multiplied by the square root of the collision energy for para-H_2^+ of $v^+ = 0$. The experiment[1] is shown by full circle and the present calculation for each rotational state N^+ is shown by curves, ——: $N^+ = 0$, - - - -: $N^+ = 2$, — · —: $N^+ = 4$, ······: $N^+ = 6$.

the experiment although the peak position is slightly shifted (about 3 meV) each other at low energies. These magnitude does not always agree with each other. However, the difference is less than or about twice. The calculation with other initial rotational states than $N^+ = 0$ can not represent the experiment. Thus the experiment is thought to be the first DR measurement specified the initial rotational state $N^+ = 0$.

The lowest peak structure in the calculation comes from the electron capture into the Rydberg state of $n = 13$ with $N = 4$ or $n = 24$ with $N = 2$, where n is the principal quantum number and N is rotational quantum number. Since the energy of this peak is at quite low energy as below 2 meV, if the peak is slightly shifted, it must be disappeared. It is open question for theory and experiment whether this structure is real or not.

We compared the calculation with the experiment of normal H_2^+ in Fig 6. The result by calculation is average over the initial rotational states $N^+ = 0$ and $N^+ = 1$ with the weight of one and tree respectively. The calculation reproduce the structure seen in the experiment although the magnitude of the peak does not always agree with each other. At the collision energies lower than 30meV, the peak positions are shifted about 3meV.

5. Concluding remarks

The off-the-energy-shell contribution to the DR of HD^+ was investigated based on the realistic electronic states obtained by ab-initio method. The dissociative processes are largely affected by the off-the-energy-shell contribution of negative energies. The DR at low energy has turned out to be induced by the indirect process with rotational excitation. The DR rate coefficient of the H_2^+ including the off-the-energy-shell contribution nicely represents the resonance structure in DR rate coefficient obtained by the state specific experiment. We have assured the effectiveness of the MQDT method for DR.

Acknowledgments

Authors wish to thank Professor W. J. van der Zande for providing us with their experimental data before publication and for having profitable discussion.

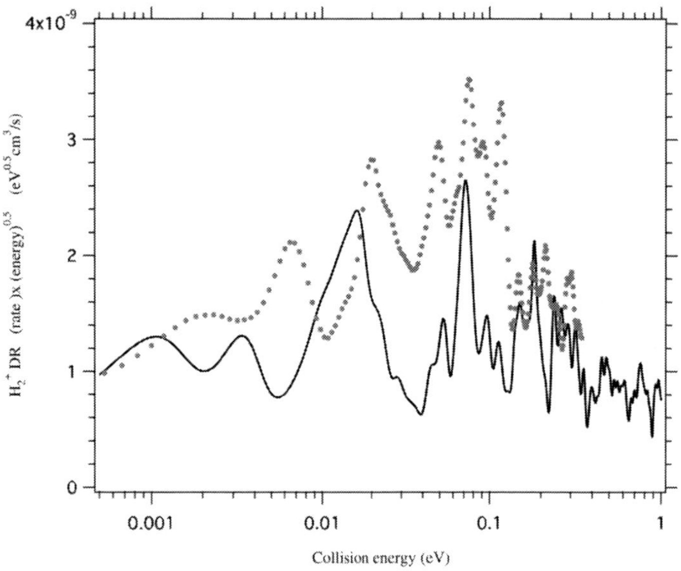

Figure 6. The rate coefficient multiplied by the square root of the collision energy for normal H_2^+ of $v^+ = 0$.The experiment[1] is shown by full circle and the present calculation is shown by solid curve.

References

[1] Zhaunerchyk V *et al.* 2007 *Phys. Rev. Lett.* **99** 013201
[2] Bates D R 1950 *Phys. Rev.* **77** 718
[3] Bardsley J N 1968 *J. Phys.* B **1** 349
[4] Jungen C and Atabek O 1977 *J. Chem. Phys.* **66** 5584
[5] Giusti A 1980 *J. Phys.* B **12** 3867
[6] Takagi H 1998 *Atomic Collision Research in Japan* vol 24 p 3
[7] Takagi H 2000 *Dissociative Recombination: Theory, Experiment, and Application iv* (World Scientific) p 180
[8] Newton R G 1982 *Scattering Theory of Wave and Particles* 2nd ed (Springer-Verlag: Springer-Verlag)
[9] Hara S and Sato H 1984 *J. Phys.* B **17** 4307
[10] Sato H and Hara S 1986 *J. Phys.* B **19** 2611
[11] Wolniewicz L and Dressler K 2077 *J. Mol. Spectrosc.* **67** 416
[12] Press W H, Teukolsky S A, Vetterling W T and Flannery B P 1992 *Numrical Recipes* 2nd ed (Cambridge Univ. Press)
[13] Takagi H, Kosugi N and Dourneuf M L 1991 *J. Phys.* B **24** 711
[14] Takagi H 1993 *J. Phys.* B **26** 4815
[15] Nakashima K, Takagi H and Nakamura H 1987 *J. Chem. Phys.* **86** 726
[16] Takagi H and Nakamura H 1988 *J. Chem. Phys.* **88** 4552
[17] Schneider I F *et al.* 1997 *J. Phys.* B **30** 2687
[18] Takagi H 2005 *J. Phys.* Conference series **4** 155

Journal of Physics: Conference Series **192** (2009) 012004 doi:10.1088/1742-6596/192/1/012004

New flowing afterglow technique for determining products of dissociative recombination: CH_5^+ and N_2H^+

Nigel G. Adams, Chris D. Molek and Jason L. McLain

Department of Chemistry, University of Georgia, Athens, Georgia, USA

adams@chem.uga.edu

Abstract. There are discrepancies in the literature for the product distributions of electron-ion (e-i) recombination when determined using different techniques. Because of this, a new technique has been developed. This is based on the flowing afterglow, with the product neutrals detected by electron impact ionization followed by mass spectrometric detection. However, in addition to the products of recombination, there are neutrals present from ion-molecule reactions and from the gases introduced into the flow tube to create the ion of interest, which often have much greater concentrations than the products. To distinguish these products, an electron attaching gas is pulsed into the flow to transiently attach the electrons, thus quenching e-i recombination. Then the difference between the attaching gas in and out yields the product distribution. Recombination products have been detected even when their signal is as much as ~10^4 less than background. Here the details of the technique are described and the possible sources of error discussed. The viability of the technique is illustrated for the recombinations of CH_5^+ and N_2H^+. The latter establishes the major product as N_2 + H (95 to 100%) correcting an error in the literature. In the former case, the major channel detected is CH_4 + H (95%) which is in disagreement with a storage ring (SR) result which gave CH_3 as the major channel (68%). Possible reasons for this are discussed.

1. Introduction

In recent years, a large number of product distributions for dissociative electron-ion recombination have been determined [1, 2]. These are critically important in the modeling of the coupled ion and neutral chemistry in many molecular plasmas (interstellar gas clouds (ISC) [3], planetary atmospheres [4-6], and with some significance to combustion [7]). The majority of the studies have been made with storage rings (SR) with a few studies using flowing afterglows (FA). Where measurements have been made using the two different techniques, (N_2H^+, H_2O^+, H_3O^+ recombinations listed in Table 1) there is a considerable degree of disagreement and this places uncertainty on all of the measurements that have been made. This is very worrisome since these data are used in models of various plasmas and the models can make very different predictions depending on the product distributions. In the case of the ISC, this can prevent the production of several important molecules. In particular, the recombinations of CH_5^+ and $CH_3OH_2^+$ could, in principle, lead to the production of interstellar methane and methanol. In practice these recombinations are considered to be dissociative to a large degree [8, 9] thus preventing the production of methanol, the only known gas phase route.

Table 1. Recombination product distributions where there is disagreement between the SR and FA techniques.

Reaction	Products	FA [10, 11]	(SR [12-14])
$H_2O^+ + e$	$\rightarrow H + OH$	55%	(22%, 30%, 20%)
	$\rightarrow O + H_2$	<22%	(10%, 13%, 9%)
	$\rightarrow O + 2H$	>23%	(67%, 57%, 71%)

Reaction	Products	FA [15]	(SR [12, 16-18])
$H_3O^+ + e$	$\rightarrow H + H_2O$	5%	(33%, 18%, 25%)
	$\rightarrow OH + H_2$	36%	(18%, 11%, 14%)
	$\rightarrow OH + 2H$	29%	(48%, 67%, 60%)
	$\rightarrow O + H + H_2$	30%	(1%, 4%, 1.3%)

Reaction	Products	FA [19]	SR [20]
$N_2H^+ + e$	$\rightarrow N_2 + H$	100%	36%
	$\rightarrow NH + N$	0%	64%

The determination of product distributions using these techniques (SR and FA) is complicated. In the FA, spectroscopic detection requires vuv absorption for atoms, and the very sensitive Laser Induced Fluorescence (LIF) for molecules such as OH. LIF can also be used for atoms by using chemical reactions to convert the atoms H [21] and O [15] to OH by the neutral reactions; $H + NO_2 \rightarrow OH + NO$ and $O + GeH_4 \rightarrow OH + GeH_3$. Note that here the spectroscopic intensities have to be converted into number densities to compare with the electron and ion number densities, i.e., using quantitative spectroscopy. Whilst this should not cause errors, it does complicate the analysis. In the case of the SR, a detailed analysis is required using grids [22] and more recently Coulomb explosion imaging [23] to determine the product distribution. Also in some cases, corrections have to be made for collisions with the background gas in the ring [8], and for the regions where the electrons and ions are merged together and later separated [24, 25]. There is also a limitation on the energy resolution in magnetic storage rings at the higher masses; for example, in previous studies it has not been possible to fully resolve individual masses in hydrocarbons such as $C_2H_3^+$, $C_4H_5^+$, and $C_4H_9^+$ [26, 27], and other hydrocarbon ions [28]. However, to some extent, this situation has been alleviated by using deuterated hydrocarbons ($C_2D_5^+$, $C_3D_7^+$ and $C_4H_9^+$) [29, 30].

In order to circumvent some of these problems, we have developed a new flowing afterglow technique in which the products of recombination are identified by electron impact ionization followed by mass analysis, amplification in an electron multiplier and ion counting (Fig. 1). This modification means that rather than using a series of spectroscopic techniques, which have to be separately calibrated, the individual ion types are straightforwardly identified by their masses (better than 1 amu mass resolution). This allows the signals at different masses to be compared relatively, removing the need for absolute calibration. This may only seem to be a simple modification, however there are some complications that need to be overcome. Firstly, after determining the composition of the plasma, these plasma ions and electrons need to be prevented from entering the detection system, allowing only neutral species to pass through the sampling orifice. This is readily achieved by biasing the potentials on the first few lens elements of the detection system, which also prevents the passage of plasma electrons. Then, only neutral species can pass into the detection system. This includes the neutral products of recombination, but also permits passage of the gases that were introduced into the plasma as well as neutral products from ion-molecule reactions producing the recombining ions of interest. These latter contributions can be several orders of magnitude larger in concentration than the products of recombination and need to be discriminated against.

Selective detection of recombination products can be achieved in two ways, both resulting in the variation in the concentration of the recombining species (either electrons or recombining ions) and is illustrated here by the electron recombination of N_2H^+.

$$N_2H^+ + e \rightarrow N_2 + H + 8.47\ eV \quad (1a)$$

$$\rightarrow NH + H + 2.25\ eV \quad (1b)$$

This recombination was studied earlier in the FA by monitoring the H-atom production by vuv absorption [19] and later in the SR where the recombination products H, N, NH and N_2H were monitored [20, 31]. In the FA, Channel (1a) was indicated to be dominant whereas the SR gave a product distribution of 36% for Channel (1a) and 64% for Channel (1b), very disparate results. Such a result is surprising since the major channel requires the breaking of the strong N≡N triple bond. However, such a distribution would be a good test of the new technique since H, N, NH and N_2 can be monitored allowing the product distribution to be over determined and thus give an indication of the errors involved in the measurement. In these studies, it was not possible to follow the H^+ monitor for H-atoms since there was considerable reverse mass discrimination in the quadrupole against such low mass ions. Thus, N^+, NH^+ and N_2^+ were used as monitors of N, NH and N_2, respectively. Note that the use of these monitors is not perfect since some fragmentation occurs in electron impact ionization; electron impact with 70 eV electrons was generally used since most data are available for this energy [32]. These fragmentations were accounted for to determine the final percentage product distribution.

To extract product distributions from the monitor ion data, N_2 was pulsed into an H_3^+/electron plasma created in the standard way [33, 34]. Then the proton transfer $H_3^+ + N_2 \rightarrow N_2H^+ + H_2$ transiently produced the recombining ion of interest. Alternatively, a CCl_4/He mix was pulsed into the afterglow causing the rapid electron attachment $e + CCl_4 \rightarrow Cl^- + CCl_3$. Thus, when the electron attaching gas is absent e-i recombination can occur and when it is present, e-i recombination is quenched allowing only a very small amount of i-i recombination. In this way, the difference in the monitor ions signals between the pulsed gas absent and present (in both methods) is a measure of the e-i recombination. Note that only the latter technique is illustrated here since this is the method of choice. The former technique is explained in a more detailed paper [35].

2. Experimental

For this purpose, a standard temperature variable FA [36, 37] was used with the addition of a pulse valve for pulsing gas addition into the flow tube and adding an axial electron impact ionizer to the detection system downstream of the nose cone and before the quadrupole mass spectrometer; see Fig. 1.

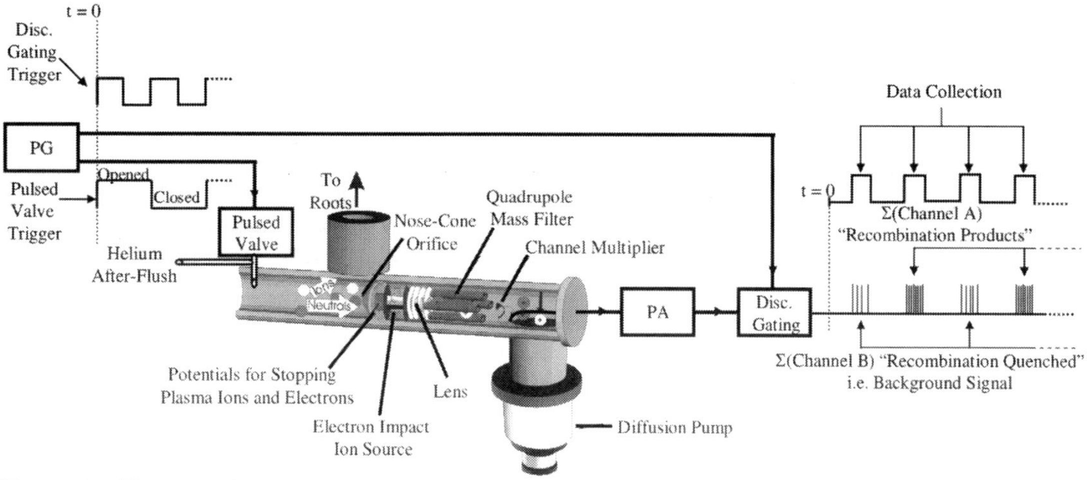

Figure 1. The detection section of the flow tube is shown with the pulse valve, the pulsing circuitry and the pulsed data collection. The pulse valve is opened and closed by a square wave (~1Hz) from the pulse generator (PG) which also sends a trigger signal at twice the frequency to the gating/discriminator. Ion counts from the channel multiplier are amplified in the preamplifier (PA) and are accumulated into a series of bins when the counting circuit is open. Because the gating/discriminator is triggered at twice the frequency of the pulse valve, alternate counting periods occur with the pulse valve closed (A) and open (B). These are separately summed together to give the integrated counts for background and signal plus background.

These may seem minor changes but they have profound significance. To enable products of recombination to be quantitatively identified requires a detailed understanding of the pulsing process and also highly stable conditions. The recombining plasma of interest was produced in the conventional way [33, 34], by sequentially adding Ar, H_2 and the reactant gas, RG, to the He flow. After ionization of He in a microwave discharge, He^+/He_2^+, Ar^+, H_3^+ and RGH^+ electron plasmas were successively produced (by a conventional reaction sequence) [33, 34]. The compositions of the plasmas were determined by sampling the plasma ions through the orifice in the nose cone in the conventional way with the electron impact ion source switched off. To sample neutral species in the flow, (including the products of recombination) the downstream lenses were biased by ~1 V, alternately positive and negative, to prevent plasma ions and electrons from entering the detection system. In this case, the ionizer is switched on to ionize the neutrals from the plasma to be sampled by the mass spectrometer. Within these ions, to separate the monitor ions for the recombination products from the background species, a rapidly attaching gas is pulsed into the plasma to quench e-i recombination; CCl_4 is usually used producing Cl^- (and CCl_3). Attachment is so rapid that only small amounts of CCl_4 are used, which are insufficient to affect the positive ion chemistry. An amount of CCl_4 is added which is sufficient to attach 80 to 90% of the electrons, thus retaining a plasma which is controlled by electron/positive ion ambipolar diffusion and therefore the positive ion diffusive loss is unchanged [38]. When the electrons are present, Counter A collects 300ms bursts of monitor ions with Counter B collecting counts when negative ions are present, see Fig. 1. The difference between these for the monitor ions, when integrated, gives counts, which are proportional to the recombination products. In order to get clean square pulsed of CCl_4 into the flow tube, a weak mixture of CCl_4 in He (~1%) is introduced into the flow tube and is continuously flushed through the pulsed valve by addition of a He flush flow. To remove the CCl_4 from the flow as quickly as possible after the pulse valve has closed, a second He flush is added just after the pulse valve. To get as square a pulse of electrons as possible, the flows were optimized while monitoring the electron signal to a positively biased Langmuir probe (collecting electrons [39]) located in the region of the forming recombining ion. The electron pulse can be seen to be quite square [35], but even so, the channels do not count until 100 ms after the pulse valve opens and are terminated 100 ms before it closes. In spite of all this, and with flows that were kept extremely stable with servo valves, the counts had to be passed through a single discriminator with a single gate width and the counts in Counters A and B were only collected after this. Within the same instrument using separate counters and discriminators for the two channels, there was a difference between the counts into Channel A and Channel B even when there was no signal. This problem was due to slight differences in the discriminators and gate widths and was eliminated using the single channel before the counters. Note that such a setup, minimizes the effect of drifts that occur on a time scale greater than that corresponding to the 1 Hz frequency. In addition, random variations on a time scale less than this will be averaged out. Only drifts on time scales corresponding to ~1 Hz will contribute a background. Tests were made by having a constant CH_4 flow into the flow tube with a pulsed flow of CH_4 10^3 times less. For this situation, the counts of CH_4^+ (a monitor of CH_4), in the A and B channels per period seemed the same, and the signal difference S (= A – B) spanned zero counts. However, when the S signal was integrated, the S built up linearly with time exactly at the rate expected for a pulsed flow of 0.1%, generating a substantial count built up in about 1 min.

3. Recombining systems studied

3.1. Electron-ion recombination of CH_5^+

With this technique the first product distribution determined was for CH_5^+/e recombination, with CH_4^+, CH_3^+, CH_2^+ and CH^+ as monitor ions of CH_4, CH_3, CH_2 and CH respectively. Note that these neutrals all fragment to some degree on ionization by 70 eV electrons ($CH_4 \rightarrow CH_4^+$ (44%), CH_3^+ (40%), CH_2^+ (9%), CH^+ (5%) and C^+ (2%); $CH_3 \rightarrow CH_3^+$ (75%), CH_2^+ (18%), and CH^+ (7%);

$CH_2 \rightarrow CH_2^+$ (~71%) and CH^+ (~29%) and CH $\rightarrow$ CH^+ (~79%) and C^+ (~21%)) [32, 35, 40] and this effect has to be accounted for. All the effects that have to be taken into account can be illustrated by considering the recombination:

$$H_3X^+ + e \rightarrow H_2X + H; \quad p_1 \tag{2a}$$
$$\rightarrow HX + H_2/2H; \quad p_2 \tag{2b}$$
$$\rightarrow X + H_2 + H; \quad p_3 \tag{2c}$$

where p_1, p_2, and p_3 represent the fractional product distribution. To detect these products, electron impact ionization is used with overall collision cross-sections of σ_{T2}, σ_{T1} and σ_{T0} for H_2X, HX and X ionization respectively. As mentioned above, electron impact ionization is considered to yield fractional monitor ion distributions of d_{22}, d_{21} and d_{20}, for H_2X ionization/ionization fragmentation to H_2X^+, HX^+ and X^+ respectively; similarly for HX and X fragmentation. It can now be determined what the relative monitor ion signals would be for recombination (2) and by rearrangement what the product distribution would be in terms of the relative monitor ion signals. This is analyzed in detail in a paper describing the technique [35]. The results are:

$$p_1 \propto \frac{[H_2X^+]}{\sigma_{T2} d_{22}} \tag{3a}$$

$$p_2 \propto \frac{1}{\sigma_{T1} d_{11}} \left([HX^+] - [H_2X^+] \frac{d_{21}}{d_{22}} \right) \tag{3b}$$

$$p_3 \propto \frac{1}{\sigma_{T0} d_{00}} \left([X^+] - [H_2X^+] \frac{d_{20}}{d_{22}} - \left\{ [HX^+] - [H_2X^+] \frac{d_{21}}{d_{22}} \right\} \frac{d_{10}}{d_{11}} \right) \tag{3c}$$

where $[H_2X^+]$, $[HX^+]$ and $[X^+]$ are the relative monitor ion signals and the constants of proportionality are all the same. In a similar manner, the neutral product distribution for CH_5^+ + e can be deduced from the monitor ion signals. The build up of the monitor ion signals with time is shown in Fig. 2 where they all increase linearly and all pass through the origin, both as expected. Note that the CH_4^+ and CH_3^+ signals are similar, but after conversion into recombination products CH_3 is small. This is an uncommon occurrence and happens here because of the high degree of fragmentation of CH_4 into CH_3^+ upon ionization. The curves in Fig. 2b are only plotted for illustrative purposes. In practice, only the values obtained over the whole integration period are used since these are the most accurate and contain all of the information in the data obtained at shorter times.

To be confident in the results, it is necessary to confirm that no increase in S occurs when there should be no signal. This was tested in several ways [35]. (i) A constant artificial signal was generated and data were taken. (ii) A situation was created where all of the reactant gases were present with the pulsing of the attaching gas, but with no plasma present. The neutral gases passing into the detection system were ionized and counts taken for all potential products, which are time invariant., (iii) A situation was set up in which CO^+ was recombining with electrons. Here no H_2 was added and the CO^+ was generated by the reaction of Ar^+ with added CO which produces only CO^+ [41]. In this case, plasma ions and electrons were prevented from entering the detection system. Since CO^+ dissociative recombination can only produce C + O and no CO, any signal and thus any build up of CO (as detected using CO^+ as the monitor ion) would be an erroneous detection of the abundant CO present in the plasma, which was not adequately discriminated against. Thus, this is a particularly stringent test. In all of these cases there was no build up of the monitor ions as expected [42]. However in this case (iii), addition of H_2 produces HCO^+ in proton transfer from H_3^+, which recombines to produce H + CO as a 92% channel [31].

Seventh International Conference on Dissociative Recombination (DR2007) IOP Publishing
Journal of Physics: Conference Series **192** (2009) 012004 doi:10.1088/1742-6596/192/1/012004

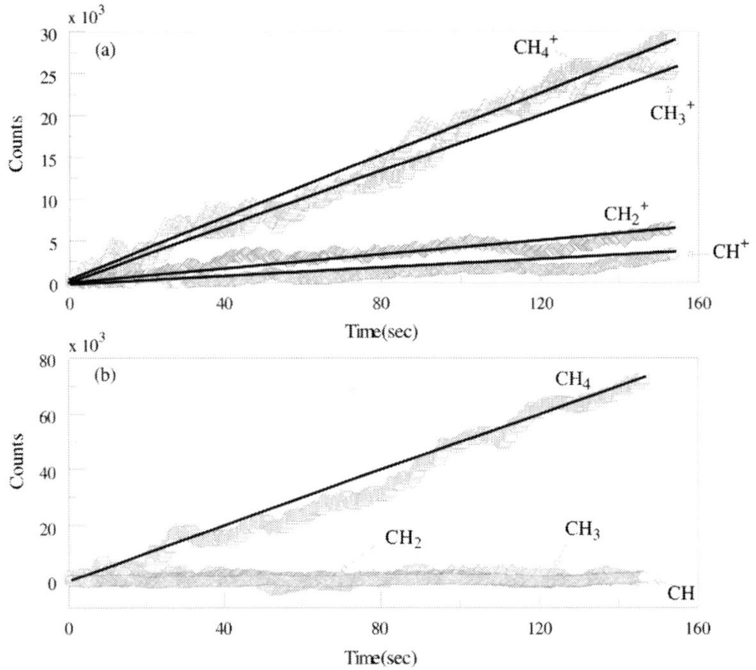

Figure 2. Build up of the products for the e-i recombination of CH_5^+. (a) The integration of the monitor ions as a function of time. (b) Build up of the products of recombination obtained from (a) using equations (3a to c). Note that the signals increase quite linearly showing good statistics.

As soon as the H_2 was added, the CO^+ monitor ion immediately started to accumulate as expected.[42] This very clearly shows that background signals, even much larger than those due to recombination, can be cleanly discriminated against.

Note that when CCl_4 is pulsed in, negative ions are produced and thus some i-i recombination involving Cl^- can occur. This is expected to make only a small contribution since this recombination, especially involving atomic ions, is slow. Such a contribution would yield negative signal counts and when it occurred at the same mass as a monitor ion for e-i recombination products, it would erroneously reduce this observed contribution. There are many energetically possible channels for this i-i recombination of $CH_5^+ + Cl^-$ including: 1) $CH_4 + HCl + 8.9$ eV, 2) $CH_4 + H + Cl + 4.4$ eV, 3) $CH_3 + H_2 + Cl + 4.4$ eV, 4) $CH_3Cl + H_2 + 8.0$ eV, 5) $CH_3 + H + HCl + 4.4$ eV, 6) $CH_3Cl + 2H + 3.5$ eV, 7) $CH_2 + H_2 + HCl + 4.0$ eV, 8) $CH_2Cl + H_2 + H + 3.7$ eV. Note that channels (2 and 3) could occur with the conventional mechanism of long-range electron transfer, but that some such recombinations have been observed which require an intimate interaction [43]. To test for any i-i recombination the monitor ions Cl^+, HCl^+, and CH_4Cl^+ were observed and the signal followed; these are reproduced in Fig. 3. Note that the signals are very small compared with CH_4^+ and CH_3^+, but that the integrated signals do still increase in an approximately linear way within the counting statistics. CH_4Cl^+ is essentially zero, but Cl^+ and HCl^+ (marginally) are slightly negative showing that a little i-i recombination maybe occurring, however it is only a few percent. This small signal is not surprising since the plasma chemistry was optimized to keep this at a minimum. Thus this would not significantly affect the product distribution determined for the CH_5^+ + e recombination. This does however introduce the possibility that product distributions for i-i recombination could be determined using this technique by maximizing this recombination process. Such recombination is increasing in importance with the detection of C_4H^- and C_6H^- in interstellar clouds [44, 45] and the indication that

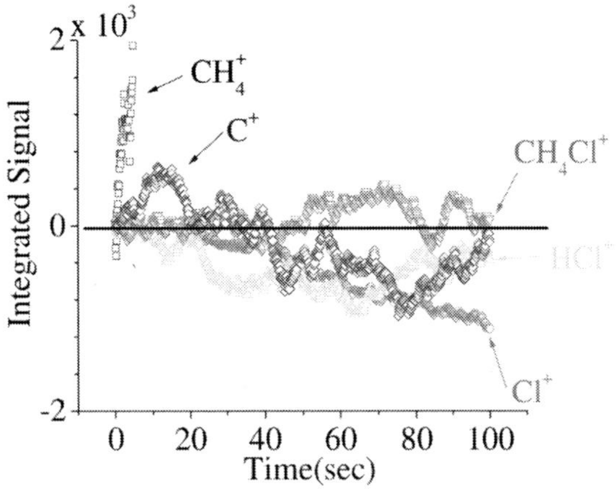

Figure 3. Build up of the monitor ions for the products of e-i and i-i recombination. Note that CH_4^+ increases very rapidly with time because of the much more sensitive scale than in Fig. 2. This is evident from the more pronounced statistical variations than in Fig. 2. Monitor signals, which are negative going, imply that they are products of i-i recombination (Cl^+ and HCl^+). C^+ and CH_4Cl^+ signals are at the limit for this count period of 100 sec showing that they are varying statistically about the zero and are not products.

there are negative ions in the atmosphere of Titan [46]. Note that since CH_4Cl^+ cannot be a product of e-i recombination; its signal represents the magnitude of the statistical errors which are very small indeed. That Cl seems to be a product is not surprising since long-range electron transfer would occur, following conventional thought on i-i recombination.

The DR data of CH_5^+ in Fig. 2 and using equation (3) determines the product distribution in Table 2. This can be compared with the distribution previously obtained by Semaniak et al. in the SR [8]. These distributions are radically different with CH_4 dominating in the FA (95%) and CH_3 dominating in the SR (68%). Sheehan and St-Maurice [47] in a single pass merged beam experiment obtained the same rate coefficient as the SR (2.9 x 10^{-7} relative to 2.8 x 10^{-7} cm^3 s^{-1}) with ions that were vibrationally excited. Based on this, they suggested that the SR ions were also excited. This seems unlikely since the CH_5^+ could relax vibrationally in the 2 sec in the SR before recombination [8], Larsson (priv. com. 2007). However, there is no radiative mechanism for completely rotationally relaxing the ions and such excitation is possible. The effect of rotational excitation can be tested by studying the recombination in the FA at elevated temperatures and such studies are planned. Another possibility that is unlikely is that the ions at mass 17 amu could have been $^{13}CH_4^+$ as well as CH_5^+, but, if this were the case, then a $^{12}CH_4^+$ signal 90 times larger (the isotope ratio of ^{12}C to ^{13}C) would be expected and this was not observed Larsson (priv. com. 2007). A further possibility is that there is some fragmentation of the CH_4 product to CH_3 in the 10^{-6} sec time scale between recombination and product detection in the SR.

Table 2. Previous and present product distributions for the e-i recombination of CH_5^+.

	Present FA	Previous SR[a)] [8]
$CH_5^+ + e \rightarrow CH_4 + H$	95 ± 5%	2 ± 2%
$\rightarrow CH_3 + H_2/2H$	≤ 8% [b)]	68 ± 1 %
$\rightarrow CH_2 + H_2 + H$	≤ 1%	25 ± 4 %
$\rightarrow CH + 2H_2$	≤ 1%	5 ± 4%

a) values at an energy of 37 meV. b) use of the earlier FA data for H-atoms (117.5%) implies that the CH_3 plus 2 H channel is dominant over the CH_3 plus H_2 channel.

In the FA, in addition to the discounted effect of i-i recombination on the product distribution the often radical products can be lost in neutral-neutral reactions or by relative diffusion. Generally the former can be quite slow with the rate coefficients: $CH_3 + CH_3 \rightarrow C_2H_6$; 4.6 x 10^{-11} cm^3 s^{-1} [48], $CH_3 + CH_2 \rightarrow C_2H_4 + H$; 2.1 x 10^{-10} cm^3 s^{-1} [48], $CH + H_2 \rightarrow CH_2 + H$; 2.3 x 10^{-11} cm^3 s^{-1} [49], $CH + CH_4 \rightarrow$ Products; 5.4 x 10^{-11} cm^3 s^{-1} [50]. To account for diffusive loss, the worst-case condition was taken with 100% loss at the walls being assumed for radicals with obviously no loss for CH_4. Modeling of this reactive and diffusive loss would modify a Semaniak et al. [8] distribution for CH_4, CH_3, CH_2 and CH from 2%, 68%, 25% and 5% to 3%, 68.3%, 27.3% and 1.4%, an insignificant change relative to the difference between the FA and SR results.

3.2. *Electron-ion recombination of N_2H^+*

Another recombination that has been included in these initial studies is that of N_2H^+ because of the drastic difference between FA and SR results as mentioned earlier. The results of this study have recently been published [42] and so will only be briefly mentioned here. The same techniques and tests were used as described above for CH_5^+ recombination and confirmed, consistent with the earlier FA study, that the NH + N channel was 0-5% with the N_2 + H channel being very dominant. These data caused a repeat of the SR experiment, which showed that in the original measurement the H_2 in the source had been depleted and the recombination of the ion at 29 amu was for $^{14}N^{15}N^+$ and not N_2H^+. A repeat of the measurement with H_2 and N_2 in the source showed that the NH + N channel is consistent with the FA, having a value of ~10% (Larsson, priv. com. 2007).

4. Summary

Detailed analysis has shown that, with some modification, the FA can be used to obtain accurate product distributions for e-i recombination. The addition of a pulse valve discriminates greatly in favor of recombination neutral products with electron impact ionization to make the products detectable by mass spectrometry. The technique has required highly stable gas flows of high purity gases, with very reproducible pulsing of the attaching gas. The need for accurate, identical and stable pulse amplitude discriminators and counting gate widths is essential for meaningful product distributions to be obtained. Initial studies have established the major product of the N_2H^+ electron recombination as being 95 to 100% N_2 + H, with the NH + N channel being minor (0 to 5%) if present at all. The CH_5^+ recombination has shown that the technique can be used with more polyatomic species, although is does reveal a discrepancy between FA and SR results. The technique can now be used with confidence to study other e-i recombinations. It has also revealed that it is possible to identify the products of i-i recombination, which is becoming increasingly important in several plasmas.

Acknowledgements
Funding from the NSF and NASA is gratefully acknowledged.

References

[1] Adams N G, Poterya V and Babcock L M 2006 Electron Molecular Ion Recombination: Product Excitation and Fragmentation *Mass Spectrom. Revs.* **25** 798-828
[2] Florescu-Mitchell A I and Mitchell J B A 2006 Dissociative Recombination *Phys Rep* **430** 277-374
[3] Herbst E 2005 Molecular Ions in Interstellar Reaction Networks *Journal of Physics: Conference Series* **4** 17-25

[4] Fox J L 2005 Effects of Dissociative Recombination on the Composition of Planetary Atmospheres *Journal of Physics: Conference Series* **4** 32-7

[5] Cravens T E 2003 *Dissociative Recombination of Molecular Ions with Electrons,* ed S L Guberman (New York: Kluwer) pp 385-400

[6] Fox J L 1993 *Dissociative Recombination Theory, Experiment and Applications II,* ed B R Rowe, *et al.* (New York: Plenum) p 219

[7] Viggiano A A and Williams S 2001 Ion-Molecule Kinetics at High Temperatures (300-1800 K): Derivation of Internal Energy Dependencies *Adv. Gas Phase Ion Chem.* **4** 85-136

[8] Semaniak J, Larson A, Le Padellec A, Stromholm C, Larsson M, Rosen S, Peverall R, Danared H, Djuric N, Dunn G H and Datz S 1998 Dissociative Recombination and Excitation of CH_5^+: Absolute Cross-Sections and Branching Fractions *Ap. J.* **498** 886-95

[9] Geppert W, Hamberg M, Thomas R D, Osterdahl F, Hellberg F, Zhaunerchyk V, Ehlerding A, Millar T J, Roberts H, Semaniak J, Af Ugglas M, Kallberg A, Simonsson A, Kaminska M and Larsson M 2006 Dissociative Recombination of Protonated Methanol *Farad. Disc.* **133** 177-90

[10] Vallee F, Gomet J C, Rowe B R, Queffelec J L and Morlais M 1987 *Astrochemistry,* ed M S Varda and S P Tarafdar (Dordrecht: Kluwer) pp 29-30

[11] Rowe B R, Vallee F, Queffelec J L, Gomet J C and Morlais M 1988 The Yield of Oxygen and Hydrogen Atoms through Dissociative Recombination of H_2O^+ with Electrons. *J. Chem. Phys.* **88** 845-50

[12] Vejby-Christensen L, Andersen L H, Heber O, Kella D, Pedersen H B, Schmidt H T and Zajfman D 1997 Complete Branching Ratios for the Dissociative Recombination of H_2O^+, H_3O^+ and CH_3^+. *Ap. J.* **483** 531-40

[13] Jensen M J, Bilodeau R C, Heber O, Pedersen H B, Safvan C P, Urbain X, Zajfman D and Andersen L H 1999 Dissociative Recombination and Excitation of H_2O^+ and HDO^+ *Phys Rev. A* **60** 2970-6

[14] Rosen S, Derkatch A M, Semaniak J, Neau A, Al-Khalili A, Le Padellec A, Vikor W S L, Thomas R, Danared H, af Ugglas M and Larsson M 2000 Recombination of Simple Molecular Ions Studied in Storage Ring: Dissociative Recombination of H_2O^+ *Farad. Discuss.* **115** 295-302

[15] Williams T L, Adams N G, Babcock L M, Herd C R and Geoghegan M 1996 Production and Loss of the Water-Related Species, H_3O^+, H_2O and OH in Dense Interstellar Clouds. *Mon. Not. Roy. Astr. Soc.* **282** 413-20

[16] Andersen L H, Heber O, Kella D, Pedersen H B, Vejby-Christensen L and Zajfman D 1996 Production of Water Molecules from Dissociative Recombination of H_3O^+ with Electrons *Phys. Rev. Letts.* **77** 4891-4

[17] Neau A, Khalili A A, Rosen S, Le Padellec A, Derkatch A M, Vikor W S L, Semaniak J, Thomas R, Nagard M B, Anderson K, Danared H and af Ugglas M 2000 Dissociative Recombination of D_3O^+ and H_3O^+: Absolute Cross Sections and Branching Ratios. *J. Chem. Phys.* **113** 1762-70

[18] Jensen M J, Bilodeau R C, Safvan C P, Seiersen K, Andersen L H, Pedersen H B and Heber O 2000 Dissociative Recombination of H_3O^+, HD_2O^+ and D_3O^+ *Ap. J.* **543** 764-74

[19] Adams N G, Herd C R, Geoghegan M, Smith D, Canosa A, Gomet J C, Rowe B R, Queffelec J L and Morlais M 1991 Laser Induced Fluorescence and Vacuum Ultraviolet Spectroscopic Studies of H-Atom Production in the Dissociative Recombination of Some Protonated Ions *J. Chem. Phys.* **94** 4852-7

[20] Geppert W, Thomas R, Semaniak J, Ehlerding A, Millar T J, Osterdahl F, Ugglas M a, Djuric N, Paal A and Larsson M 2004 Dissociative Recombination of N_2H^+ : Evidence for Fracture of the N-N Bond *Ap. J.* **609** 459

[21] Herd C R, Adams N G and Smith D 1990 OH Production in the Dissociative Recombination of H_3O^+, HCO_2^+ and N_2OH^+: Comparison with Theory and Interstellar Implications *Ap. J.* **349** 388-92

[22] Datz S, Larsson M, Stromholm C, Sundstrom G, Zengin V, Danared H, Kallberg A and af Ugglas M 1995 Dissociative Recombination of H_2D^+: Cross Sections, Branching Fractions, and Isotope Effects *Phys. Rev. A* **52** 2901-9

[23] Krohn S, Amitay Z, Baer A, Zajfman D, Lange M, Knoll L, Levin J, Schwalm D, Wester R and Wolf A 2000 Electron Induced Vibrational Deexcitation of H_2^+ *Phys Rev. A* **62** 032713-8

[24] Amitay Z, Zaijman D, Forck P, Hechtfischer U, Seidel B, Grieser M, Habs D, Repnow R, Schwalm D and Wolf A 1996 Dissociative Recombination of CH^+: Cross-Section and Final States *Phys. Rev. A.* **54** 4032-50

[25] Kokoouline V and Greene C H 2005 Theoretical study of the H_3^+ ion dissociative recombination process *Journal of Physics: Conference Series* **4** 74-82

[26] Kalhori S, Viggiano A A, Arnold S T, Rosen S, Semaniak J, Derkatch A M, af Ugglas M and Larsson M 2002 Dissociative Recombination of $C_2H_3^+$ *Astron. Astrophys.* **391** 1159-65

[27] Mitchell J B A, Rebrion-Rowe C, Le Garrec J L, Angelova G, Bluhme H, Seiersen K and Andersen L H 2003 Branching Ratios for the Dissociative Recombination of Hydrocarbon Ions I: The Cases of $C_4H_9^+$ and $C_4H_5^+$ *Int. J. Mass Spectrom.* **227** 273-9

[28] Mitchell J B A, Angelova G, Rebrion-Rowe C, Novotny O, Le Garrec J L, Bluhme H, Seiersen K, Svendsen A and Andersen L H 2005 Branching ratios for the dissociative recombination of hydrocarbon ions *Journal of Physics: Conference Series* **4** 198-204

[29] Geppert W, Ehlerding A, Hellberg F, Kalhori S, Thomas R, Novotny O, Arnold S T, Miller T M, Viggiano A A and Larsson M 2004 First Observation of Four Body Break-Up in Electron Recombination: $C_2D_5^+$ *Phys. Rev. Lett.* **93** 153201-4

[30] Larsson M, Ehlerding A, Geppert W, Hellberg F, Kalhori S, Osterdahl F, Angelova G, Semaniak J, Novotny O, Arnold S T and Viggiano A A 2005 Rate Constants and Branching Ratios for the Dissociative Recombination of $C_3D_7^+$ and $C_4D_9^+$ *J. Chem. Phys.* **122** 156101-3

[31] Geppert W D, Thomas R D, Ehlerding A, Hellberg F, Osterdahl F, Hamberg M, Semaniak J, Zhaunerchyk V, Kaminska M, Kallberg A, Paal A and Larsson M 2005 *Dissociative Recombination: Theory, Experiments and Applications,* ed W Wolf, *et al.* (London: Inst. Phys. J. Phys. Conf. Ser. 4) pp 26-31

[32] NIST Chemistry WebBook *NIST Standard Database 69* http://webbook.nist.gov/chemistry/ **June 2005**

[33] Poterya V, McLain J L, Adams N G and Babcock L M 2005 Mechanisms of Electron-Ion Recombination of $N_2H^+/(N_2D^+)$ and $HCO^+/(DCO^+)$ Ions: Temperature Dependence and Isotopic Effect *J. Phys. Chem. A* **109** 7181-6

[34] Adams N G and Babcock L M 1994 Optical Emissions from the Dissociative Electron Recombination of N_2H^+ and HCO^+ *J. Phys. Chem.* **98** 4564-9

[35] Molek C D, Poterya V, Adams N G and McLain J L 2009 Development of a Novel Technique for Quantitatively Determining the Products of Electron-Ion Dissociative Recombination *Int. J Mass Spec* **289** 1-11.

[36] Smith D and Adams N G 1984 *Swarms of Ions and Electrons in Gases,* ed W Lindinger, *et al.* (Vienna: Springer-Verlag) pp 284-306

[37] Adams N G and Smith A D 1988 *Techniques for the Study of Ion-Neutral Reactions,* ed. J M Farrar and J W H Saunders (New York: Wiley-Interscience) p 165

[38] Smith D, Dean A G and Adams N G 1974 Space Charge Fields in Afterglow Plasmas. *J.Phys.D.* **7** 1944-62

[39] Swift J D and Schwar M J R 1970 *Electrical Probes for Plasma Diagnostics.* (London: Iliffe)

[40] Tarnovsky V, Levin A, Deutsch H and Becker K 1996 Electron impact ionization of CD_x (x=1-4) *J Phys B-at Mol Opt* **29** 139-52

[41] Anicich V 2003 *An Index of the Literature for Bimolecular Gas Phase Cation-Molecule Reaction Kinetics: JPL Publication 03-19* (Pasadena: Jet Propulsion Laboratory)

[42] Molek C D, McLain J L, Poterya V and Adams N G 2007 A Remeasurement of the Products for Dissociative Electron Recombination of N_2H^+: No Significant NH + N Production *J. Phys. Chem.* **A** 6760-5

[43] Adams N G, Babcock L M and Molek C D 2003 *Encyclopedia of Mass Spectrometry: Theory and Ion Chemistry, Vol.1,* ed P Armentrout (Amsterdam: Elsevier) pp 555-61

[44] Brunken S, Gupta H, Gottlieb C A, McCarthy M C and Thaddeus P 2007 Detection of the carbon chain negative ion C8H- in TMC-1 *Ap. J.* **664** L43-L6

[45] Gupta H, Brunken S, Tamassia F, Gottlieb C A, McCarthy M C and Thaddeus P 2007 Rotational spectra of the carbon chain negative ions C4H- and C8H *Ap. J.* **655** L57-L60

[46] Waite J H, Jr., Young D T, Cravens T E, Coates A J, Crary F J, Magee B and Westlake J 2007 The process of tholin formation in Titan's upper atmosphere *Science FIELD Full Journal Title:Science (New York, N.Y.)* **316** 870-5

[47] Sheehan C H and St.-Maurice J-P 2004 Dissociative recombination of the methane family ions: rate coefficients and implications *Adv. Space Res.* **33** 216-20

[48] Wang B and Fockenberg C 2001 Direct Measurement of the Rate Constant for the $CH_2(X^3B_1)$ + CH_3 Reaction at 300K *J. Phys. Chem. A* **105** 8449-55

[49] Becker K H, Kurtenback R and Wiesen P 1991 Temperature and Pressure-Dependence of the Reaction CH + H_2 *J. Phys. Chem.* **95** 2390-4

[50] Anderson S M, Freeman A and Kolb C E 1987 Fast-Flow Studies of CH Radical Kinetics at 290K *J. Phys. Chem.* **91** 6272-7

Multicollision character of recombination of H_3^+ ions in afterglow plasma

J Glosik, R Plasil, I Korolov, O Novotny, T Kotrik

Charles University, Faculty of Mathematics and Physics, Department of Surface and Plasma Science, Prague, Czech Republic

E-mail: juraj.glosik@mff.cuni.cz

Abstract. Reported are studies of recombination of H_3^+ ions with electrons in flowing afterglow (FALP) experiment. Reported is observation of dependence of overall recombination process on He pressure and hydrogen partial pressure. The obtained data are in good agreement with previous stationary afterglow (AISA) results and clearly indicate the multicollision and multistep character of the recombination process. Proposed is scheme of overall recombination process.

1. Introduction

Recombination of H_3^+ ions with electrons is key process playing role in evolution of hydrogen containing laboratory and natural plasmas. Because of its importance in plasma physics, technology, astrophysics and theory, it is attracting broad attention for nearly 50 years. Over these years recombination of H_3^+ has become the "fundamental problem". The results of H_3^+ recombination studies were for a long time very controversial [1,2,3]. The problems were also with interpretation of astronomical observations of H_3^+ [4,5]. The "enigma" of H_3^+ has been solved only very recently, with the inclusion of the non-Born-Oppenheimer Jahn-Teller coupling, which gave high rate of dissociative recombination of this ion [6,7]. The theoretical predictions were confirmed by very recent storage ring experiments. These experiments have used new type of ion sources, where ions prior to injection into the ring were rotationally and vibrationally cooled [8,9]. In contradiction with mentioned theory was just the measured increase [9,10] of the low energy recombination rate coefficient when the ion beam was enriched by para H_3^+ (the component with total nuclear spin ½). This was solved in the recently published theoretical work with higher accuracy of calculations [11]. Now the qualitative agreement of theory and storage ring experiments has been achieved. The theory now predicts that at temperatures below 100 K the recombination of para H_3^+ will be at low energies substantially faster than the recombination of ortho H_3^+.

The theory is in agreement with the storage ring experiments, but there are still some unsolved problems with recombination of H_3^+ ions. The main problem is that up to now there is no reliable interpretation of many plasmatic experiments which give very different values of recombination rate coefficient (see compilation in refs. [1,2,12]). In recent years we have studied in our laboratory the recombination of H_3^+ ions using stationary afterglow experiment AISA and we have observed the dependence of the recombination rate coefficients on partial pressure of hydrogen [12,13,14,15,16,17]. AISA was build to study slow recombination processes, nevertheless it has the main disadvantage of stationary afterglow experiments: the mixture of gases, here He, Ar and H_2, are exposed to discharge and long-lived excited species and radicals can be produced. These particles can influence the decay of the plasma and the apparent recombination rate coefficient determined from the decay of electron number density. We were trying to eliminate such uncertainties in our experiments. To identify recombining ions we build second stationary afterglow experiment, Test Discharge Tube (TDT). In TDT the density of recombining ions ($H_3^+(v=0)$) was measured by NIR cavity ring-down absorption spectroscopy (CRDS) [18,19]. The TDT-CRDS studies were carried out only at relatively high hydrogen densities. The dependence on partial pressure of hydrogen was not measured. In the covered range of hydrogen partial pressures there is good agreement between TDT-CRDS and AISA results.

The dependence of recombination rate coefficient on partial pressure of H_2 observed in the AISA is clear proof that the recombination process is not pure binary dissociative recombination but it is a multi-collision process. Nevertheless the data from AISA and TDT-CRDS did not give enough information for the characterisation of this multicollisional process; the results were left without interpretation. We also didn't any have theoretical prediction of a lifetime of a neutral H_3* formed in collision of H_3^+ with electron. Nevertheless it was obvious that the recombination rate coefficient measured in AISA is an effective recombination rate coefficient and not the rate coefficient corresponding to the binary dissociative recombination (DR), which is measured in colliding beams experiments and which is calculated by theory of DR. In another words the recombination of H_3^+ in plasma differs from DR. Here we have in mind afterglow plasma in He buffer used in recombination studies.

The common feature of many experiments with H_3^+ dominated plasma was, that the plot of the reciprocal electron density ($1/n_e$) did not show linear dependence on decay time as it follows from the solution of the balance equation with the assumption of binary recombination process, typical e.g. for decay of O_2^+ dominated plasma [12]. In several experiments "evolution" of the recombination rate coefficient during an afterglow was observed [2,20]. First obvious explanation is vibrational excitation of recombining ions and relaxation during the afterglow [2]. On many occasions three-body character of the recombination process was discussed, but there was lack of clear experimental evidences. The problem of complex formation in electron-ion recombination of molecular ions including H_3^+ was discussed in paper of R. Johnsen and J.B.A. Mitchell [21]. In their paper many experimental evidences for formation of long-lived H_3* were discussed, nevertheless no quantitative results were given.

As was already mentioned the problem of stationary afterglow technique is, that in the process of plasma formation H atoms and some excited particles can be formed and they can later influence the process of plasma decay and hence the determination of apparent recombination rate coefficients. Even spectroscopic identification of recombining ions cannot exclude this "fingerprints of formation". The flowing afterglow technique (Flowing Afterglow with Langmuir Probe - FALP) can solve the problem of formation by ignition of discharge in pure helium and formation of H_3^+ ions by sequential introduction of gases to already cold afterglow plasma. `

2. Experiments

To create in the FALP experiment conditions at which the dependence on $[H_2]$ was observed in AISA we have changed construction of flow tube, see Fig. 1. In FALP fast flow of He buffer gas is carrying decaying plasma along the flow tube. To obtain a long decay time of the afterglow plasma the used pressure of buffer gas is 1000-2000 Pa. With the new FALP [22] recombination rate coefficients as small as 5×10^{-9} cm^{3}s^{-1} can be measured. To obtain very low partial pressures of hydrogen (densities down to 10^{11} cm^{-3}) the FALP and the gas handling systems were constructed with UHV technology. The plasma in the FALP is generated upstream by a microwave discharge in the flow of helium. Downstream from the discharge (~3 ms in time scale) Ar is added and in sequence of ion-molecule reactions Ar^+ dominated plasma is formed. Further downstream (~35 ms from the discharge) hydrogen is added and H_3^+ dominated plasma is formed. The measured velocity of the helium flow is used for conversion of the probe position to the decay time. In figures we arbitrarily use $t = 0$ at hydrogen inlet, upstream from this port $t < 0$ and downstream $t > 0$. The kinetics of formation and plasma decay is clear and can be calculated [16,23]. The time necessary for the transition from Ar^+ dominated plasma to H_3^+ dominated plasma depends on hydrogen partial density, [H_2]. The advantage of FALP over the AISA is that measurements are in continuous regime and hydrogen is added to already relaxed cold plasma with electron temperature (T_e < 300 K) given by the buffer gas temperature.

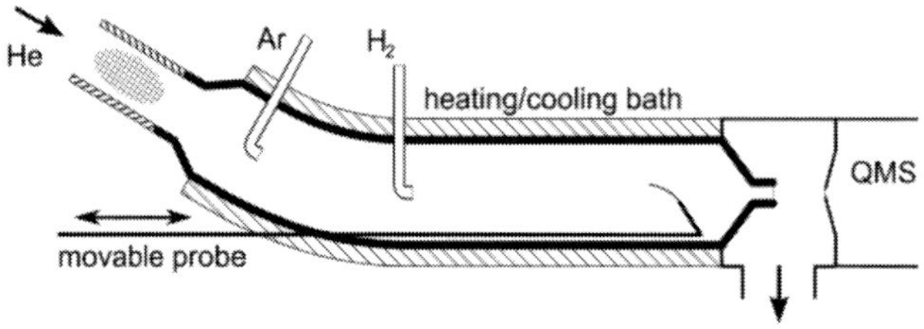

Figure 1. The FALP with long decay of afterglow plasma (not in scale). Indicated are the He input and the position of Ar and H_2 ports. The Langmuir probe is movable from the hydrogen port up to the end of the flow tube. The decay time in the range 0-60 ms is given by the position of the probe.

Fig. 2. shows the calculated plasma evolution along the flow tube. For the calculations we have deliberately used relatively high density of hydrogen ([H_2] = 5×10^{12} cm^{-3}). Below we will discuss the results of measurements at this particular density. Even if hydrogen density will be lower by factor of 20 (the formation will be longer by factor of 20) the plasma decay for t > 10 ms will be governed by the recombination of H_3^+ ions. This is because Ar^+ and ArH^+ recombine very slowly and H_2^+ recombines slowly in comparison with H_3^+ [24]. For study of the recombination of H_3^+ in the mixture with these ions it is enough if the relative density of H_3^+ is higher than 90 %.

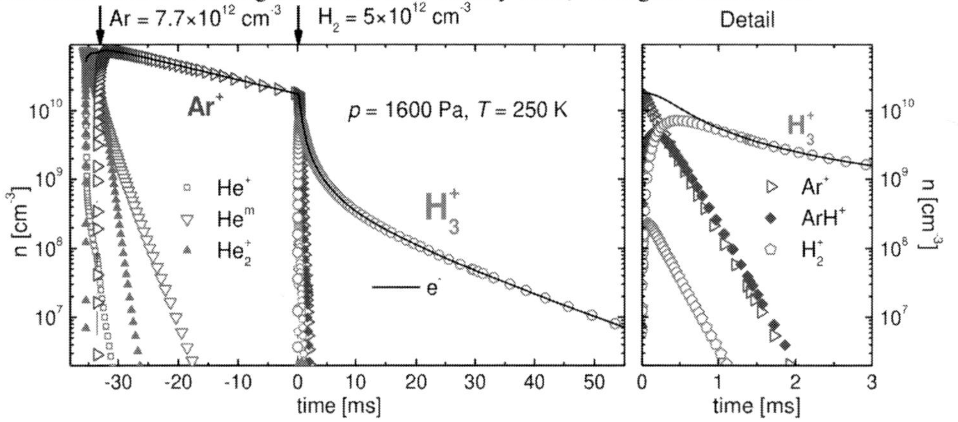

Figure 2. The calculated plasma formation and decay along the flow tube after the addition of Ar and H_2. In the right panel the detail of transition from Ar^+ dominated to H_3^+ dominated plasma after addition of hydrogen is shown.

3. Results

The examples of measured decreases of the electron density along the flow tube ("decay curves") are plotted in the left panel of Fig. 3. The dependence of the decay rate on the hydrogen partial pressure is evident. The apparent recombination rate coefficients (α_{eff}) were calculated from the electron density decays. In the data analyses the formation of H_3^+ dominated plasma was considered (see details in ref. [25]). For the comparison the decay curves obtained in AISA experiment are also shown in the right panel of Fig. 3. Without going into further details of the measurements we will discuss the results obtained in both FALP and AISA.

3.1. Dependence on the hydrogen partial pressure

Dependence of α_{eff} on the hydrogen partial density measured in the FALP experiment is plotted in Fig. 4. The dependence is similar to one obtained in AISA experiment.

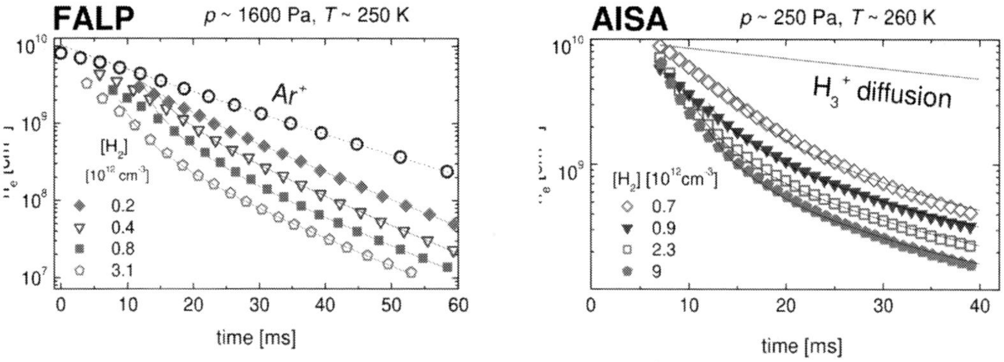

Figure 3. The measured decay of the electron density in H_3^+ dominated plasma. **Left panel**: the FALP experiment. **Right panel**: the AISA experiment.

The value of measured α_{eff} is increasing with $[H_2]$ (up to $[H_2] \sim 10^{12}$ cm^{-3}) and for higher densities of hydrogen (up to $[H_2] \sim 10^{14}$ cm^{-3}) the value of α_{eff} is constant. It is clear that the deionisation mechanism cannot be a pure binary dissociative recombination process. Because the conditions in the FALP are very well defined we can conclude, that the process of the recombination in the H_3^+ dominated low temperature plasma in He/Ar/H_2 mixture is a multicolisional process in which hydrogen molecules are involved. For even higher hydrogen densities, $[H_2] > 10^{14}$ cm^{-3}, we have observed further increase of α_{eff}, but this is given by the formation of H_5^+ ions and their fast recombination. Using the FALP technique we have already studied the influence of H_5^+ formation on overall recombination process, for details see ref. [26]. We will see below that it is substantial to divide the measured dependence to the "low hydrogen density region" with $[H_2] \leq 10^{12}$ cm^{-3} and to the "saturated region" with 2×10^{12} cm^{-3} $\leq [H_2] \leq 1\times10^{14}$ cm^{-3}. In the low H_2 density region α_{eff} is increasing with $[H_2]$, in saturated region α_{eff} is constant independent on $[H_2]$.

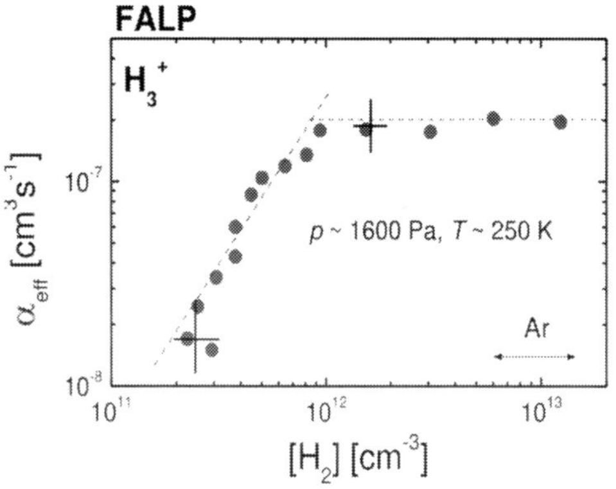

Figure 4. The dependence of the measured effective recombination rate coefficient in H_3^+ dominated plasma on hydrogen density.

3.2. Dependence on the temperature

Three body recombination processes, e.g. colisionall radiative recombination (CRR) [27], have usually strong temperature dependence, $\alpha_{CRR} \sim T_e^{-4.5}$. Having indication of multicollisional character of the recombination process in low temperature plasma we measured the dependence of α_{eff} on the buffer gas temperature. In this study we assume that decaying afterglow plasma is in thermodynamic equilibrium (or very close to equilibrium) with temperature given by buffer gas temperature (also discussion in accompanying paper by Plasil et al. [23]). The measurements of temperature dependence were carried out using the AISA [17]. When He density was kept nearly constant, the measured value of α_{eff} was decreasing with increasing buffer gas temperature. The observed dependence was very small, close to $\alpha_{eff} \sim T_e^{-0.5}$. An exponential factor was not 4.5 as it is expected for three body CRR recombination.

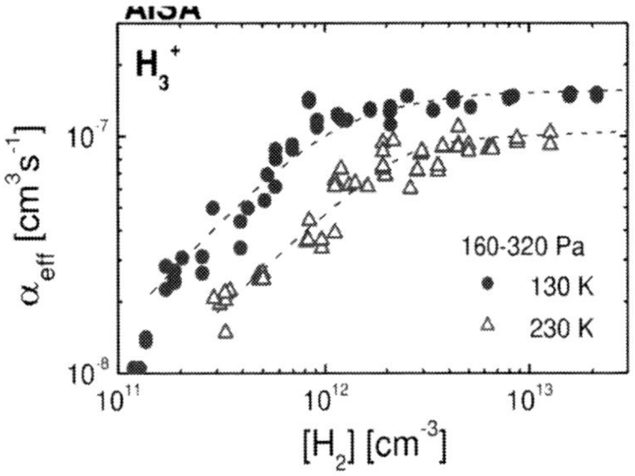

Figure 5. The temperature dependence of the effective recombination rate coefficient measured in AISA experiment.

The form of the dependence on hydrogen density did not change with the temperature of the buffer gas. The dependencies of α_{eff} on hydrogen density measured at two different temperatures are plotted in Fig. 5. Note that the position of the transition to the saturated region is shifted at higher temperature towards higher [H$_2$].

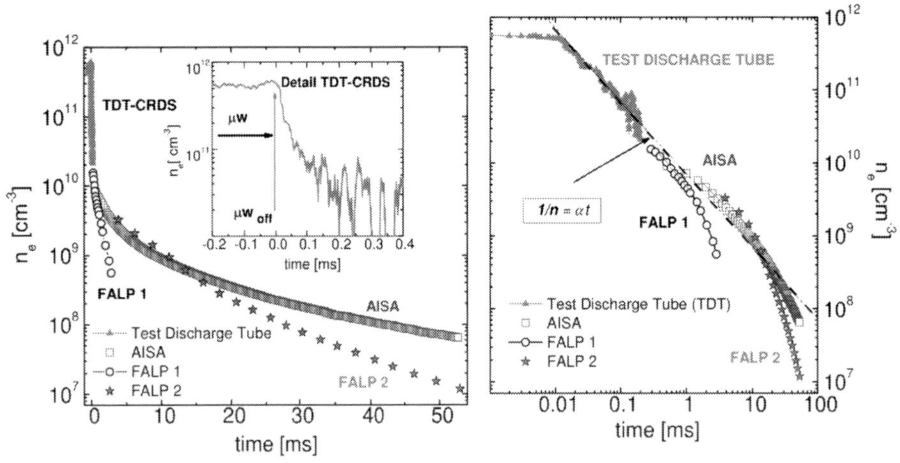

Figure 6. Left panel: The decay of H_3^+ dominated recombination-governed plasma. The used experimental techniques are indicated. In the insert the decay curve measured in TDT-CRDS is plotted. **Right panel**: log-log plot of the measured decay curves. The deviations from linearity at the end of the decays are due to diffusion losses. The diffusion losses are considered in the data analyse.

3.3. Dependence on the electron density

In our studies of H_3^+ recombination we have used three different experiments in several different configurations. In these experiments the electron density was varying over four orders of magnitude, from $n_e = 10^7$ up to $n_e = 5\times10^{11}$ cm^{-3}. When comparing obtained α_{eff} at fixed hydrogen density ([H$_2$] ~ 10^{13} cm^{-3}) we cannot see substantial dependence on n_e. In the left panel of Fig. 6 the decay curves obtained in the different experiments are plotted. The differences between the decay curves are given by the differences in rates of diffusion losses, which are given by differences in pressures (from 200 to 1600 Pa) and diffusion lengths. If the plasma decay is governed by the recombination, then for electron density can be written: $1/n_e = 1/n_{e0} + \alpha_{eff}\, t$. In log-log scale for large t, where $1/n_{e0}$ can be neglected, we should obtain linear dependence: $\ln(n_e) \sim \{-\ln(t) - \ln(\alpha_{eff})\}$. The values $\ln(\alpha_{eff})$ give the vertical shift. In the right panel of Fig. 6 there are such plots of the measured decay curves (at hydrogen density $[H_2] \sim 10^{13}$ cm^{-3}). As was already stated, we cannot see substantial dependence of α_{eff} on electron density.

3.4. Dependence on the He density

The experimental conditions in the AISA and in the FALP differ in used pressures of the buffer gas. The obtained dependencies on hydrogen density measured at different pressures have similar shape but they differ in absolute value of the rate coefficients. The recombination rate coefficients measured in different experiments are in the Fig. 7 plotted versus helium buffer number density [He]. The values from "saturated region" are used in the plot, i.e. measured for 2×10^{12} cm$^{-3} \leq [H_2] \leq 1\times10^{14}$ cm^{-3}. The data obtained in previous studies using He/Ar/H_2 afterglow are also included in the plot [28,29]. The binary recombination rate coefficients obtained in the storage ring experiments with the cold ion sources [8,9,10] and the value obtained in the theoretical calculation [11] is also indicated in the Fig. 7. The linear dependence on the helium density is obvious. The dependence can be expressed in the form: $\alpha_{eff} = \alpha_{eff\,0} + K_{He}\cdot[He]$, where K_{He} is constant given by the slope of the plot. The value obtained from the plot is $K_{He}(260\text{ K}) = 2.8\times10^{-25}$ cm^{6}s^{-1}. The extrapolation for [He] → 0 gives the value $\alpha_{eff\,0}(260\text{ K}) = 7.5\times10^{-8}$ cm^{3}s^{-1}. The proposed mechanism of the multicollision recombination process is described in ref. [30] and will only shortly be described here.

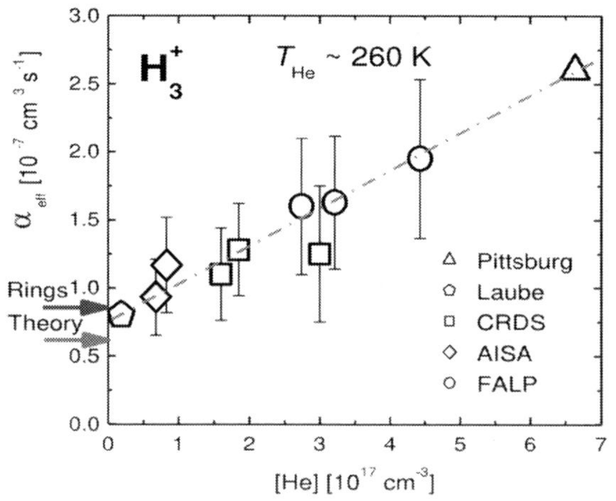

Figure 7. The effective recombination rate coefficients measured in He/Ar/H_2 afterglow experiments. The corresponding references are: Pittsburg - [28], Laube - [29]. The line is connecting just the points indicating data obtained by Pittsburg group and by Laube et al.

Calculations show [30] that at thermal energies in a collision of H_3^+ with an electron highly excited neutral Rydberg molecule with lifetime 10-100 ps is formed. The collisions of these molecules with present He atoms are influencing the process of recombination in the plasma. These collisions can form long living highly excited molecular H_3*, which can autoionise or dissociate depending on the hydrogen partial density. If the density of H_2 is small H_3* autoionise and H_3^+ are formed, the electron density then decays slowly. If the density of H_2 is large (>2×10^{12} cm^{-3}) H_3^* dissociate in collisions with H_2 and apparent recombination get faster. Assuming this process the linear dependence of α_{eff} on [He] can be obtained (for more details see [30]). Further studies on elucidation of role of H_2 are in progress.

4. Conclusions

We studied recombination of H_3^+ ions in low temperature He/Ar/H_2 afterglow plasma in three different experiments, AISA, TDT CRDS and FALP. In all three experiments He was used as a buffer gas. From the measured recombination rate coefficient we obtained dependence on hydrogen partial pressure and on helium pressure. This clearly indicates, that the recombination of H_3^+ in afterglow plasma in He buffer is a multicollisional multistep process.

Acknowledgements: This work is a part of the research plan MSM 0021620834 financed by the Ministry of Education of the Czech Republic and partly was supported by GACR (202/05/P095, 205/05/0390, 202/07/0495) by GAUK 53607 and GAUK 124707.

References

[1] Johnsen R 2005 *J. Phys.: Conf. Series* **4** 83
[2] Smith D and Spanel P 1993 *Int. J. Mass Spectrom. Ion Processes* **129** 163
[3] Oka T 2003 *Dissociative Recombination of Molecular Ions with Electrons* ed S L Guberman (New York: Kluwer, Plenum) pp 209–220
[4] McCall B J, Huneycutt A J, Saykally R J, Geballe T R, Djuric N, Dunn G H, Semaniak J, Novotny O, Al-Khalili A, Ehlerding A, Hellberg F, Kalhori S, Neau A, Thomas R, Osterdahl F and Larsson M 2003 *Nature* **422** 500
[5] Oka T 2000 *Phil. Trans. R. Soc. Lond.* **A358** 2363
[6] Kokoouline V, Greene C H, Esry B D 2001 *Nature* **412** 891–894
[7] Kokoouline V and Greene C H 2003 *Phys. Rev.* A **68** 012703
[8] McCall B J, Huneycutt A J, Saykally R J, Djuric N, Dunn G H, Semaniak J, Novotny O, Al-Khalili A, Ehlerding A, Hellberg F, Kalhori S, Neau A, Thomas R D, Paal A, Österdahl F and Larsson M 2004 *Phys. Rev.* A **70** 052716
[9] Kreckel H, Motsch M, Mikosch J, Glosik J, Plasil R, Altevogt S, Andrianarijaona V, Buhr H, Hoffmann J, Lammich L, Lestinsky M, Nevo Y, Novotny S, Orlov D A, Pedersen H B, Sprenger F, Toker J, Wester R, Gerlich D, Schwalm D, Wolf A and Zajfman D 2005 *Phys. Rev. Lett.* **95** 263201
[10] Wolf A, Kreckel H, Lammich L, Strasser D, Mikosch J, Glosik J, Plasil R, Altevogt S, Andrianarijaona V, Buhr H, Hoffmann J, Lestinsky M, Nevo I, Novotny S, Orlov D A, Pedersen H B, Terekhov A S, Toker J, Wester R, Gerlich D, Schwalm D and Zajfman D 2006 *Phil. Trans. R. Soc. Lond.* **A364** 2981
[11] Fonseca dos Santos S, Kokoouline V and Greene C H 2007 *J. Chem. Phys.* **127** 124309
[12] Plasil R, Glosik J, Poterya V, Kudrna P, Rusz J, Tichy M and Pysanenko A 2002 *Int. J. Mass Spectrom.* **218** 105–130
[13] Glosik J, Plasil R, Poterya V, Kurdna P and Tichy M 2000 *Chem. Phys. Lett.* **331** 209
[14] Glosík J, Plašil R, Poterya V, Kudrna P, Tichý M, Pysanenko A 2001 *J. Phys.* B: *At. Mol. Opt. Phys.* **34** L485
[15] Poterya V, Glosík J, Plašil R, Tichý M, Kudrna P, Pysanenko A 2002 *Phys. Rev. Lett.* **88** 044802
[16] Plašil R, Glosík J, Poterya V, Kudrna P, Vicher M, Pysanenko A 2003 *Dissociative Recombination of Molecular Ions with Electrons* ed S L Guberman (New York: Kluwer, Plenum) pp 249–263
[17] Glosik J, Plašil R, Pysanenko A, Novotny O, Hlavenka P, Macko P, Bánó G 2005 *J. Phys.: Conf. Series* **4** 104
[18] Macko P, Bánó G, Hlavenka P, Plašil R, Poterya V, Pysanenko A, Votava O, Johnsen R and Glosík J 2004 *Int. J. Mass Spectr.* **233** 299–304
[19] Macko P, Bánó G, Hlavenka P, Plašil R, Poterya V, Pysanenko A, Dryahina K, Votava O, Glosík J 2004 *Acta Physica Slovaca* **54** 263–271
[20] Gougousi T, Johnsen R, Golde M F 1995 *Int. J. Mass Spectrom. Ion Processes* **149/150** 131

[21] Johnsen R and Mitchell J B A 1998 *Advances in Gas-Phase Ion Chemistry* **3** (JAI Press Inc.)
[22] Novotny O, Plasil R, Pysanenko A, Korolov I, Glosik J 2006 *J. Phys.* B: *At. Mol. Opt. Phys.* **39** 2561
[23] Plasil R, Korolov I, Kotrik T, Varju J, Dohnal P, Donko Z, Bano G, Glosik J 2008 Measurements of EEDF in recombination dominated afterglow plasma *J.Phys.: Conf. Series* in this volume
[24] Florescu-Mitchell A I, Mitchell J B A 2006 *Physics Reports* **430** 277–374
[25] Korolov I, Novotný O, Kotrík T, Plašil R, Varju J, Hejduk M, Glosík J 2008 *Contrib. Plasma Phys.* **48** 521-526
[26] Glosík J, Novotný O, Pysanenko A, Zakouřil P, Plašil R, Kudrna P, Poterya V 2003 *Plasma Source Sci. Technol.* **12(4)** S117
[27] Flannery M R 1990 *Molecular Processes in Space* (New York: Plenum)
[28] Leu M T, Biondy M A and Johnsen R 1973 *Phys. Rev.* A **8** 413–419
[29] Laube S, Le Padellec A, Sidko O, Rebrion-Rowe C, Mitchell J B A and Rowe B R 1998 *J. Phys.* B: *At. Mol. Opt. Phys.* **31** 2111–2128
[30] Glosik J, Plasil R, Korolov I, Kotrik T, Novotny O, Hlavenka P, Varju J, Dohnal P, Kokoouline V and Greene C H 2009 Phys. Rev. A **79** 052707

Resonances in dissociative recombination: trends and patterns

A. E. Orel[1], V. Ngassam[1], J. B. Roos[2], J. Royal[1] and Å. Larson[2]

[1] Department of Applied Science, University of California, Davis, USA
[2] Department of Theoretical Chemistry, Royal Institute of Technology, Stockholm, Sweden

E-mail: aeorel@ucdavis.edu

Abstract. In dissociative recombination, the kinetic energy of the incident electron is transferred into excitation of the electrons of the target ion and then into kinetic energy of the fragments. In general, this proceeds via a resonance where the electron is temporarily trapped by the ion, leading to efficient energy transfer. The study of dissociative recombination is the study of these resonances, Rydberg states converging to the ground and excited states of the ion. For a number of systems, we have studied the electronic states involved in dissociative recombination, including the ground and excited states of the ion, the resonant states and the bound Rydberg states of the system, by combining electron scattering calculations with multi-reference configuration interaction quantum chemistry calculations. We will report on trends and patterns in these resonance states. We will discuss studies of dissociative recombination of the rare-gas ions, moving down the periodic table from He_2^+ to Ne_2^+ to Ar_2^+, where the ground electronic state of the ion is constant, but its polarizability increases. We will also present results on isoelectronic polyatomic systems, such as HCO^+ and $HCNH^+$, as well as the effects of changing the electronic structure slightly such as HCN^+/HNC^+ and H_2CO^+.

1. Introduction

Electronic energy must be transferred into the kinetic energy of the nuclei in order for dissociative recombination to occur. The mass of the electron is more than 1800 times lighter than the mass of the hydrogen atom, and this mass ratio is even larger when heavier diatomics or polyatomics are considered. Therefore, the time scale for the electron collision is very different than the time scale for molecular motion, such as vibration. So in the case of non-resonant electron collisions with molecular ions, there would be inefficient transfer of energy. However, in the case of resonant collisions, the electron is temporarily attached to the molecular ion, changing the forces that the atoms in the molecule feel. This leads to an electron collision time which is commensurate with molecular vibration, allowing the electron collision to drive dissociation and vibration. Therefore the study of dissociative recombination is the study of the resonant states of the system.

Unlike the case of dissociative attachment, where there are only a handful of anion states, often only one that are involved in the process, dissociative recombination is dominated by series of Rydberg states. These resonances can include repulsive states, usually a Rydberg series converging to the first excited state of the ion. These are, in many cases, the resonances which are responsible for direct dissociative recombination. There is also an infinite series of Rydberg states converging to the ground state of the molecular ion. This leads to structure in the dissociative recombination cross section as a function of initial electron energy. These structures are the signature of the indirect recombination process. In general, they can be assigned to excited rovibrational levels of bound Rydberg states with a ground-state ion core, which couple to the initial ground-state electronic continuum by non-adiabatic interactions only. In addition, these Rydberg states play a critical role in the non-crossing mechanism of dissociative recombination, seen in systems such as HeH^+[1], H_3^+ [2, 3] and LiH^+[4]. In addition, there exists another class of resonances. These are members of a Rydberg series converging to higher excited states of the ion, generally referred to as core-excited states. These states are bound with respect to dissociation, if the ion to which the series converges is bound. They are coupled both to the electronic initial continuum and to the dissociative resonance states and tend to produce broader and more prominent structures in the dissociative recombination cross section[5, 6]. Since the energy dependence of the Rydberg states mirror that of the state of the ion to which they converge, the shape and postion of the excited states of the ion can yield information about resonant states.

Our process for studying dissociative recombination divides the calculation into two parts. First we use electron scattering at fixed nuclear geometries to calculate the position and lifetime of the Feshbach resonances by fitting the eigenphase sum to a Breit-Wigner form. We then use standard quantum chemistry techniques at fixed nuclear geometries to calculate the potential energy curves for the bound Rydberg states and the resonance states (after they have crossed the ion ground state so they are electronically bound) and finally to diabatize the resonances. These calculations yield potential energy curves, quantum defects, and the couplings between the states, and between the states and the continuum (autoionization widths). Using this information as input we then calculate the quantum molecular dynamics leading to dissociative recombination. These calculations are carried out using Multi-Channel Quantum Defect Theory (MQDT)[7], wave packet methods[8], and for polyatomics, the Multi-Configuration Time Dependent Hartree (MCTDH) method of Meyer *et. al.*[9]

In the following sections, we will briefly describe a series of examples of such calculations and

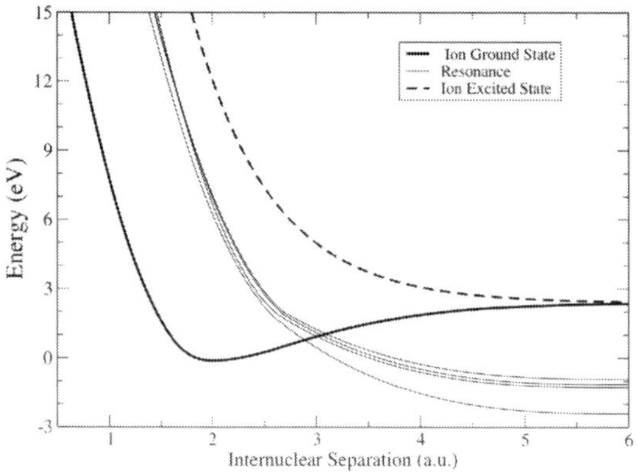

Figure 1. Potential energy curves for He_2^+ ground and low-lying excited states and lower He_2 resonance states. Ground state of ion heavy solid line ——, excited states of ion dashed line - - - -, resonance states light solid line ——.

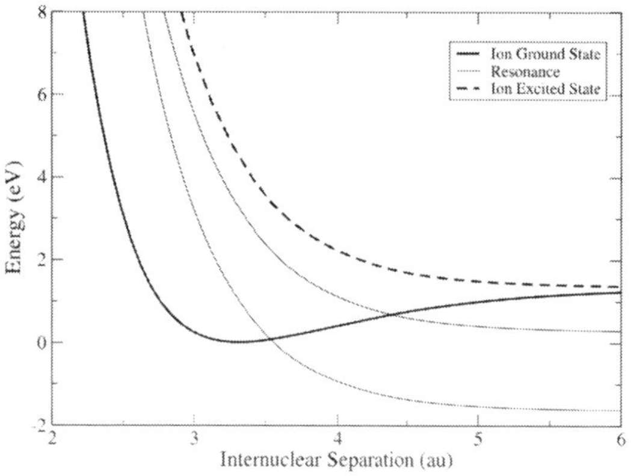

Figure 2. Potential energy curves for Ne_2^+ ground and low-lying excited states and lower Ne_2 resonance states. Ground state of ion heavy solid line ——, excited states of ion dashed line - - - -, resonance states light solid line ——.

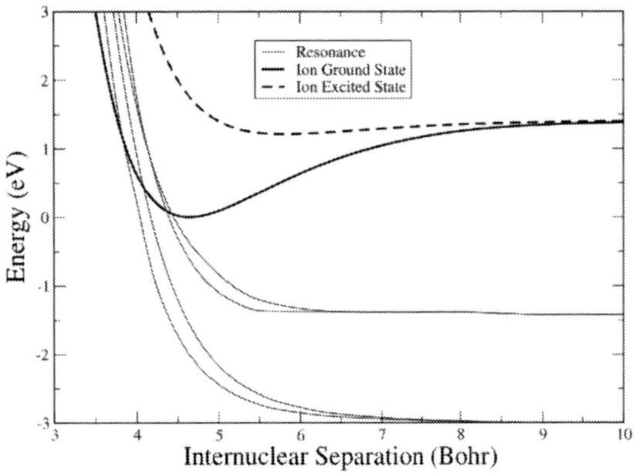

Figure 3. Potential energy curves for Ar_2^+ ground and low-lying excited states and lower Ar_2 resonance states. Ground state of ion heavy solid line ——, excited states of ion dashed line - - - -, resonance states light solid line ——.

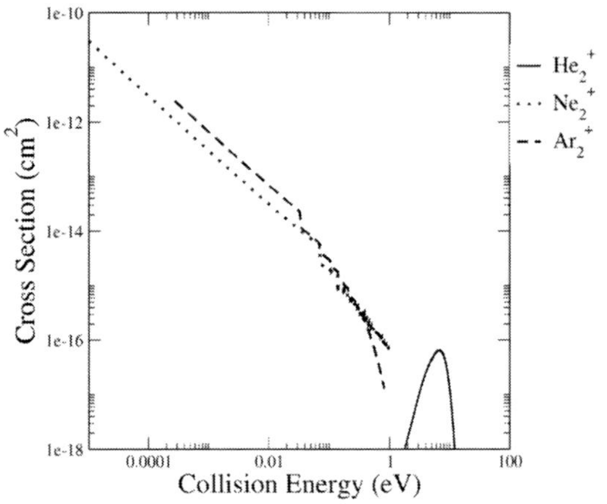

Figure 4. Cross section for dissociative recombination. Solid line —— He_2^+, dashed line - - - -Ne_2^+, and dotted line $\cdots\cdots$$Ar_2^+$.

In our first example, we consider dissociative recombination of electrons with the rare gas ions, He_2^+, Ne_2^+ and Ar_2^+. These systems have been previously studied[10, 11, 12], the results are summarized here. The ground state, for these ions is:

$$[core](n\sigma_g)^2(n\sigma_u)$$

For He_2^+, there is no core, for Ne_2^+, there are 16 electrons in the core, and and Ar_2^+ has 32 core electrons. The lowest excited ion state is:

$$[core](n\sigma_g)(n\sigma_u)^2$$

for He_2^+ and Ne_2^+. In Ar_2^+, this is one of the low-lying excited states, but the lowest excited state is:

$$[core](n\pi_g)^3(n\sigma_u)^2$$

Moving from He_2^+, Ne_2^+ to Ar_2^+, the polarizability of the molecule increases and the excited states drop in energy. This causes the resonance states to lower in energy also. This can be seen in Figures 1-3, where the ground state and first excited state of each ion and the low lying resonant states are plotted. Each curve is shifted such that the zero of energy for each system is the energy of the ground state of the ion at its equilibrium separation. Note that the first resonance of He_2^+ crosses at much larger internuclear distance, outside of the Frank-Condon region, while in the case of Ne_2^+, the state crosses just to the right of the equilibrium position, but in the Frank-Condon region. In the case of Ar_2^+, the curve drops further in energy, and now crosses at smaller distances. This effects the dissociative recombination cross section dramatically as can be seen in Figure 4, where the three systems are compared. He_2^+ has almost zero cross section at low energy, while Ne_2^+ and Ar_2^+ have a significant value even at the lowest energy studied. The cross section at the lowest energy studied for Ne_2^+ is higher than that of Ar_2^+ due to the more favorable crossing.

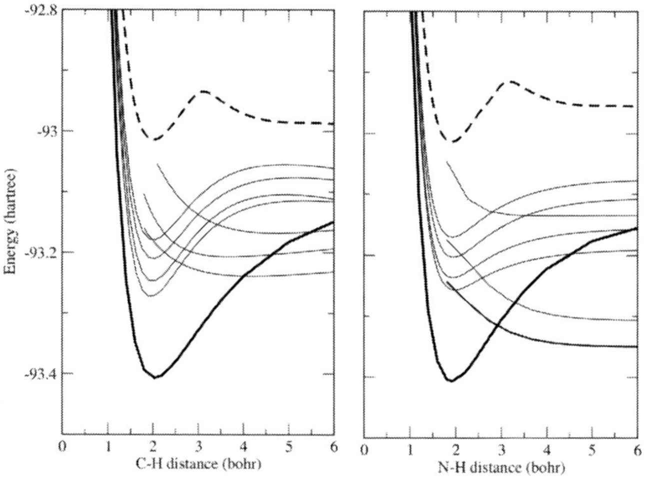

Figure 5. Potential energy curves for $HCNH^+$ Ground and low-lying excited states and HCNH resonance states of $^1\Sigma$ symmetry. Ground state of ion heavy solid line ——, excited states of ion dashed line - - - -, resonance states light solid line ——.

3. $HCNH^+$

The ground state of $HCNH^+$ is a closed shell with $^1\Sigma$ symmetry,

$$(1\sigma)^2(2\sigma)^2(3\sigma)^2(4\sigma)^2(5\sigma)(1\pi)^4$$

The low-lying excited states involve the excitation of an electron from the highest lying orbital (1π) to higher orbitals,

$$(1\sigma)^2(2\sigma)^2(3\sigma)^2(4\sigma)^2(5\sigma)(1\pi)^3(2\pi)$$

and

$$(1\sigma)^2(2\sigma)^2(3\sigma)^2(4\sigma)^2(5\sigma)(1\pi)^3(6\sigma)$$

with symmetries $^{1,3}\Sigma$,$^{1,3}\Pi$ and $^{1,3}\Delta$.

Details of this calculation have been published[13]. The target wave function was constructed by first carrying out a self-consistent-field (SCF) calculation on the ion using a contracted set of triple zeta plus polarization (TZP) basis set. These orbitals were then used in a multi-reference Configuration Interaction (CI) calculation, where the lowest three orbitals were frozen, and all single and double excitations from a reference space generated by a full CI for eight electrons in the 4σ,5σ,6σ,7σ,$1\pi_{x,y}$,$2\pi_{x,y}$ space. Natural orbitals were generated from averaging the density matrices for the lowest two states of the ion. The five lowest orbitals were frozen and the target states were determined from a full CI calculation in the natural orbital space of eleven orbitals with the largest occupation number. The basis set was expanded to include more diffuse orbitals. The Complex Kohn variational method[14]was used to study the electron scatttering at fixed geometries. The position and lifetime of the Feshbach resonances was obtained by fitting the eigenphase sum to a Breit-Wigner form. Additional structure calculations were carried out to determine the dominant character of the states and to obtain the resonant curves after crossing the ion.

An example of the results are shown in Figure 5. We have plotted the ground and first excited state of the ion as a function of the C–H bond distance, keeping all other bonds frozen at their equilibrium geometry and in the second panel, the variation with N–H bond distance. Only the states with $^1\Sigma$ symmetry are shown. The states have been diabatized using information from the configuration-interaction coefficients of the wave function. There are no crossings at low energy, but there are multiple resonances starting at approximately 3 eV. The resonance curves reflect the ion states to which they converge. The ion states change character and have several curve crossings as the bonds of the system are stretched. Future calculations will study the structure and dissociative recombination mechanisms of the H_2CN^+ isomer, which has a similar geometry to the H_2CO^+ ion.

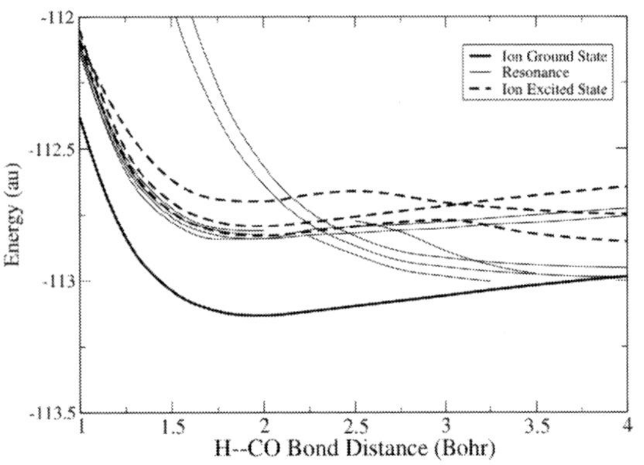

Figure 6. Potential energy curves for HCO^+ ground and low-lying excited states and HCO resonance states of $^1\Sigma$ symmetry. Ground state of ion heavy solid line ——, excited states of ion dashed line - - - -, resonance states light solid line ——.

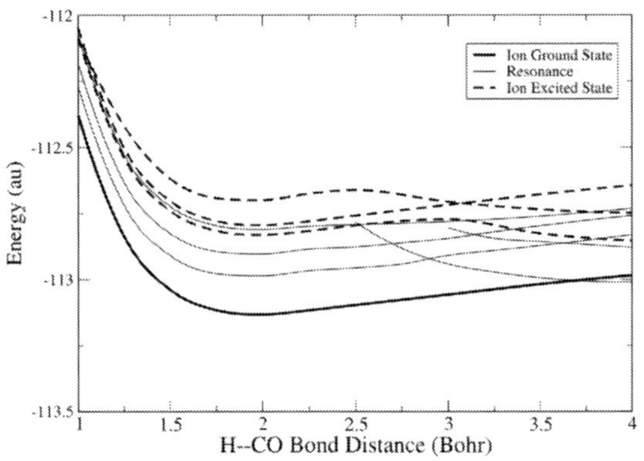

Figure 7. Potential energy curves for HCO^+ ground and low-lying excited states, and HCO resonance states of $^1\Pi$ symmetry. Ground state of ion heavy solid line ——, excited states of ion dashed line - - - -, resonance states light solid line ——.

4. HCO^+

HCO^+ is isoelectronic to $HCNH^+$ and therefore has similar ground and low-lying excited states. Our level of calculation in this system is the same as in $HCNH^+$. Again, as in $HCNH^+$, the excited states of the ion lie approximately 10.5eV above the ground state of the ion, and exhibit curve crossings between the lower energy states that are bound in the Frank-Condon region and higher lying dissociative states. These results are shown in Figures 6 and 7.

5. HCN^+/HNC^+

HCN^+ and HNC^+ have one fewer electron than HCO^+ and $HCNH^+$. In these systems, the (1π) and the (5σ) orbitals are very close in energy and switch positions between the two systems. The ground state of HNC^+ is:

$$(1\sigma)^2(2\sigma)^2(3\sigma)^2(4\sigma)^2(1\pi)^4(5\sigma)$$

with a low-lying excited state

$$(1\sigma)^2(2\sigma)^2(3\sigma)^2(4\sigma)^2(5\sigma)^2(1\pi)^3$$

There exists a series of Rydberg states converging to this excited state of the ion with configurations

$$(1\sigma)^2(2\sigma)^2(3\sigma)^2(4\sigma)^2(5\sigma)^2(1\pi)^3(n\sigma)$$

$$(1\sigma)^2(2\sigma)^2(3\sigma)^2(4\sigma)^2(5\sigma)^2(1\pi)^3(n\pi)$$

with symmetries $^{1,3}\Sigma, ^{1,3}\Pi$ and $^{1,3}\Delta$. In contrast, the ground state of HCN^+ is:

$$(1\sigma)^2(2\sigma)^2(3\sigma)^2(4\sigma)^2(5\sigma)^2(1\pi)^3$$

Journal of Physics: Conference Series **192** (2009) 012006 doi:10.1088/1742-6596/192/1/012006

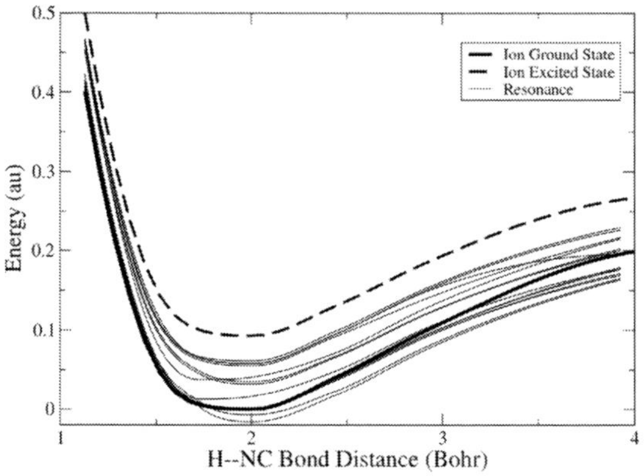

Figure 8. Potential energy curves for HNC^+ and HNC resonance states. Ground state of ion heavy solid line ——, excited state of ion dashed line - - - -, resonance states light solid line ——.

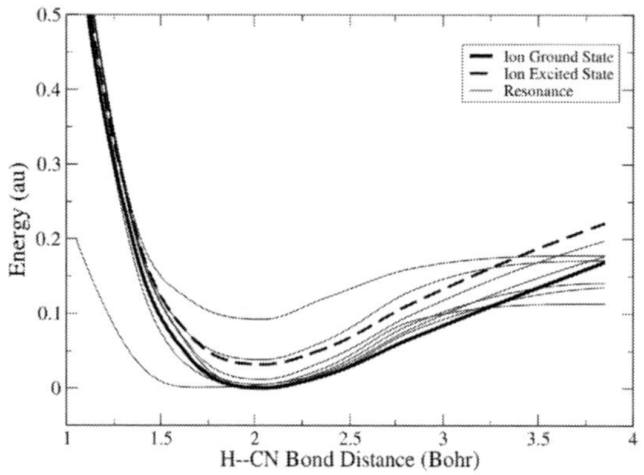

Figure 9. Potential energy curves for HCN^+ HNC resonance states. Ground state of ion heavy solid line ——, excited state of ion dashed line - - - -, resonance states light solid line ——.

with a low-lying excited state

$$(1\sigma)^2(2\sigma)^2(3\sigma)^2(4\sigma)^2(1\pi)^4(5\sigma)$$

and a series of Rydberg states converging to this state,

$$(1\sigma)^2(2\sigma)^2(3\sigma)^2(4\sigma)^2(1\pi)^4(5\sigma)(n\sigma)$$

$$(1\sigma)^2(2\sigma)^2(3\sigma)^2(4\sigma)^2(1\pi)^4(5\sigma)(n\pi)$$

with symmetries $^{1,3}\Sigma, ^{1,3}\Pi$ and $^{1,3}\Delta$. Since the first excited state of the ion is low-lying, the resonances are also expected to be low in energy. Again, the level of calculation was similar to that of HCNH. The results of the calculation are shown in Figures 8 and 9.

These calculations were carried out in linear geometry and fixed C-N bond distance. Calculations which considered the variation in energy with the H—-CN (or NC) distance fixed but for varying C–N bond distances showed no qualitative differences. Further calculations will study the effect of bend on the system, breaking the degeneracy of the Π states and allow the interaction of Σ states with one component of the Π states. It is interesting to note that although HCN and HNC are different chemical species, they represent two stable points on a three dimensional potential energy surface. The effect of the bend must be considered to understand the dynamics of dissociation.

6. H_2CO^+

H_2CO^+ has one more electron than HCO^+ and $HCNH^+$. This system is no longer linear, but bent, leading to C_{2v} not $C_{\infty v}$ symmetry. The ground state of the ion is:

$$(1a_1)^2(2a_1)^2(3a_1)^2(4a_1)^2(1b_2)^2(1b_1)^2(2b_2)^1$$

The ($1b_1$ and $2b_2$) which are roughly π orbitals, either oriented in or perpendicular to the molecular plane are close in energy, leading to a low lying excited state,

$$(1a_1)^2(2a_1)^2(3a_1)^2(4a_1)^2(1b_2)^2(1b_1)(2b_2)^2$$

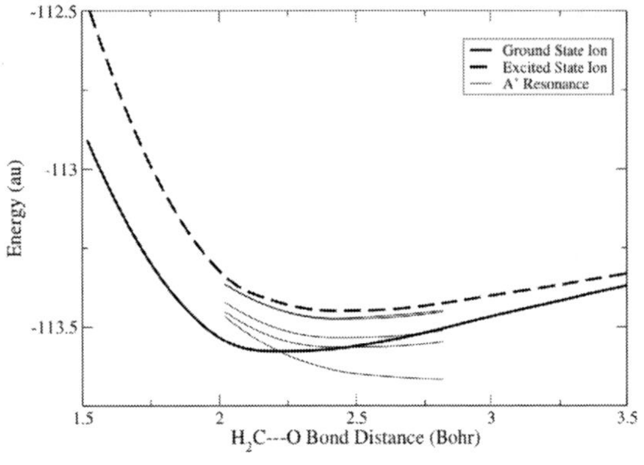

Figure 10. Potential energy curves for H_2CO^+ and H_2CO resonance states of A' symmetry. Ground state of ion heavy solid line ———, excited state of ion dashed line - - - -, resonance states light solid line ———

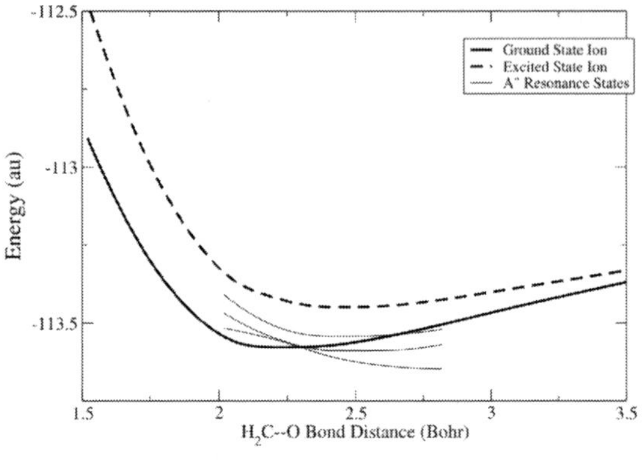

Figure 11. Potential energy curves for H_2CO^+ and H_2CO resonance states of A" symmetry.Ground state of ion heavy solid line ———, excited state of ion dashed line - - - -, resonance states light solid line ———

The low-lying resonances in this system, consist of the Rydberg series converging up to this excited ionic state. These are shown in Figures 10 and 11.

The initial calculations on this system held all angles and bond distances fixed at the equilibrium geometry of the ground state of the ion and varied the C–O bond distance. Only the singlet resonant states have been calculated so far. These studies show a number of resonances, that as in HCO^+ and $HCNH^+$ interact, showing evidence of curve crossings as the C–O bond is stretched. Further calculations on this system are necessary to both understand how the triplet resonance states vary with C–O stretch, but also, how the resonances change as other bond distances and angles are modified.

7. Conclusions

Electron resonances are ubiquitous in the scattering of low-energy electrons from molecules and molecular ions. These resonances provide an efficient path for channeling energy from the electron collision into nuclear motion leading to dissociative recombination, vibrational excitation or resonant dissociative excitation. Dissociative recombination is driven by Rydberg states, either converging to the ground or excited states of the ion. Therefore, calculations on the ground and excited states of the ion can yield information about the resonance states. It is clear much more work remains in the study of both diatomic and polyatomic systems, particularly studies leading to information not just on total dissociative recombination cross sections, but branching ratios into the various excited fragment states.

7.1. Acknowledgments

Å. Larson acknowledges support from The Swedish Research Council. A. E. Orel acknowledges support form the National Science Foundation Grant No. PHY-05-55401.

References

[1] S. Guberman, Phys. Rev. A **49**, R4277 (1994).
[2] V. Kokoouline, C. H. Greene and B. D. Esry, *Nature*, **412**, 891 (2001).
[3] V. Kokoouline and C. H. Greene, *Phys. Rev. A*, **68**, 012703 (2003).
[4] C. Greene, this volume.

[5] L. Carata, M. Raoult, I. F. Schneider, A. Suzor-Weiner and A. E. Orel, Phys. Rev. A **62** 0527711 (2000).
[6] S. Guberman, , J. Chem. Phys. **102**, 1699 (1995).
[7] Giusti A 1980 J. Phys. B: At. Mol. Phys. **13**, 3867 (1980).
[8] C. W. McCurdy and J. Turner, *J. Chem. Phys.*, **78**, 6773 (1983)
[9] G. A. Worth, M. H. Beck, A. Jäckle and H. -D. Meyer. The MCTDH Package, Version 8.2 (2000), University of Heidelberg, Heidelberg, Germany. H. -D. Meyer, Version 8.3 (2002). See http://www.pci.uni-heidelberg.de/tc/usr/mctdh/
For a review see: M. H. Beck, A. Jäckle, G. H. Worth and H. -D. Meyer. *Phys. Rep.*, **324:1**, 1 (2000).
[10] J. Royal and A. E. Orel, Phys. Rev. A **72**, 022719 (2005).
[11] V. Ngassam and A. E. Orel, Phys. Rev. A. **73** 32720 (2006).
[12] J. Royal and A. E. Orel, Phys. Rev. A. **73** 042706 (2006).
[13] V. Ngassam and A. E. Orel, Phys. Rev. A **75** 062702 (2007).
[14] T. N. Rescigno, C. W. McCurdy, A. E. Orel and B. H. Lengsfield III, *"The Complex Kohn Variational Method in Computational Methods for Electron-Molecule Scattering"* eds. W. H. Huo and F. A. Gianturco (Plenum, New York (1995)).

Dissociation dynamics of highly excited molecules produced by charge exchange: Two-body dynamics of CH_5 and three-body dynamics of *sym*-triazine

John D. Savee, Jennifer E. Mann, and Robert E. Continetti[1]

Department of Chemistry and Biochemistry, University of California, San Diego, 9500 Gilman Drive, La Jolla, CA 92093-0340, USA

E-mail: rcontinetti@ucsd.edu

Abstract. Translational spectroscopy and coincidence detection of the neutral photofragments have been used to observe the dissociation dynamics of highly excited neutrals produced by charge exchange between keV cation beams with Cs, and the results from two novel systems are presented. CH_5 is formed slightly above the 3s Rydberg level and dissociates into two possible fragmentation channels, H loss and H_2 loss. The kinetic energy release distributions of the two products are presented and the branching ratio between the two is found to be 11.4 ± 1.5 : 1 with the H loss being the dominant channel. Production of the highly symmetric azabenzene *sym*-triazine in its 3s Rydberg state has been shown to induce dissociation to $3\,\mathrm{HCN}(\tilde{X}\,^1\Sigma^+)$. Examination of momentum correlation in the dissociation products shows that this dissociation occurs by two distinct mechanisms. Evidence from Monte Carlo simulations suggest a sequential mechanism occurs creating products accompanied by a kinetic energy release of ~1.5-5 eV. A symmetric concerted mechanism is also observed and is associated with products receiving a 2-4 eV kinetic energy release.

1. Introduction

Translational spectroscopy coupled with coincidence detection of neutral molecules allows a full kinematic description of many-body dissociation processes occurring from highly-excited neutral molecules to be obtained. This full kinematic description provides a complete measure of momentum correlation among dissociation products, bypassing many of the assumptions required by typical time-of-flight (TOF) techniques.[1] While momentum matching can be avoided when determining the kinetic energy release (KER) in the case of a two-body dissociation, the true strength of this technique lies in its ability to provide full three-dimensional information about higher-order dissociation processes. A prime example of the level of detail obtained in this type of experiment can be seen in recent studies of the three-body dissociation dynamics of H_3.[2, 3]

In the present experiments, charge exchange (CE) between a keV cation beam and Cs vapor is used to create excited neutral molecules which then dissociate. This is a well-studied technique and has been previously described in detail, so only key elements will be presented here.[4-6] The amount of internal energy deposited in the neutral upon CE is not discrete, and can occur via what has been described as resonant or near-resonant electron capture. Resonant CE excites the neutral to a state lying above the ground state by an energy equal to the adiabatic ionization potential of the neutral minus the ionization potential of Cs (3.89 eV). There exists a probability for populating neutral states

[1] To whom any correspondence should be addressed

that lie off resonance that has been shown to increase non-linearly with shorter interaction time (i.e. faster beam velocity). Reaction (1) illustrates several ways in which a polyatomic molecule (ABC) can dissociate following CE with Cs. In the absence of decay by fluorescence, excess energy is distributed to the fragments as internal and translational energy, the latter of which is measured in translational spectroscopy experiments. Product internal energies are inferred by translational (KER) measurements, and are often difficult to interpret in the case of molecular fragments.

$$\begin{aligned} ABC^{+} + Cs \xrightarrow{-Cs^{+}} ABC^{*} &\rightarrow A + BC \\ &\rightarrow B + AC \\ &\rightarrow C + AB \\ &\rightarrow A + B + C \end{aligned} \quad (1)$$

The application of this technique to the examination of the product branching ratios in the dissociative CE of CH_5^+ and *sym*-triazine cation ($H_3C_3N_3^+$) is presented here. In the case of CH_5^+, the two low-lying product channels H + CH_4 and H_2 + CH_3 are observed, while in the case of *sym*-triazine, competing two- and three-body dissociation pathways are observed, with a focus in this manuscript on the three-body dissociation pathway $H_3C_3N_3 \rightarrow 3$ HCN.

2. Experimental

The fast cation translational spectrometer used to carry out the current CE experiments is a modified version of an existing apparatus.[7] Cations are formed in an electric discharge on a pulsed (1 kHz) supersonic expansion of a gas mixture and then skimmed, accelerated, and focused using electrostatic optics. Cations of different mass-to-charge ratios are then separated by time-of-flight and the cation of interest is selected using a mass-gate prior to entering a Cs oven with a ~1 mm^3 interaction region. Any neutral fragments or stable neutrals created in the interaction region propagate 110 cm forward to the neutral particle detector. Unreacted cations are deflected out of the beam path and monitored using a time-sensitive microchannel plate detector. The time- and-position sensitive neutral particle detector consists of three microchannel plates in a z-stack arrangement above a four-quadrant crossed delay line anode.[8, 9] Each quadrant is capable of detecting two particles in coincidence, thus allowing coincidence detection of up to eight particles in a single event. Once empirical N(KER) distributions are obtained, Monte Carlo (MC) simulations of the detector's geometric collection efficiency can be used to convert measured N(KER) distributions to true probability distributions, denoted by P(KER).[10]

3. Results and discussion

3.1 CH_5

CH_5^+ is the simplest non-classical carbonium ion.[11] It has importance in interstellar chemistry where it is formed via the radiative association of CH_3^+ and H_2.[12] There have been many studies, both theoretical and experimental of the cation, but very few of its corresponding neutral, due to its transient nature. CH_5^+ has a very small barrier to isomerization between C_s (I), C_s (II) and C_{2v} geometries [13], resulting in free exchange of all H atoms within the molecule. Although the existence of CH_5^+ has been known since its discovery via mass spectrometry in the 1950s, it has only been within the last 10 years that a high resolution IR spectrum was taken. [14]

Neutralization reionization mass spectrometry (NRMS) experiments have studied the dissociative CE of this neutral using a variety of electron donors.[15-18] More recently, the dissociative recombination branching ratios have been measured using two methods, flowing afterglow[19] and ion storage rings.[20] These two methods have produced conflicting results; in the flowing afterglow, the H loss 2-body channel is dominant, whereas the ion storage rings have found the dominant channel to be the CH_3 + H + H channel. In the present experiment, fragments are detected in coincidence, and translational KER distributions and branching ratios between different product channels are measured. Two possible dissociation channels are energetically accessible for dissociative CE with Cs.

$$CH_5 \rightarrow CH_4 + H \tag{2}$$

$$CH_5 \rightarrow CH_3 + H_2 \tag{3}$$

Upon resonant CE of CH_5^+ with Cs, CH_5 is formed ~ 0.1 eV above the predicted energy of the 3s Rydberg state.[21] There are two dissociation channels that are energetically available, reactions (2) and (3) above and they are energetically nearly degenerate as shown in figure 1. Although H_2 is kinematically more favorable to detect, the dominant channel in the dissociation is the atomic H loss.

If CE with Cs were resonant, and CH_5^+ is assumed to be in its ground vibrational state, the maximum KER would be 4.2 eV for both channels, however, the experimentally observed maximum KER is ~5 eV. As shown in previous studies, near-resonant charge exchange (NRCE) can occur, increasing the probability of forming a neutral above the resonant level, resulting in a larger maximum KER.[2]

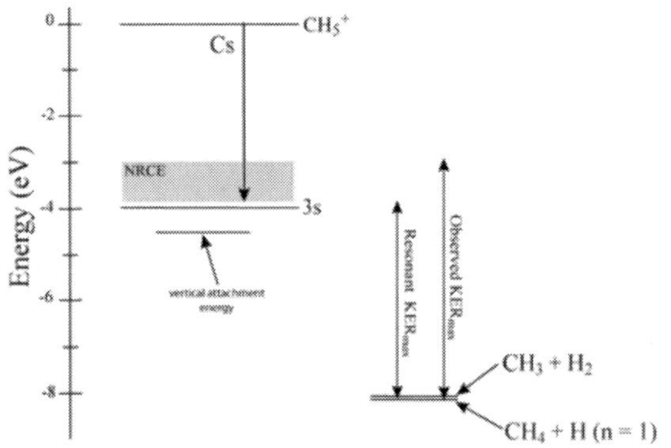

Figure 1. Energy level diagram showing the two energetically accessible dissociation limits of CH_5 and the location on the neutral surface where CH_5 is formed upon neutralization with Cs. The three-body dissociation limit is nearly resonant with the 3s Rydberg state and was not observed in the present experiment.

The P(KER) distribution for the dominant H-loss channel is shown in figure 2. It has a main peak with a shoulder at lower energy. The two Gaussians fit to the experimental data are also shown and peak at 2.0 eV and 3.1 eV. Both Gaussians extend to the maximum KER, indicating that the transient neutral for both features is formed in the same initial state. Most of the energy goes into internal excitation (vibration and rotation) of CH_4 as indicated by the low probability of products at the maximum KER.

The P(KER) distribution for the minor channel is shown in figure 3. Its distribution is bimodal with Gaussians peaking at 1.2 eV and 2.9 eV. The distribution of energy is more complicated for this channel because internal energy can go into both the CH_3 and H_2 products. Again, as in the case of the dominant channel, there is a low probability of products at the maximum KER, with most of the energy going into the internal excitation of the products. The branching ratios between the two channels can be calculated by integrating the P(KER) distributions over the given KER and dividing. The branching ratio for H loss to H_2 loss is 11.4 ± 1.5:1.

At first glance, it is surprising that the branching ratio shows a dominance for the H loss channel. The CH_5^+ cation is generally thought of as a CH_3 and H_2^+ moiety, and it would be expected that loss of the H_2 would have a more significant role in the dissociation. This may be a result of the fluxional nature of CH_5^+, with the protons essentially delocalized around the carbon center. Dynamical calculations on the *ab initio* potential energy surfaces of Bowman and co-workers may provide further insights into this phenomenon.[22]

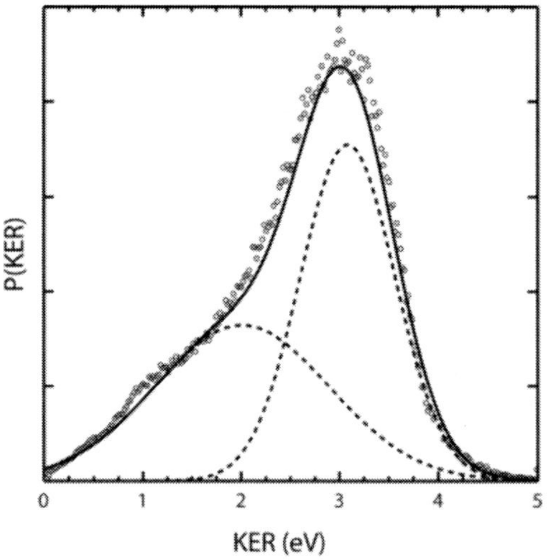

Figure 2. The P(KER) distribution for $CH_5 \rightarrow CH_4 + H$. The dotted lines represent Gaussian fits to the data. The solid line represents the sum of the two Gaussians. The experimental P(KER) is shown by open circles.

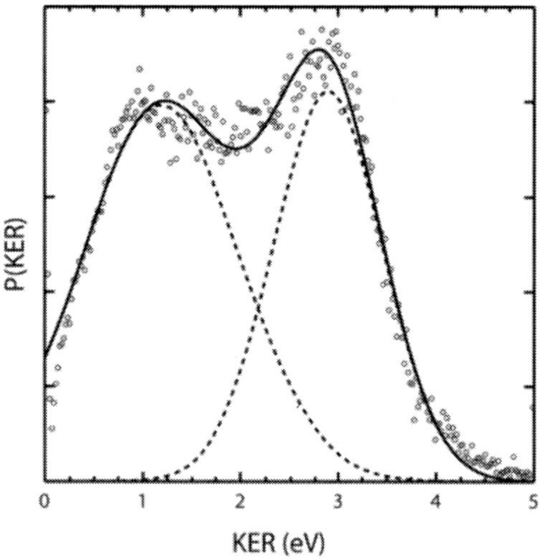

Figure 3. The P(KER) distribution for $CH_5 \rightarrow CH_3 + H_2$. The dotted lines represent Gaussian fits to the data. The solid line represents the sum of the two Gaussians. The experimental P(KER) is shown by open circles.

3.2 Sym-triazine

Sym-triazine (1,3,5-triazine) is the most symmetric member of the azabenzene family and is a common backbone for a wide variety of organic compounds (e.g. RDX, melamine, atrazine). *Sym*-triazine is of particular interest in the present work as it has long been known to photodissociate to three $HCN(\tilde{X}\,^1\Sigma^+)$ products by internal conversion from low-lying electronic states back to the $^1A_1'$ ground electronic state.[23-26] Early spectroscopic work probing infrared emission of HCN produced by photolysis of *sym*-triazine showed evidence for strong bending excitation and little evidence for the production of HNC fragments.[24] Previous photofragment translational spectroscopy (PTS) experiments have yielded several different interpretations of the three-body dynamics occurring from photoexcitation by 248 and 193 nm photons ($^1A_2''$ or $^1E'' \leftarrow {}^1A_1'$ and $^1A_2' \leftarrow {}^1A_1'$ respectively). Low-lying regions of the *sym*-triazine potential energy surface (PES) are densely populated by several electronic states which have been shown to have strong coupling, making the assignment of an exact mechanism extremely difficult.

While the first PTS experiment on the three-body dissociation of *sym*-triazine assumed HCN products were dissociating in a symmetric concerted manner, a similar experiment carried out later presented evidence for an asymmetric partitioning of momentum among the fragments.[23, 26] This asymmetric partitioning of momentum was described as being the result of a sequential process during which a short-lived $(HCN)_2$ intermediate is formed prior to complete dissociation to three HCN products, and is illustrated in reaction (4).

$$H_3C_3N_3^* \rightarrow (HCN)_2 + HCN \rightarrow 3HCN \tag{4}$$

A series of later *ab initio* studies found that a concerted mechanism with asymmetric partitioning of momentum among the products was a more likely reaction pathway.[27-29]

In the present experiment, the two- and three-body dynamics resulting from CE of the *sym*-triazine radical cation with Cs were examined using 12 and 16 keV cation beams. The adiabatic ionization potential for the lowest energy *sym*-triazine radical cation has been reported as 10.01 eV and corresponds to removal of an e' electron from a lone-pair orbital localized on a nitrogen atom in the ring structure.[30] The resulting doubly degenerate $^2E'$ state of the molecule is subject to the E⊗e Jahn-Teller effect and distorts from D_{3h} to C_{2v} equilibrium geometry.[31] While no direct experiments have been carried out on this cation in the gas-phase, inference from an investigation into the corresponding 3s $^1E'$ Rydberg state ($\sigma^* \leftarrow n$) of the neutral suggests the ground vibronic state of the cation undergoes hindered pseudorotation along a single ring-breathing mode analogous to ν_6 in benzene.[32, 33] Upon resonant CE, neutral *sym*-triazine is expected to be formed 6.12 eV above its ground state (4.26 eV above the ground state 3 HCN limit).[34] The competing two-body dissociation channel forming $(HCN)_2$ + HCN products will not be further discussed in this work.

The empirically measured N(KER) distributions and corrected P(KER) probability distributions for the three-body dissociation at both beam energies are presented in figure 4. The resulting P(KER) distributions at both beam energies are extremely broad and exhibit two features, one peaked at ~2.6 eV (Feature I) and another peaked at ~0.5 eV (Feature II). The 12 and 16 keV distributions are very similar, varying only slightly on the rising edge of Feature I. The maximum observed KER in the P(KER) distributions extends ~0.75 eV past the theoretical limit for resonant CE and indicates the initially populated state of the neutral lies ~6.9 eV above the ground state. Based on these energetics, it is likely that the neutral is being formed initially in the 3s Rydberg state which has favorable Franck-Condon (FC) overlap with the cation and preserves the cation core (i.e. requires no electronic rearrangement). The peak of Feature I at 2.6 eV suggests that on average ~48% of the available energy is distributed to internal modes of HCN, the majority of which is presumably deposited in the HCN bending mode.[24]

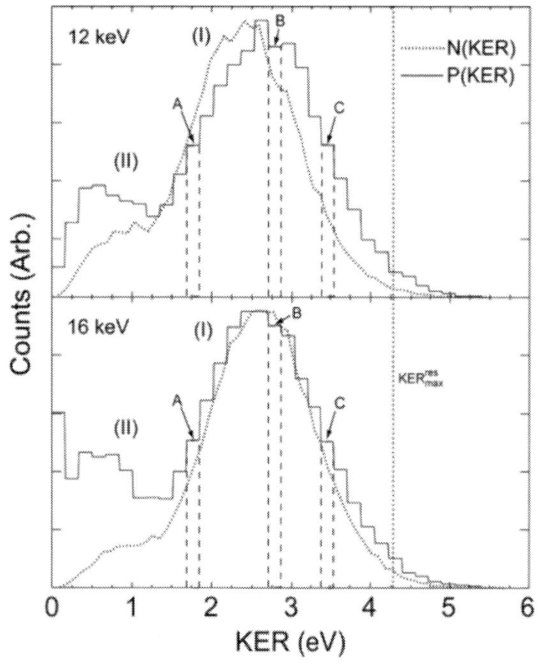

Figure 4. The resulting N(KER) and P(KER) distributions for *sym*-triazine → 3HCN upon CE with Cs at 12 and 16 keV cation beam energies. The dotted line labeled KER^{res}_{max} marks the theoretical limit for the maximum observed KER upon resonant CE. P(KER) intervals A-C are described in the text.

In order to illustrate momentum correlation dynamics among the fragments, Dalitz representations of the momentum partitioning in each event were constructed over small P(KER) intervals. The Dalitz plot is a histogram with each axis corresponding to a single fragment's fractional square of the total momentum ($f_i = p_i^2/\Sigma p_i^2$).[35] In practical use a single point on the Dalitz plot represents the partitioning of momentum between the three fragments in a single event, as illustrated in figure 5.

The intervals over which Dalitz plots were constructed, labeled in figure 4, correspond to the following KER ranges: A(1.69 – 1.86 eV), B(2.70 – 2.87 eV), and C(3.38 – 3.54 eV). Figure 6 shows the resulting Dalitz plots over these intervals. In the 12 keV distributions, a feature corresponding to two slow and one fast fragment is apparent in intervals A-C and will be referred to as the 'acute' feature. This acute feature is also present in the corresponding 16 keV Dalitz distributions, although a symmetric (equal momentum partitioning) feature is also prevalent for intervals B and C. This velocity-dependent change suggests that dissociation proceeds from two different initial neutral states. Examination of the momentum correlation over small KER ranges reveals the KER range over which

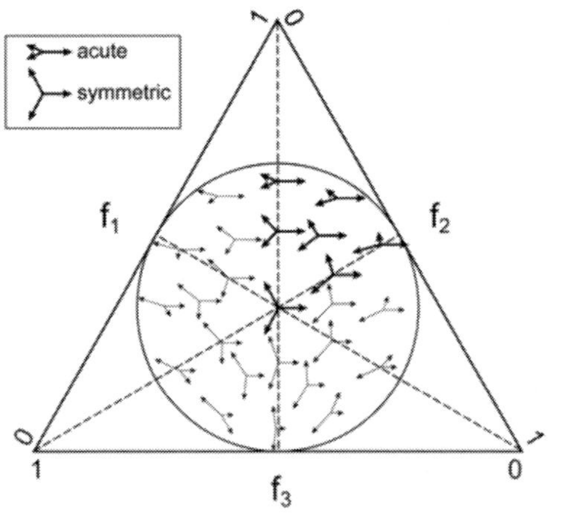

Figure 5. Illustrative map of momentum partitioning for select points in the Dalitz representation of three-body dissociation with three equal-mass products. The specific partitioning of momentum that forms an acute isosceles triangle between the heads of momentum vectors is referred to as 'acute' in the text. Equal partitioning of momentum occurs at the center of the circle and is referred to as 'symmetric' dissociation.

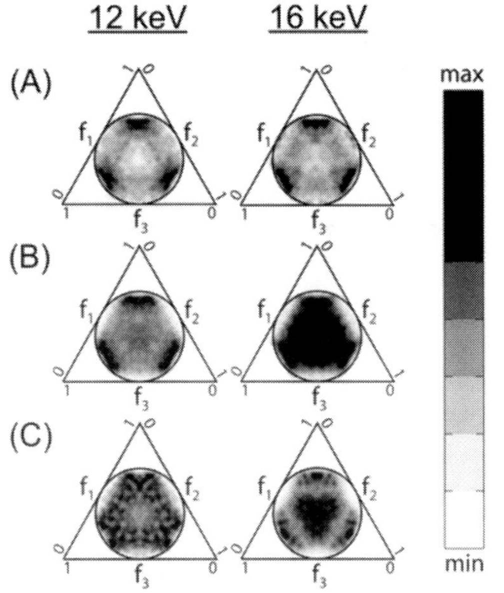

Figure 6. Dalitz distributions resulting from *sym*-triazine → 3HCN over the P(KER) intervals shown in figure 4.

each feature is present, and is not readily apparent from P(KER) distributions alone. The acute feature occurs over a KER range of ~1.5-5 eV while the symmetric feature occurs from ~2-4 eV.

The observed acute feature shows a nearly two-fold symmetry in the slow HCN products when interpreted as being the result of a concerted process. Nuclear motion on the lower sheet of the 3s $^1E'$ surface is largely dictated by motion along the ν_6 coordinate.[32, 33] A simple model using *ab initio* predicted ν_6 displacement vectors for the D_{3h} ground state of *sym*-triazine shows that while distortion along ν_6 preserves a two-fold (C_{2v}) symmetry in the molecule as a whole, the geometry never passes through a configuration with two identical HCN fragments constructed from adjacent atoms.[27] Investigations employing MC techniques show that interpretation of the Dalitz plot is not intuitive in the case of a sequential dissociation. If the $(HCN)_2$ intermediate is assumed to have a lifetime greater than one rotational period (represented by random rotation of the intermediate), the simulations are dependent on only one parameter describing the partitioning of the total (measured) KER into each step of the sequential process. Simulations with a small primary KER (25%) and a larger secondary KER from the intermediate (75%) result in Dalitz distributions that are notably similar to the acute feature observed in the present experiment (figure 7).

Symmetry correlation predicts that the three-body dissociation of *sym*-triazine must ultimately occur by coupling of the initial excited state to the $^1A_1'$ ground state.[23] The observed energetics suggest that the acute mechanism is occurring from an initial neutral state ~6.9 eV above the ground state, most likely in the 3s Rydberg manifold which is then coupled to the ground state. The symmetric mechanism is presumably the result of a near-concerted process from a three-fold symmetric geometry, although the origin is somewhat ambiguous due to uncertainty in the energetics of near-resonant CE processes. The velocity dependent nature of this feature's intensity relative to the acute feature qualitatively suggest that it arises from a state lying further off resonant excitation. The following two mechanisms are proposed, and would benefit greatly from further theoretical investigation. The splitting between the

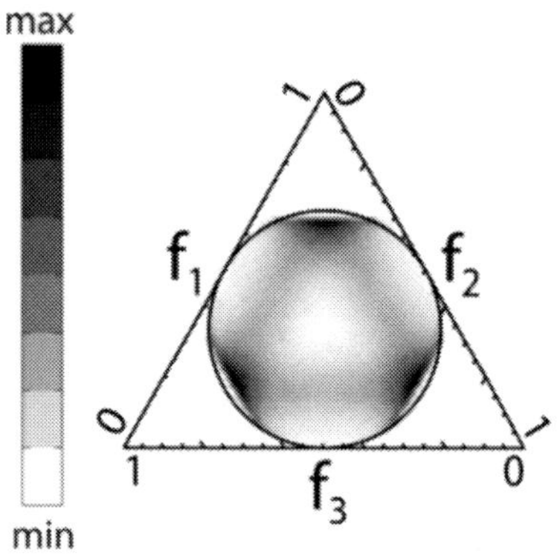

Figure 7. Dalitz distribution resulting from MC simulations of a sequential dissociation over interval A with 25% of the total KER partitioned to the first step and 75% to the second step.

zero-point level on the lower sheet and the conical intersection on the 3s $^1E'$ surface has been reported as 0.13 eV, and states lying above the conical intersection are expected to have D_{3h} geometries.[32, 33] Based on the approximate maximum KER observed for this feature (~4 eV), it is possible that population of a vibronic state on the upper sheet of the 3s surface followed by non-radiative decay to a dissociative state is giving rise to this feature. Alternatively, the lower-lying $^1A_2'$ valence state of *sym*-triazine ($\pi^* \leftarrow n$) has D_{3h} equilibrium geometry and excitation of the neutral to this highly symmetric region of the PES could lead directly to the dissociation mechanism.

4. Conclusion

The study presented here reveals some of the insights studies of dissociative charge exchange can provide to understanding the structure and dynamics of energized neutral species. The dissociative CE of CH_5^+ is particularly interesting as the H + CH_4 product channel is found to be strongly favored in spite of the fact that the CH_5^+ equilibrium geometry is more easily classified as a CH_3-H_2^+ molecule. In the case of the three-body dissociation of *sym*-triazine following CE we see a beautiful example of competing two- and three-body dissociation with both symmetric and asymmetric momentum partitioning occurring in the dissociation.

Acknowledgements

This work was supported by the U.S. Air Force Office of Scientific Research under Grant No. FA9550-04-1-0035. We thank Dr. Morton A. Fineman for his contributions to this research.

References

[1] Continetti R E 2001 *Annu. Rev. Phys. Chem.* **52** 165
[2] Laperle C M, Mann J E, Clements T G and Continetti R E 2004 *Phys. Rev. Lett.* **93** 153202
[3] Galster U, Baumgartner F, Muller U, Helm H and Jungen M 2005 *Phys. Rev. A* **72** 062506
[4] Vanderkamp A B, Los J, Vanderzande W J and Sidis V 1994 *Chem. Phys.* **179** 165
[5] Sidis V 1989 *J. Phys. Chem.* **93** 8128
[6] Debruijn D P, Neuteboom J, Sidis V and Los J 1984 *Chem. Phys.* **85** 215
[7] Hanold K A, Luong A K, Clements T G and Continetti R E 1999 *Rev. Sci. Instrum.* **70** 2268
[8] Lampton M, Siegmund O and Raffanti R 1987 *Rev. Sci. Instrum.* **58** 2298
[9] Friedman P G, Cuza R A, Fleischman J R, Martin C, Schiminovich D and Doyle D J 1996 *Rev. Sci. Instrum.* **67** 596
[10] Continetti R E, Cyr D R, Osborn D L, Leahy D J and Neumark D M 1993 *J. Chem. Phys.* **99** 2616
[11] Olah G A, Klopman G and Schlosberg R H 1969 *J. Am. Chem. Soc.* **91** 3261
[12] Herbst E and Bates D R 1988 *Astrophys. J.* **329** 410
[13] Huang X C, McCoy A B, Bowman J M, Johnson L M, Savage C, Dong F and Nesbitt D J 2006 *Science* **311** 60
[14] White E T, Tang J and Oka T 1999 *Science* **284** 135
[15] Williams B W and Porter R F 1980 *J. Chem. Phys.* **73** 5598
[16] Griffiths W J, Harris F M, Brenton A G and Beynon J H 1986 *Int. J. Mass. Spectrom. Ion Proc.* **74** 317
[17] Selgren S F and Gellene G I 1987 *J. Chem. Phys.* **87** 5804
[18] Bordas-Nagy J, Holmes J L and Hop C E C A 1988 *Int. J. Mass. Spectrom. Ion Proc.* **85** 241
[19] Adams N G, Herd C R, Geoghegan M, Smith D, Canosa A, Gomet J C, Rowe B R, Queffelec J L and Morlais M 1991 *J. Chem. Phys.* **94** 4852

[20] Semaniak J, Larson A, Le Padellec A, Stromholm C, Larsson M, Rosen S, Peverall R, Danared H, Djuric N, Dunn G H and Datz S 1998 *Astrophys. J.* **498** 886
[21] Raynor S and Herschbach D R 1982 *J. Phys. Chem.* **86** 3592
[22] Jin Z, Braams B J and Bowman J M 2006 *J. Phys. Chem. A* **110**
[23] Ondrey G S and Bersohn R 1984 *J. Chem. Phys.* **81** 4517
[24] Goates S R, Chu J O and Flynn G W 1984 *J. Chem. Phys.* **81** 4521
[25] Kim J H and Kim H L 2001 *Chem. Phys. Lett.* **333** 45
[26] Gejo T, Harrison J A and Huber J R 1996 *J. Phys. Chem.* **100** 13941
[27] Pai S V, Chabalowski C F and Rice B M 1996 *J. Phys. Chem.* **100** 5681
[28] Song K Y and Collins M A 2001 *Chem. Phys. Lett.* **335** 481
[29] Lee J, Dong E J, Jin D S, Song K Y and Collins M A 2004 *Phys. Chem. Chem. Phys* **6** 945
[30] C. Fridh L A, B. O. Jonsson, E. Lindholm 1972 *International Journal of Mass Spectrometry and Ion Physics* **8**
[31] Kato H, Hirao K and Yamashita K 1982 *Theochem-Journal of Molecular Structure* **5** 265
[32] Whetten R L and Grant E R 1984 *J. Chem. Phys.* **81** 691
[33] Whetten R L, Haber K S and Grant E R 1986 *J. Chem. Phys.* **84** 1270
[34] Bystrom K 1982 *Journal of Chemical Thermodynamics* **14** 865
[35] Dalitz R H 1953 *Philosophical Magazine* **44** 1068

Molecular Ions in Astrophysics

Evelyne Roueff[1], Eric Herbst[2]
[1]LUTH, Observatoire de Paris, Place J. Janssen, 92190 Meudon, France
[2]Dept. of Physics, The Ohio State University, 191 W. Woodruff Ave., Physics Research Building, Columbus, OH 43210 USA

E-mail: evelyne.roueff@obspm.fr

Abstract. The chemical content of the interstellar medium has been drastically revised in the last two years with the discovery of negative ions at a significant level of abundance. We emphasize the chemical processes involved in the formation and depletion of anions and the need for new laboratory work to understand them better. In addition, we comment on the detection of the first fluorine ion, CF^+, and on hydrocarbon chemistry, specifically the recent detection of propene and a study of deuterated methylacetylene isotopomers.

1. Introduction

General overviews of interstellar molecules and their chemistry were given at the last DR meeting [1, 2] and we refer readers to these contributions for a general introduction to astrochemical modelling. There have, however, been important updates to our knowledge of the molecular content during the last two years with the detections of:

- new organic molecules such as acetamide (CH_3CONH_2) towards star-forming regions in the galactic center
- propene (C_3H_6) in the "standard" cold molecular cloud TMC 1 [3], although this stable molecule was not included in any chemical network
- the fluorine-bearing molecular ion CF^+ in a photon-dominated region
- negative ions in a variety of sources, following their laboratory spectroscopic identification.

We display in Table 1 an up-to-date list of interstellar and circumstellar molecules without their isotopic variants. The most dramatic change from previous lists is the inclusion of negative molecular ions. The italic symbols refer to detections via absorption in electronic transitions in the UV or visible wavelength range, which are possible in diffuse or so-called translucent clouds. Molecular ions, very important tracers of the ionization state of the environment, appear in **boldface**. Molecules detected only in circumstellar environments are labelled with an asterisk whereas molecules detected both in the interstellar medium and circumstellar environments are labelled with two asterisks. In addition, molecules detected in external galaxies are underlined. The present inventory includes 133 neutral molecules, and 17 molecular ions including three negative species: C_4H^-, C_6H^-, and C_8H^-.

Table 1. Interstellar and Circumstellar Molecules as of October 2007.

Molecules with Hydrogen only				
$\underline{H_2}^{**}$	**$\mathrm{H_3^+}$**			
Molecules with Carbon and Hydrogen				
$\underline{CH}$	**CH^+**	C_2^{**}	$\mathrm{CH_2}$	$\underline{\mathrm{CCH}}^{**}$
C_3^{**}	$\mathrm{CH_3}$	$\mathrm{C_2H_2}^{**}$	l-$\mathrm{C_3H}^{**}$	c-$\mathrm{C_3H}^{**}$
$\mathrm{CH_4}^{**}$	c-$\mathrm{C_3H_2}^{**}$	l-$\mathrm{C_3H_2}^{**}$	$\mathrm{C_4H}^{**}$	**$\mathrm{C_4H^-}$** *
$\mathrm{C_5}^{*}$	$\mathrm{C_2H_4}^{*}$	$\mathrm{C_5H}^{**}$	l-$\mathrm{H_2C_4}^{**}$	$\mathrm{HC_4H}^{*}$
$\underline{\mathrm{CH_3CCH}}^{**}$	$\mathrm{C_6H}^{**}$	**$\mathrm{C_6H^-}$** **	$\mathrm{C_6H_2}^{**}$	$\mathrm{HC_6H}^{*}$
$\mathrm{C_7H}^{*}$	$\mathrm{C_2H_6}$	$\mathrm{CH_3C_4H}^{**}$	$\mathrm{C_8H}^{**}$	**$\mathrm{C_8H^-}$****
$\mathrm{CH_3CHCH_2}$	$\mathrm{CH_3C_6H}$	$\mathrm{C_6H_6}^{*}$		
Molecules with Oxygen and/or Hydrogen and/or Carbon				
$\mathrm{O_2}$	$\underline{OH}^{**}$	$\underline{CO}^{**}$	**$\mathrm{CO^+}$**	$\underline{\mathrm{H_2O}}^{**}$
$\underline{\mathrm{HCO}}$	**$\underline{\mathrm{HCO^+}}$****	**$\mathrm{HOC^+}$**	$\mathrm{C_2O}$	$\underline{\mathrm{CO_2}}$
$\mathrm{H_3O^+}$	**$\mathrm{HOCO^+}$**	$\underline{\mathrm{H_2CO}}^{**}$	$\mathrm{C_3O}^{**}$	HCOOH
$\mathrm{H_2CCO}$	**$\mathrm{H_2COH^+}$**	$\underline{\mathrm{CH_3OH}}^{**}$	$\mathrm{CH_2CHO}$	$\mathrm{HC_2CHO}$
c-$\mathrm{H_2C_3O}$	$\mathrm{CH_3CHO}$	c-$\mathrm{CH_2OCH_2}$	$\mathrm{CH_2CHOH}$	$\mathrm{CH_2OHCHO}$
$\mathrm{CH_3COOH}$	$\mathrm{HCOOCH_3}$	$\mathrm{CH_2CHCHO}$	$\mathrm{(CH_3)_2O}$	$\mathrm{CH_3CH_2OH}$
$\mathrm{CH_3CH_2CHO}$	$\mathrm{CH_3COCH_3}$	$\mathrm{HOCH_2CH_2OH}$	$\mathrm{C_2H_5OCH_3}$	
Molecules with Nitrogen and/or Hydrogen and/or Carbon				
NH	$\underline{CN}^{**}$	N_2	$\mathrm{NH_2}$	$\underline{\mathrm{HCN}}^{**}$
$\underline{\mathrm{HNC}}^{**}$	**$\underline{\mathrm{N_2H^+}}$**	$\underline{\mathrm{NH_3}}^{**}$	**$\mathrm{HCNH^+}$**	$\mathrm{H_2CN}$
HCCN^{**}	$\underline{\mathrm{C_3N}}^{**}$	$\mathrm{CH_2CN}$	$\mathrm{CH_2NH}$	$\underline{\mathrm{HC_3N}}^{**}$
HNCCC^{**}	$\mathrm{HC_2NC}^{**}$	$\mathrm{NH_2CN}$	$\underline{\mathrm{CH_3CN}}^{**}$	$\mathrm{CH_3NC}$
$\mathrm{CH_2CNH}$	**$\mathrm{HC_3NH^+}$**	$\mathrm{HC_4N}^{*}$	$\mathrm{C_5N}$ **	$\mathrm{CH_3NH_2}$
$\mathrm{HC_5N}^{**}$	$\mathrm{C_2H_3CN}$	$\mathrm{CH_3C_3N}$	$\mathrm{CH_2CCHCN}$	$\mathrm{CH_3CH_2CN}$
$\mathrm{HC_7N}^{**}$	$\mathrm{CH_3C_5N}$	$\mathrm{HC_9N}^{**}$	$\mathrm{HC_{11}N}$	
Molecules with Nitrogen, Oxygen				
$\underline{\mathrm{NO}}$	HNO	$\mathrm{N_2O}$	$\underline{\mathrm{HNCO}}$	$\mathrm{NH_2CHO}$
$\mathrm{CH_3CONH_2}$				
Molecules with Phosphorus, Sulfur, Silicon, Halogen or a Metal				
SH	$\underline{\mathrm{CS}}^{**}$	$\underline{\mathrm{SO}}^{**}$	**$\mathrm{SO^+}$**	$\underline{\mathrm{NS}}$
SiC	SiN^{**}	$\underline{\mathrm{SiO}}^{**}$	SiS^{**}	HCl
NaCl^{*}	AlCl^{*}	KCl^{*}	HF	AlF^{*}
$\mathrm{CF^+}$	CP^{*}	PN^{**}	PO^{*}	FeO
$\mathrm{H_2S}^{**}$	$\mathrm{C_2S}^{**}$	$\mathrm{SO_2}$	OCS	**$\mathrm{HCS^+}$**
c-$\mathrm{SiC_2}^{*}$	SiCN^{*}	SiNC^{*}	NaCN^{*}	MgCN^{*}
MgNC^{*}	AlNC^{*}	HCP^{*}	$\mathrm{H_2CS}$	HNCS
$\mathrm{C_3S}^{**}$	c-$\mathrm{SiC_3}^{*}$	$\mathrm{SiH_4}^{*}$	$\mathrm{SiC_4}^{*}$	$\mathrm{CH_3SH}$
$\mathrm{C_5S}^{*}$				

The remainder of the paper is organized as follows. In Section 2 we discuss fluorine chemistry, a potentially important subtopic in interstellar chemistry, interest in which has been raised by the detection of CF^+ [4]. Molecular anions are then discussed in Section 3. Finally, we discuss interstellar hydrocarbon chemistry including the detection of propene in the cold cloud TMC 1 and the presence in the same cloud of different isomers of deuterated methyl acetylene.

2. Fluorine chemistry

Three fluorine-bearing molecules have been detected so far. Of these, AlF has only been found in the circumstellar envelope IRC+10216 [5, 6], whereas both HF [7] and CF^+ [4] have been detected in the Orion Bar, a well-known photon dominated region (PDR). A preliminary chemical scheme for the chemistry of fluorine-bearing species has been suggested by Neufeld, Wolfire and Schilke [8]. Theoretical and experimental work show, rather remarkably, that the chemistry can be accounted for with a relatively restricted chemical network. Because the ionization energy of atomic fluorine is larger than 23 eV, it is safe to assume that the atom is mainly neutral, even in the high-photon environment of PDR's. Atomic fluorine (F) can react with H_2 to produce HF and H. Although this well-studied reaction is exothermic, a small activation barrier in the potential energy surface moderates its reaction rate coefficient. The main destruction channels for HF in the Orion bar environment are photodissociation and reaction with C^+. Photodissociation cross sections have been computed by Brown and Balint-Kurti [9], while the reaction of C^+ and HF is supposed to occur at the Langevin rate and lead to CF^+, a very stable molecular ion with the same electronic structure as CO. Finally, dissociative recombination and possible photodissociation are the main destruction channels of the CF^+ molecular ion. The chemical pathways leading to and destroying HF and CF^+ are as follows:

$$F + H_2 \rightarrow HF + H, \tag{1}$$

$$C^+ + HF \rightarrow CF^+ + H, \tag{2}$$

$$HF + h\nu \rightarrow H + F, \tag{3}$$

$$CF^+ + e \rightarrow C + F, \tag{4}$$

$$CF^+ + h\nu \rightarrow C^+ + F. \tag{5}$$

In their model calculation, Neufeld et al. [4, 8] estimated the DR recombination rate coefficient of CF^+ to be 2×10^{-7} $(T/300)^{-0.5}$ cm^3 s^{-1} and obtained satisfactory agreement between their observations of CF^+ and their predictions. The CF^+ dissociative recombination (DR) rate coefficient has now been measured by Novotny et al. [10] to be $5.2 \pm 1.0 \times 10^{-8}$ $(T/300)^{-0.8}$ cm^3 s^{-1}, which is rather close to the estimated value.

In addition to CF^+, other fluorine-bearing molecular ions may be of astrophysical interest. Neufeld and et al. [8] suggested that $F + H_3^+$ could lead to $FH_2^+ + H$, based on a study of the analogous reaction between N and H_3^+ [11]. However, this latter reaction has now been shown *not* to proceed [12]. Whether the $F + H_3^+$ reaction occurs is not clear, but it is true that this process satisfies spin conservation, as opposed to $N + H_3^+$. In any case, other formation channels exist, such as:

$$F^+ + H_2 \rightarrow HF^+ + H, \tag{6}$$

$$HF^+ + H_2 \rightarrow H_2F^+ + H, \tag{7}$$

$$HF + H_3^+ \rightarrow H_2F^+ + H_2. \tag{8}$$

The H_2F^+ molecular ion may be of potential astrophysical interest, as it is quite stable and isoelectronic with H_2O. Further studies on H_2F^+ would will be quite relevant in the astrophysical context.

Journal of Physics: Conference Series 192 (2009) 012008 doi:10.1088/1742-6596/192/1/012008

3. Negative molecular ions

The presence of anions has been demonstrated in the last year thanks to spectroscopic studies performed in the laboratory of P. Thaddeus [13, 14, 15]. The molecular anions C_6H^- and C_8H^- have been found in the "standard" molecular cloud TMC 1 [15, 16] while these as well as C_4H^- have been detected in the circumstellar envelope IRC+10216 [15, 16, 17, 18, 19]. Most recently, Sakai et al [20] reported the detection of C_6H^- in L1527, a low-mass star-forming region. Table 2 displays the ground-state electronic structures, permanent dipole moments (μ), and derived column densities (cm^{-2}) of the anions and their neutral counterparts in TMC-1 and IRC+10216. Whereas the neutral C_nH species have an open shell structure, leading to either $^2\Sigma$ or $^2\Pi$ ground states, electron attachment collapses the structure and reduces the spectrum to that of a simple $^1\Sigma$ molecule. Together the smaller partition function and the larger dipole moment enhance the lines of the anions relative to those of the neutral radicals by almost one order of magnitude.

Table 2. Summary of negative ion properties and their column densities (cm^{-2}).

Ref	Electronic symmetry	μ D	TMC 1 [16]	IRC+10216 [17, 18, 19]
C_4H^-	$^1\Sigma+$	6.2	$< 8.5 \times 10^{10}$	$7.1 \pm 2.0 \times 10^{11}$
C_4H	$^2\Sigma+$	0.9	$(6.1 \pm .5) \times 10^{13}$	3×10^{15}
C_4H^- / C_4H				1/4400
C_6H^-	$^1\Sigma+$	8.2	$(12 \pm 2) \times 10^{10}$	$4.1 \pm 1.5 \times 10^{12}$
C_6H	$^2\Pi$	5.5	$(750 \pm 80) \times 10^{10}$	$6.6 \pm 2.0 \times 10^{13}$
C_6H^- / C_6H			$(1.6 \pm .3) \times 10^{-2}$	1/16
C_8H^-	$^1\Sigma+$	11.9	$(2.1 \pm 0.4) \times 10^{10}$	$1.5 \pm .6 \times 10^{12}$
C_8H	$^2\Pi$	6.3	$(46 \pm 4) \times 10^{10}$	$8 \pm 3 \times 10^{12}$
C_8H^- / C_8H			$(5 \pm 1) \times 10^{-2}$	1/3.8

The chemistry associated with the formation of these molecular anions is not well known. Herbst [21] suggested more than 25 years ago that negative ions could be produced at low densities by radiative attachment while Petrie and Herbst [22] and Terzieva and Herbst [23] specifically addressed the formation of C_3N^- and carbon cluster negative ions (C_n^-), respectively, in interstellar space, using phase-space theory. Petrie [24] also considered dissociative attachment of weakly-bound neutral precursors as an exothermic formation mechanism. For the detected anions, both radiative attachment; e.g,

$$C_6H + e \rightarrow C_6H^- + h\nu, \tag{9}$$

and dissociative attachment of a precursor carbene; e.g.,

$$C_6H_2 + e \rightarrow C_6H^- + H \tag{10}$$

are likely formation processes [20].

The destruction of anions is better understood. The chemical reactions of assorted C_n^- and C_nH^- ions with H, H_2, N, and O have been studied by Barkholtz et al. [25] and Eichelberger et al. [26], who showed that these ions are not destroyed by H_2 but that reactions with atomic hydrogen and other atoms can be rapid. In their high-density experiment, three-body association (not applicable to the interstellar medium) is the dominant product channel for the C_n^- + H reactions while the C_nH^- + H reactions proceed via associative detachment, leading to HC_nH neutrals. Mutual ionization between cations and anions also plays a role in L1527 [20]; additional studies of this process are urgently required.

The discovery of molecular anions has generated the need to include their formation and destruction in chemical models. With estimates of the radiative attachment reaction rate coefficients for C_n and C_nH radicals and measured and estimated anion destruction rates, Millar et al. [27] computed column densities for molecular anions in a variety of sources. They found, in agreement with observation, that the larger molecular anions detected (n = 6, 8) have higher abundances relative to their neutral precursors (see Table 2) because the radiative attachment rate increases with the number of degrees of freedom of the anion. However, their rate estimates are quite uncertain and experimental studies are highly welcome. Flower et al. [28] reported a more restricted anion chemistry to derive astrophysical constraints from the ratios of the column densities of the negative ions to their neutral counterparts. Whether their deductions concerning the actual value of the cosmic ionization rate in TMC 1 are correct remains to be demonstrated.

4. Hydrocarbon chemistry

4.1. Propene, a new interstellar molecule

Propene, $CH_3CH{=}CH_2$ (C_3H_6), a closed shell molecule with a low dipole moment of 0.36 D, has been found in TMC 1 by Marcelino et al. [3] with an unexpectedly large fractional abundance relative to molecular hydrogen of about 10^{-8}, which is comparable with the main carbon-containing molecules such as C_2H, C_3H_2, and HCN. This molecule has not been considered, up to now, in any interstellar chemical network, because the gas-phase production of molecules with this many hydrogen atoms is often inefficient. One possibility is to first produce the precursor ion $C_3H_7^+$. Normal and ternary associative gas phase ion-molecule reactions of $C_nH_m^+$ ions with H_2 have been studied by McEwan et al. [29] at 300 K for various values of n and m. Although $C_3H_5^+$ has been found not to react associatively with H_2 to form $C_3H_7^+$, the upper limit of the association channel may still allow a reasonable radiative association rate in the cold interstellar medium. The DR of $C_3H_7^+$ has been studied with the CRYRING and ASTRID storage rings [30, 31]. Sixteen product channels are energetically possible, and the branching ratio leading to propene + H is a matter of debate. A value of 0.67, found by Angelova et al. [31], is used in our preliminary model calculations.

In addition to ion-molecule reactions, selected neutral-neutral reactions can be rapid at low temperatures. Canosa et al. [32, 33] have shown that CH and C_2 react with ethane to form propene:

$$CH + C_2H_6 \rightarrow C_3H_6 + H, \tag{11}$$

$$C_2 + C_2H_6 \rightarrow C_3H_6 + C, \tag{12}$$

with rate coefficients of 2×10^{-11} and 2×10^{-10} cm^3 s^{-1} respectively at 23 K. Once formed, propene is destroyed by ions and, as has been recently found, by atomic oxygen [34]. Detection of this molecule in the TMC 1 environment, where many carbon chain molecules have already been found, may be a signature of a large C/O elemental ratio. We have performed model calculations including propene, using the latest information described above, for physical conditions relevant to TMC 1. The chemical network comprises 169 species, including carbon chains up to C_8H_2, the negative ions H^-, C^-, CH^-, OH^-, C_2H^-, C_4H^-, C_5H^-, C_6H^-, C_7H^-, C_8H^-, and more than 2000 chemical reactions. Figure 1 displays the time dependence of the fractional abundances of several hydrocarbon molecules relative to total hydrogen nuclei. We used two different sets of elemental abundances with differing C/O ratios. Propene peaks at about 0.1 Myr, corresponding with the decrease of C_2 and CH (not shown). Its maximum fractional abundance reaches $1 - 8 \times 10^{-9}$, depending strongly on the C/O ratio. More work on the corresponding chemistry is clearly needed.

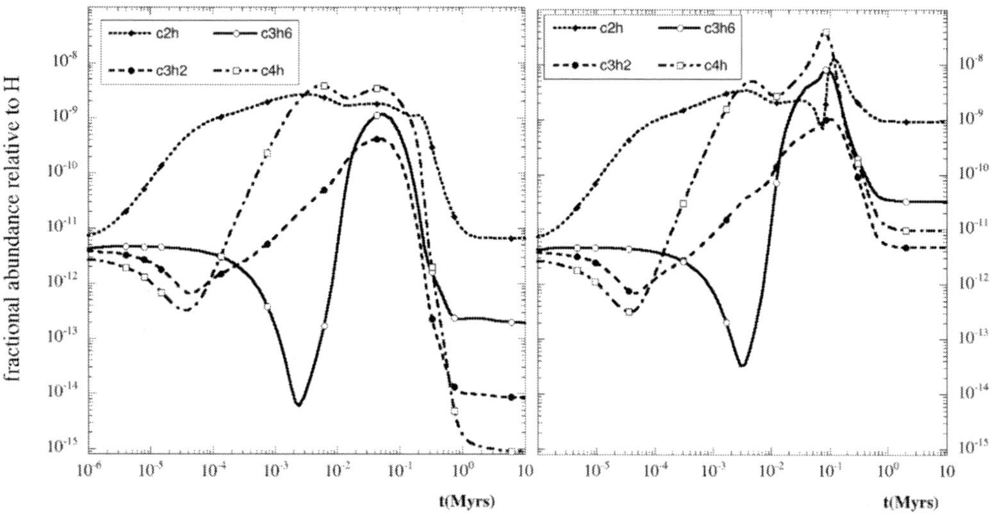

Figure 1. Time evolution of different carbon-containing species with an H_2 density of 10^4 cm^{-3} and a temperature T = 10 K. Left: standard low-metal elemental abundances, C/O = 0.4; Right: C/O = 0.73 and same elemental abundance of C as in the left panel.

4.2. Deuterated isomers

Despite a long history, deuterated molecules still represent an active topic, both for observations and modelling studies. We display detected deuterated isotopologues in Table 3; these include seven doubly-deuterated and two triply-deuterated species. As previously described [1] in

Table 3. Detected deuterated molecules in the interstellar medium

HD*	HDCO	$CH_3D^\diamond$	**D_2H^+**
HDO†	HDCS	**H_2D^+**	D_2O
DCN†	DC_3N	**DCO^+**	ND_2H
DNC†	CH_2DOH	**N_2D^+**	D_2CO
HDS	CH_3OD		CHD_2OH
C_2D	DC_5N		D_2S
C_4D	CH_2DCCH		D_2CS
C_3HD	CH_3CCD		ND_3
NH_2D	CH_2DCN		CD_3OH

* detected in planetary atmospheres, comets and the interstellar medium via the rotational transition at 112 μm and through UV absorption in diffuse and translucent interstellar clouds as well as in front of a distant quasar where the UV Lyman transitions of HD are shifted into the visible range.
† also detected in comets
◇ only detected in planetary atmospheres

Table 4. Chemical differentiation between deuterated isomers.

H_3^+	$+ CH_3CCH$	$\rightarrow$	$CH_3CCH_2^+$	$+ H_2$	H_2D^+	$+ CH_3CCH$	$\rightarrow$	CH_3CCHD^+	$+ H_2$
CH_3^+	$+ C_2H_4$	$\rightarrow$	$CH_3CCH_2^+$	$+ H_2$	CH_2D^+	$+ C_2H_4$	$\rightarrow$	CH_2DCCH^+	$+ H_2$
$C_2H_3^+$	$+ CH_4$	$\rightarrow$	$CH_3CCH_2^+$	$+ H_2$	$C_2H_2D^+$	$+ CH_4$	$\rightarrow$	CH_3CCHD^+	$+ H_2$

the DR2005 conference, the abundance ratio of deuterated species to their hydrogenated counterparts is much larger than the elemental D/H ratio, which is on the order of 1.5×10^{-5}. Deuterium enhancement is readily explained from gas phase chemical fractionation reactions involving deuteron exchange, such as $H_3{}^+ + HD \rightleftharpoons H_2D^+ + H_2$. The backward reaction is significantly slowed at low temperatures due to its endothermicity, which results from differences in the zero point energies and the extra rotational energy due to the absence of the 0_0 level in H_3^+. Similar reaction pairs can occur with H_2D^+ and HD and D_2H^+ and HD, producing D_2H^+ and finally D_3^+. These deuterated ions then react with neutrals to form high abundances of other deuterated species, both ionic and neutral. Other efficient deuteration vectors are provided by the deuterated isotopologues of CH_3^+ and $C_2H_3^+$ ions, which do not react rapidly with H_2 but can exchange a deuteron with HD.

The study of the abundance ratio between abundances of different deuterated versions of polyatomic molecules, such as the isotopomers CH_3CCD and CH_2DCCH of methyl acetylene, is most probably linked to subtle chemical processes, where care has to be taken with the various mechanisms. Marckwick et al. [35] suggest the occurence of the chemical reactions shown in Table 4, under the assumption that deuterons cannot transfer between functional groups. Experimental studies by Angelova et al [31] on the DR reaction of protonated methyl acetylene have shown that the product channel leaving the carbon skeleton intact occurs with an 87% probability. If, in addition, ejection of H is more likely than ejection of D for the deuterated precursor ions, the abundances of the deuterated isotopologues in TMC 1 can reasonably be accounted for. But deuterium fractionation can also occur on grain mantles with the products subsequently released into the gas phase by thermal and non-thermal processes. Whether gas-phase mechanisms or surface processes dominate is still an open question, and more work is needed on these topics, such as the extent to which non-equilibrium effects reduce the gas-phase fractionation.

5. Conclusions

We have given some examples showing the relevance of dissociative recombination and other ion processes for astrophysical modelling. The recent discovery of molecular anions demonstrates the need to improve our knowledge of assorted classes of reactions such as electron attachment, dissociative attachment, associative detachment, and mutual neutralization, all processes for which the theoretical and experimental DR community has the appropriate expertise. Finally, we recall the need for studying specific isotopic processes. For example, whether H or D atoms are more easily ejected during DR has a profound influence on the model results.

References

[1] Roueff E 2005 *J. Phys.: Conf. Series* **4**, 1
[2] Herbst E. 2005, *J. Phys.: Conf. Series* **4**, 17
[3] Marcelino N., Cernicharo J., Agúndez M., Roueff E., Gerin M., Martin-Píntado J., Mauersberger R., Thum C. 2007, *Astrophys. J.* **655**, L127

[4] Neufeld D et al. 2006, *Astron. Astrophys.* **454**, L37
[5] Cernicharo J., Guelin M 1987, *Astron. Astrophys.* **183**, L10
[6] Ziurys L.L., Apponi A.J., Phillips T.G. 1994, *Astrophys. J.* **433**, 729
[7] Neufeld D, Zmuidzinas J., Schilke P. and Phillips T.G. 1997, *Astrophys. J.* **488**, L141
[8] Neufeld D, Wolfire M. and Schilke P. 2005, *Astrophys. J.* **628**, 260
[9] Brown A., Balint-Kurti G.G. 2000, *J. Chem. Phys.* **113**, 1879
[10] Novotny O. et al. 2006, *J. Phys. B: At. Mol. Opt. Phys.* **38**, 1471
[11] Scott G.B., Fairley D.A., Freeman G., McEwan M.J. 1997 *Chem. Phys. Lett.* **269**, 88
[12] Scott G.B., Fairley D.A., Freeman G., McEwan M.J., Anicich G.A. 1998 *J. Chem. Phys.* **109**, 9010
[13] Bruenken S., Gottlieb C.A., Gupta H., McCarthy M.C., Thaddeus P. 2007 *Astron. Astrophys.* **464**, L33
[14] Gupta H., Bruenken S., Tamassia F., Gottlieb C.A., McCarthy M.C., Thaddeus P. 2007 *Astrophys. J.* **655**, L57
[15] McCarthy M.C., Gottlieb C.A., Gupta H., Thaddeus P. 2006 *Astrophys. J.* **652**, L141
[16] Bruenken S., Gupta H., Gottlieb C.A., McCarthy M.C., Thaddeus P. 2007 *Astrophys. J.* **664**, L43
[17] Cernicharo J., Guélin M., Agúndez M.Kawaguchi K., McCarthy M.C., Thaddeus P. 2007 *Astron. Astrophys.* **467**, L37
[18] Kasai Y., Kagi E., Kawaguchi K. 2007, *Astrophys. J.* **661**, L61
[19] Remijan A.J., Hollis J.M., Lovas F.J., Cordiner M.A., Millar T.J., Marckwick-Kemper A.J., Jewell P.R. 2007, *Astrophys. J.* **664**, L50
[20] Sakai N., Sakai T., Osamura Y., Yamamoto S. 2007, *Astrophys. J.* **667**, L65
[21] Herbst E. 1981, *Nature* **289**, 656
[22] Petrie S, Herbst E. 1997, *Astrophys. J.* **491**, 210
[23] Terzieva R, Herbst E. 2000, *Int. J. Mass Spectr.* **201**, 135
[24] Petrie S. 1996, *Mon. Not. R. Astr. Soc.* **282**, 807
[25] Barcholtz C., Snow T.P., Birnbaum V. 2001, *Astrophys. J.* **547**, L171
[26] Eichelberger B., Snow T.P., Barckholtz C., Bierbaum V.M. 2007, Astrophys. J. **667**, 1283
[27] Millar T., Walsh C., Cordiner M.A., Chuimín R.Ní, Herbst E. 2007, *Astrophys. J.* **662**, L87
[28] Flower D.R., Pineau des Forêts G., Walmsley M.M., 2007, *Astron. Astrophys.*,**462**, 221
[29] McEwan M.J., Scott G. B., Adams N. G., Babcock L. M., Terzieva R., Herbst E. 1999 *Astrophys. J.* **513**, 287
[30] Ehlerding A. et al. 2003, *J. Phys. Chem. A* **107**, 2179
[31] Angelova G. et al. 2004, *Int. J. Mass Spectr.* **235**, 7
[32] Canosa A., et al. 1997 *Astron. Astrophys.* **323**, 644
[33] Canosa A., et al. 1997 *Icarus* **187**, 558 **667**, 1283
[34] Sabbah H. et al. 2007 *Science* **317**, 102
[35] Marckwick A.J., Charnley S.B., Butner H.M., Millar T.J. 2005, *Astrophys. J.* **627**, L17

Journal of Physics: Conference Series **192** (2009) 012009 doi:10.1088/1742-6596/192/1/012009

Optical emissions from the dissociative recombination of N_2H^+, HCO^+, HOC^+, and HNC^+

R Johnsen[1], M F Golde[2], R E Rosati[3], D Pappas[4], and M P Skrzypkowski[5]

[1]Department of Physics and Astronomy University of Pittsburgh, Pittsburgh, PA 15260, USA
[2]Department of Chemistry, University of Pittsburgh, Pittsburgh, PA 15260, USA
[3]Smithsonian Astrophysical Observatory, 60 Garden St. MS-50 Cambridge, MA, 02138, USA
[4]Army Research Laboratory, 4600 Deer Creek Loop, Aberdeen Proving Ground, MD, USA
[5]Prometheus Energy Company, 3311 S. 120th Place Suite 100, Seattle, WA 98168, USA

E-mail rj@pitt.edu

Abstract. We present recent flowing-afterglow measurements of branching fractions for electronically and vibrationally excited products arising from the dissociative recombination of N_2H^+, HCO^+, HOC^+, and HNC^+ ions with thermal electrons. State-specific yields were derived by fitting the observed, spatially resolved emission band intensities to models that simulate all ion-chemical processes, recombination, diffusion, and gas mixing.

1. Introduction

Absolute emission spectroscopy in recombining afterglow plasmas is currently the only method to determine branching fractions for radiating products of dissociative recombination (DR). This contribution summarizes the results of such measurements on the protonated species N_2H^+, HCO^+, HOC^+, and HNC^+. Emissions from recombining CO_2^+ plasmas were used as an intensity reference, as has been described in an earlier publication [1,2]. Additional details on specific systems, and more complete references to sources of data, are given in several of our recent publications [3,4,5].

2. Experimental methods

Our flowing afterglow system consists of a small flow tube (diameter of 6 cm, length of 36 cm), equipped with a Langmuir probe to measure electron densities, two long spectroscopic windows, and a downstream quadrupole ion mass filter. The spectroscopic equipment, a movable monochromator/PMT system, covers wavelengths from 200 nm to 900 nm with bandwidths from 0.2 nm to 1.3 nm. Its spectral response was calibrated with deuterium and tungsten lamps. Almost all measurements were done in a spatially resolved fashion, i.e., emissions were recorded at several axial positions, and were compared to models that simulate the spatial/temporal evolution of the plasma. This procedure provides an essential consistency check of the kinetic models that are used in the data analysis. N_2H^+, HCO^+, HOC^+, and HNC^+ ions were derived *via* a sequence of ion-chemical steps from metastable helium atoms (He*) and a few He^+ ions that flow out of the microwave discharge in pure helium (see figure 1), followed by Penning ionization of argon gas, $He^* + Ar \rightarrow He + Ar^+ + e^-$. The initial upstream plasma contains mainly Ar^+ ions and electrons with number densities of typically 4×10^{10} cm^{-3}. Subsequent reactions of Ar^+ with added N_2, H_2, CO, or HCN or suitable blends of those gases then produce the recombining luminous plasma (the "flame").

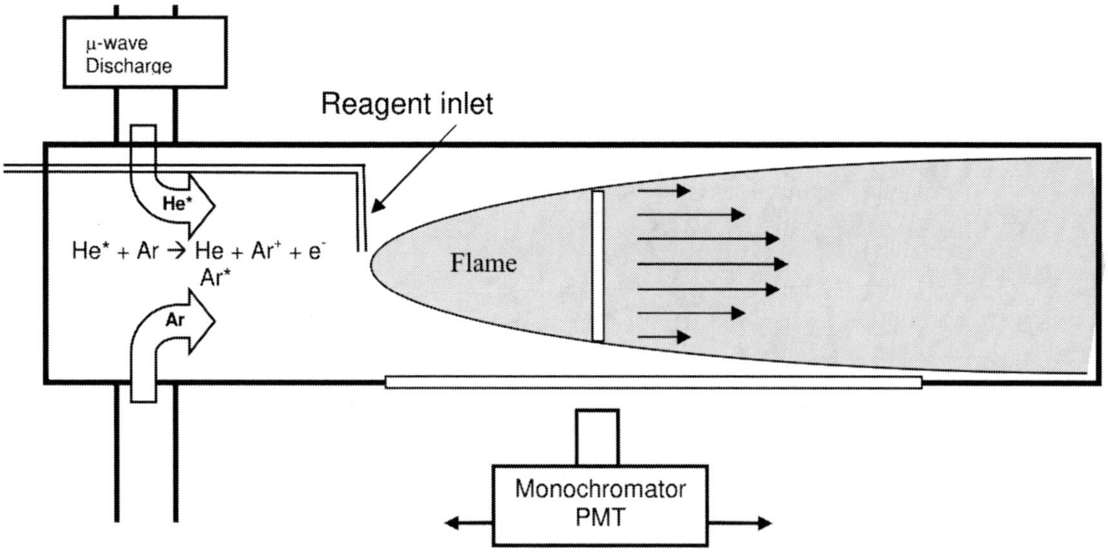

Figure 1. Diagram showing the formation of the afterglow plasma in the case of a point-like gas inlet. The white, column-shaped region of the "flame" indicates the volume seen by the monochromator.

Some metastable argon atoms (Ar*) enter the reaction zone and can give rise to background molecular emissions that have to be separated from those due to recombination; this can be done by adding an attaching gas (SF_6) that removes free electrons and thus eliminates DR. The total pressure in the flow tube in these experiments was on the order of 0.9 Torr (80% He, 20% Ar). The on-axis flow velocity of the buffer gases was typically $\approx$ 5900 cm/s. The observable "flame" extends about 10 cm downstream from the reagent inlet, corresponding roughly to 2 ms of afterglow time.

3. Afterglow models

To derive product yields from the observed emission intensities, the concentrations of the recombining ions and electrons at each point in the afterglow plasma must be obtained by calculation. Our computer models simulate the diffusion of the reagents from the reagent inlet, transport of particles in the flow field, ambipolar diffusion of ions and electrons, destruction of active particles at the flow-tube walls, recombination, and radiation. The models generate the radial and axial density profiles of all relevant particles and predict the z-dependence of the recombination radiation (integrated over the radial coordinate along the line of sight of the monochromator) for trial yields of a particular electronic state. We insisted that the models adequately reproduce both the shape and the magnitude of the observed emissions for several reagent concentrations (see figure 2). To test the validity of the models, several experimental tests were made, and two independent computer models were developed. The first model uses an implicit-explicit algorithm for solving the coupled differential equations while the second uses simpler algorithms that can be implemented on a spreadsheet program and produces a graphical output in a time of a few seconds. Except for the short region a few millimeters downstream from the reagent inlet both models gave essentially identical results. The models also showed, somewhat surprisingly at first, that the volume emission rate of the recombination radiation downstream from the gas inlet has a minimum on the flow-tube axis. When a point-like gas inlet is used, the flame resembles a hollow rather than a solid cone. In the case of a ring-shaped inlet, a "windsock" might be a better descriptive term.

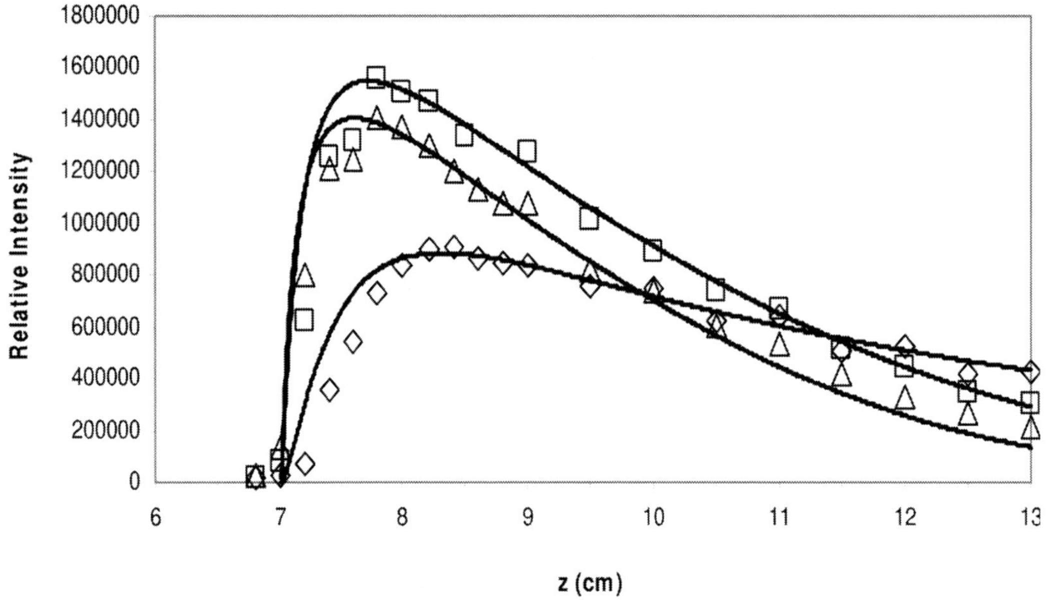

Figure 2: Example of computer fits (lines) to the measured axial dependence (data symbols) of an observed emission intensity. In the example, the CN(B – X) intensities due to HNC^+ recombination were fitted for three HCN concentrations of 2.5×10^{11} cm^{-3} (diamonds), 3×10^{12} cm^{-3} (squares), and 5×10^{12} cm^{-3} (triangles). The HCN gas inlet was located at z = 7 cm. The only adjustable parameter in the fits was a single normalizing factor in the absolute intensity.

To determine absolute yields of for a particular electronic state, the axial (z) dependence of the intensity of all observable transitions from that state was compared to one of the CO bands arising from recombination of CO_2^+, for instance CO(a-X), for which the yield has been determined previously [2] to be 0.29 ± 0.10. In practical work, to reduce errors due to possible drifts of the photometric detection efficiency, all intensity measurements were preceded and followed by reference measurements in recombining CO_2^+ afterglows.

4. Measurements and results

4.1 Emissions from N_2H^+ recombination

The dissociative recombination $N_2H^+ \rightarrow N_2 + H$ with thermal electrons releases sufficient energy to populate N_2 in the ground state and in the following electronically excited states:

$$\begin{aligned} N_2H^+ + e^- &\rightarrow N_2(X\,^1\Sigma_g^+) + H(^2S) + 8.49\text{eV} \\ &\rightarrow N_2(A\,^3\Sigma_u^+) + H(^2S) + 2.32\text{eV} \\ &\rightarrow N_2(B\,^3\Pi_g) + H(^2S) + 1.14\text{eV} \\ &\rightarrow N_2(W\,^3\Delta_u) + H(^2S) + 1.13\text{eV} \\ &\rightarrow N_2(B'\,^3\Sigma_u^-) + H(^2S) + 0.33\text{eV} \\ &\rightarrow N_2(a'\,^1\Sigma_u^-) + H(^2S) + 0.09\text{eV} \end{aligned} \quad (1)$$

Accessible vibrational states are: $A\ ^3\Sigma_u^+$ (v ≤ 15), $B\ ^3\Pi_g$ (v ≤ 5), $W\ ^3\Delta_u$ (v ≤ 6), $B'\ ^3\Sigma_u^-$ (v ≤ 2), and $a'\ ^1\Sigma_u^-$ (v = 0). Afterglow measurements of the total H-atom yield [6] have shown that dissociation into N_2 and H is the only important product channel. An earlier, disagreeing ion-storage ring measurement [7] has been traced to an error and has been retracted. As is discussed by Adams [8], new measurements again show that the H-atom yield is close to unity. The total recombination coefficient of N_2H^+ at 300 K (2.4×10^{-7} cm^3/s) [9] is typical for a simple triatomic ion.

The principal kinetic processes in the N_2H^+ experiments, in which H_2 and N_2 were the molecular reagents, consist of the following fast ion-molecule reactions, followed by DR and radiation:

Ion formation:

$$\mathrm{Ar^+ + H_2 \rightarrow ArH^+ + H} \qquad \mathrm{ArH^+ + H_2 \rightarrow H_3^+ + Ar} \qquad \mathrm{H_3^+ + N_2 \rightarrow N_2H^+ + H_2}$$

$$\mathrm{Ar^+ + N_2 \rightarrow N_2^+ + Ar} \qquad \mathrm{N_2^+ + H_2 \rightarrow N_2H^+ + H} \qquad (2)$$

Recombination: $\mathrm{N_2H^+ + e^- \rightarrow N_2}(X, A, B, W, B') + \mathrm{H}$

Radiation: $\mathrm{N_2}(B\ ^3\Pi_g) \rightarrow \mathrm{N_2}(A\ ^3\Sigma_u^+) + h\nu$

The H_2 concentrations were 4.16×10^{12} cm^{-3} and 1.22×10^{13} cm^{-3} while the N_2 concentrations were 3.66×10^{13} cm^{-3} and 1.17×10^{14} cm^{-3}. All combinations of these concentrations were used.

In an earlier afterglow experiment, Adams and Babcock [10,11] observed strong $N_2(B\ ^3\Pi_g - A\ ^3\Sigma_u^+)(v', v'')$ emission in a recombining N_2H^+/e^- afterglow plasma. Emission from $v' > 5$ was seen and ascribed to recombination of vibrationally excited N_2H^+, while the more relaxed vibrational distribution obtained at a higher N_2 concentration was thought to arise from N_2-induced vibrational relaxation of the N_2H^+ precursor.

Our analysis of the *B*-*A* spectra showed that the vibrational distribution of the *B*-state shifted to lower states in the later afterglow. The effect depended only very weakly on the experimental H_2 and N_2 concentrations.

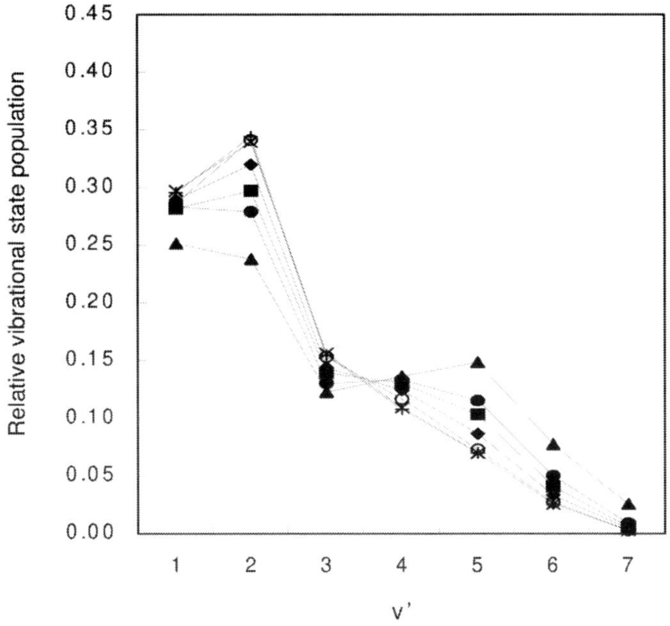

Figure 3.
$N_2(B)$ vibrational distributions from DR of N_2H^+ at seven different positions *z* [cm]:
▲ 5.4, ● 6.2, ■ 6.6, ◆ 7.0, ○ 8.0, + 9.0, × 10.0. The gas inlet was at 5 cm.

Concentrations:
$[N_2] = 1.17\times10^{14}$ cm^{-3}, $[H_2] = 1.22\times10^{13}$ cm^{-3}

Vibrational relaxation of the recombining N_2H^+ in collisions with N_2 was invoked by Adams and Babcock, who observed a similar effect. However, since we did not find any significant changes when the concentrations of H_2 and N_2 were varied, we suggest as a second, perhaps more important mechanism, namely that recombination of N_2H^+ also produces N_2 in the nearby long-lived *W*-state and that collisionally induced and radiative transitions from *W* to *B* populate $N_2(B)$ in low vibrational states. This fits well with recent theoretical work by Talbi [12,13], who finds that dissociation into the *W*-state should be the dominant dissociation path in linear geometry. In a more realistic bent geometry, both the *B* and *W* states would be formed.

Using the absolute branching fraction for CO(*d*) obtained in our previous work [1,2] , we derived an effective spectroscopic yield for formation of $N_2(B)$ of 0.19±0.08. The combined yield of all triplet states ($B\,^3\Pi_g$, $B'\,^3\Sigma_u^-$, $A\,^3\Sigma_u^+$, $W\,^3\Delta_u$) of N_2 could be significantly larger.

4.2 Emissions from HCO^+, HOC^+ recombination

Analyzing the recombination spectra emitted by the recombining formyl ion HCO^+ and its isomer HOC^+ presented a more challenging task than that encountered in the studies of N_2H^+. Recombination of HCO^+ and HOC^+ share the following three CO-forming branches, namely

$$\begin{aligned} HCO^+ + e^- &\rightarrow CO(X\,^1\Sigma^+) + H(^2S) \;+ 7.31eV \\ &\rightarrow CO(a\,^3\Pi_r) + H(^2S) \;+ 1.30eV \\ &\rightarrow CO(a'\,^3\Sigma^+) + H(^2S) \;+ 0.44eV \end{aligned} \tag{3}$$

but HOC^+ (1.74 eV above HCO^+) recombination can form three additional states:

$$\begin{aligned} HOC^+ + e^- &\rightarrow CO(d\,^3\Delta_i) + H(^2S) + 1.53eV \\ &\rightarrow CO(e\,^3\Sigma^-) + H(^2S) + 1.29eV \\ &\rightarrow CO(A\,^1\Pi) + H(^2S) + 1.02eV \end{aligned} \tag{4}$$

Recombination of HCO^+ in the vibrational ground state with 300 K electrons provides sufficient energy to populate the following vibrational levels of electronically excited states of CO: $a(v' < 8)$, and $a'(v' < 4)$. Recombination of HOC^+ ions may populate $a(v' < 16)$, $a'(v' < 18)$, $d(v' < 13)$, and $e(v' < 11)$.

Extensive laboratory measurements [14,15,16,17,18,19,20,21,22] have shown that the recombination rate of HCO^+ with 300 K electrons is on the order of 2×10^{-7} cm^3s^{-1}. However, theoretical calculations [23,24,25,26] so far have not reproduced this value and the question if HCO^+ recombines by a direct or indirect process has not been entirely resolved. The H-atom yield is known to be near unity [6,21]. Observations by Adams and Babcock [10] and Butler et al.[27] of $CO(a\,^3\Pi - X\,^1\Sigma^+)$ band emissions from recombining HCO^+ afterglows have shown that a fraction of CO is formed in the long-lived $a^3\Pi$ state.

Our experimental procedures were similar to those used in the N_2H^+ work, but the molecular reagents were H_2 and CO. Most of the kinetic processes that convert the initial Ar^+ ions to HCO^+ and HOC^+ in reactions with H_2 and CO are quite well known, with the unfortunate exception of some steps that determine the relative abundance of the two isomers. However, since the abundance of HOC^+ always remains at the level of only a few % of that of HCO^+, the uncertainty in the HOC^+ abundance has a fairly small effect on the determination of product yields of HCO^+, but a larger effect on the yields of HOC^+ recombination. Here, we have to omit the rather lengthy discussion of the ion chemistry that is described in more detail in the paper by Rosati et al. [4].

The observed emission spectra consisted of the ultraviolet CO(a – X) bands and the visible/near infrared CO(a' – a), CO(d – a) and CO(e – a) band systems. CO(a) can be formed by recombination of both isomers, but all levels $v' \geq 5$ of CO(a') and all levels of CO(d) and CO(e) must be ascribed to recombination of the HOC^+ since the upper levels are not energetically accessible to recombination of HCO^+. We do not believe that vibrationally excited HCO^+ was a significant source of the emissions that we ascribe to HOC^+, since we found no evidence for vibrational quenching of $HCO^+(v>0)$ with any of the added gases. Formation of CO(a',v' = 4) from HCO^+ is close to thermoneutral and this state may be formed by both HCO^+ and HOC^+.

4.2.1 Formation of CO(a) in recombination of HCO^+. At low concentrations of CO ($\sim 1\times10^{13}$cm^{-3}), the vibrational distribution for levels v = 0 to 4 of CO(a) was found to be 0.45±0.01, 0.21±0.01, 0.13±0.01, 0.10±0.01, 0.10±0.01. Very weak emission from v = 5 was also detected. Further downstream the distribution shifted slowly towards lower vibrational levels, consistent with electronic quenching of the long-lived CO(a, v') by CO. The observed vibrational distributions for this state are quite close to those found by Butler et al. [27]. Both distributions are slightly colder than those calculated by Tomashevsky et al [26].

The spectroscopic CO(a) yield from $HCO^+ + e^-$ recombination was found to be 0.26 ± 0.12, where the uncertainty reflects our assessment of a possible systematic uncertainty in the CO(a) yield from $CO_2^+ + e^-$. Random errors were far smaller. We estimate that indirect formation of CO(a) via radiative transitions from the CO(a', d, e) states contributes about 10% to the raw spectroscopic yield. Hence, the direct CO(a) recombination yield from HCO^+ recombination should be 0.23 ± 0.12.

4.2.2 Formation of CO(a′, d, e) in recombination of HOC^+. The relative vibrational distribution for levels v' = 5 – 8 of CO(a') was: 0.59 ± 0.03: 0.22 ± 0.03: 0.12 ± 0.01: 0.07 ± 0.01. Close to the reagent inlet, the population of CO(a', v'=4) was about 80% of that of v'= 5. For CO(d, v'= 0–5), the relative populations were: 0.21 ± 0.10: 0.16 ± 0.05: 0.17 ± 0.02: 0.28 ± 0.01: 0.12 ± 0.01: 0.05 ± 0.01. The weaker emission between 380 and 500 nm could be ascribed mainly to CO(d – a)(v',0) and (v',1) bands. Approximate analysis of the band intensity in this region showed a gradual decrease in population in CO(d) from v'=5 to v'=9. The ratio of the total observed population in CO(a', v'=4–8) to that in CO(d, v'=0–5) close to the reagent inlet was 1.12 ± 0.03. The CO(e – a) emission occurs in a congested region between 490 and 590 nm. The (v', 0) bands for v'= 2 – 5 can be tentatively assigned, with the (3,0) band being the strongest. Populations of CO(e, v' = 3) and of CO(e, v' = 2–5) are estimated to be respectively about 2% and 5% of the total population of the CO(a', d, e) states.

To summarize the HOC^+ yields: HOC^+ seems to recombine largely into the higher triplet states a', d, and e. The combined yield of the three states is large, probably larger than 0.4, of which 20% go into a', 18% into d, and 2 to 4% into e.

4.3 Emissions from HNC^+ *recombination*

Recombination of HNC^+ with thermal electrons has the following CN producing branches:

$$\begin{aligned} HNC^+ + e^- &\rightarrow CN(B\,^2\Sigma^+) + H \quad + \; 3.82\ \text{eV} \\ &\rightarrow CN(A\,^2\Pi) + H \quad + \; 5.89\,\text{eV} \qquad (5) \\ &\rightarrow CN(X\,^2\Sigma^+) + H \quad + \; 7.02\,\text{eV} \end{aligned}$$

The following vibrational states are energetically accessible: $B\,^2\Sigma^+$ ($v \leq 18$), $A\,^2\Pi$ ($v \leq 34$), and $X\,^2\Sigma^+$ ($v \leq 35$). While most needed ion-molecule reactions involving HCN^+ and HNC^+ were known [28,29, 30,31,32], the rate and isomer branching of the $Ar^+ + HCN \rightarrow HCN^+ (HNC^+) + Ar$ reaction had to be measured by us in a separate drift tube experiment. By comparing the reactivities of the Ar^+ + HCN product ions with xenon we found that HNC^+, the lower-energy isomer, is by far the dominant product. Small amounts of HCN^+ may have been present in the afterglow measurements, but there was no positive indication of their presence. The recombination coefficient of HNC^+ of 2×10^{-7} cm^3/s was obtained as part of this experiment, in good agreement with the result of a merged-beam measurement of Sheehan et al.[33].

Spectral scans of the emissions from the region downstream of the HCN inlet showed the well characterized [34] *B*-*X* violet bands as well as the *A* – *X* red bands of CN. Because of spectral congestion in some wavelength ranges, the analysis of the spectra was done, in part, by fitting them to synthetic spectra. In some measurements, xenon was added to destroy argon metastables that were found to produce some background emissions. Also, addition of hydrogen, which converts HNC^+ ions to $HCNH^+$, was found to strongly reduce all emissions, indicating that recombination of $HCNH^+$ does not produce significant emissions.

4.3.1 B-state vibrational distribution. The form of the *B* – *X* violet recombination spectrum was essentially independent of HCN concentration and reaction time (with the exception of anomalous v'= 11) suggesting that the vibrational distribution of the recombining HNC^+ ions is dominated by a single vibrational state, presumably the lowest state. Good fits by synthetic spectra required a rotational temperature of 2500 ± 500 K. For v'= 0 to 4, fairly reliable relative vibrational populations could be obtained which were 100: 58±6: 40±7: 25±5: 27±3. The relative populations of v'= 5 – 10 are less certain, but the fitting of the spectra suggested an average population per level of at least 25% of that of v'=0. Emission from v'= 11 – 15 was too weak to allow useful comparisons with synthetic spectra, but the integrated emission intensity of these bands (between 390 and 410 nm) is one-half of the integrated intensity from v'= 0 – 10 (370 – 390nm), indicates very significant population of levels v'= 11 – 15. In summary, a very wide range of *B*-state vibrational levels, v'= 0 to 15 and beyond, is strongly populated in this reaction. The vibrational distribution is probably bimodal, with peaks at v'= 0 and at high v'. The lower levels are characterized also by a high rotational temperature.

4.3.2 A-state vibrational distribution. A-*X* spectra measured at different distances from the reactant inlet yielded consistent vibrational distributions. The relative populations of levels *A*, v'= 2 – 15 were: 11: 57: 80: 91: 91: 100: 80: 68: 45: 57: 41: 37: 34: 23, indicating an inverted distribution, peaking at levels v'= 5 – 8. We estimate that the derived populations of most levels have uncertainties of about 10%, but for the lowest and highest levels (v'= 2 - 4 and 12 – 15) the uncertainties are larger, about 30%. Unlike the CN(*B*) state, the radiative lifetime of CN(*A*) is sufficiently long (4 – 9 μs) so that rotational relaxation of CN(*A*) is expected to be complete.

4.3.3 B-state and A-state yields. The *B*-state yield, defined as the fraction of recombination events producing CN(*B*) in any of the observed vibrational states, was found to be 0.22±0.08, while the yield of CN(*A*) is somewhat smaller, namely 0.14±0.05. Since the *B*-*A* radiative transition probabilities are only of the order of 1% of those for *B*-*X*, radiative *B*-*A* transitions make a minute contribution to the observed CN(*A*) yield, much smaller than the quoted uncertainties, and have little effect on the vibrational distribution.

5. Conclusions

Afterglow emission spectroscopy, in conjunction with extensive model calculations, can provide fairly accurate absolute yields for radiating states formed by recombination. The experimental uncertainties in the absolute yields largely reflect the systematic uncertainty in the reference source intensity. The ratios of yields for different states of the same product are better determined. Hence, if later work shows that one of the measured yields is incorrect, all other yields would have to be corrected by the same factor. The spectroscopic yields of particular electronic states were found to be as large as ~ 20% of the total yield into all branches. Since usually not all optical transitions are observable, the true yields may be larger. We find that the vibrational distributions in some cases extend to high quantum numbers, and that they do not always follow a monotonic pattern. The observed vibrational distributions may contain valuable hints concerning dissociation mechanism. In some cases, vibrational excitation may come about because the equilibrium bond distance in the final state differs from that in the parent ion. It is also possible that some recombination processes proceed *via* long-lived complexes, in which case vibrational excitation could result from intramolecular energy transfer.

Rotational excitation of products was seen in only one case; nascent rotational excitation may be more common but is frequently not observable because rotational thermalization occurs before radiation.

We did not find clear evidence that the recombining ions had supra-thermal vibrational distributions. This should have been detectable by slow changes in the vibrational populations of the products, but was not seen. Perhaps, one should be more careful in invoking vibrational excitation to rationalize discrepancies in DR measurements. In one case ($N_2(B)$ form N_2H^+), changes in vibrational product distributions were seen, but they were probably caused by mixing of the *W* and *B* states.

It is not possible at this time to compare our experimental findings to results of *ab-initio* theoretical calculations. Triatomic ions still pose a serious challenge to theory, and, although much progress has been made, there are no firm predictions of product branching yields of specified electronic and vibrational levels.

Acknowledgement: This work was, in part, supported by the NASA Planetary Atmospheres Program

References

[1] Rosati R E, Johnsen R and Golde M F 2003 *J. Chem. Phys.* **119** 11630
[2] Skrzypkowski M P, Gougousi T, Johnsen R and Golde M F 1998 *J. Chem. Phys.* **108** 8400
[3] Rosati R E , Johnsen R and Golde M F 2004 *J. Chem. Phys.* **120**, 8025
[4] Rosati R E , Skrzypkowski M P, Johnsen R and Golde M F 2007 *J. Chem. Phys.* **126**, 154302
[5] Rosati R E, Pappas D, Johnsen R and Golde M F *J. Chem. Phys 2007* **126**, 154303
[6] Adams N G, Herd C R, Geoghegan M, Smith D, Canosa A, Gomet J C, Rowe B R , Queffelec J L , and Morlais M 1991 *J. Chem. Phys.* **94** 4852
[7] Geppert W D, Thomas R, Semaniak J, Ehlerding A, Millar T J, Österdahl F, af Ugglas M, Djuric N, Paal A and Larsson M 2004 *Astrophysical J.* **609** 459
[8] Adams N G et al. , *this volume: New flowing afterglow technique for determining products of. . .*
[9] Smith D and Spanel P 1993 *Int. J. Mass Spectrom. Ion Processes* **129** 163
[10] Adams N G and Babcock L M 1994 *J. Phys. Chem.* **98** 4564
[11] Adams N G and Babcock L M 1994 *Astrophysical J.* **434** 184
[12] Talbi D 2007 *Chemical Physics* **332** 298
[13] Talbi D et al. , this volume: *A comparative study of the DR reactions of c-C3H3+ and l-..*
[14] Gougousi T, Golde M F and Johnsen R 1996 *Chem. Phys. Lett.* **265** 399
[15] Leu M T, Biondi M A and Johnsen R, 1973 *Phys. Rev.* A **8** 420
[16] Adams N G, Smith D and Alge E J 1984 *Chem. Phys.* **81** 1778
[17] Ganguli B, Biondi M A and Johnsen R 1988 *Phys. Rev.* A **37** 2543
[18] Amano T 1990 *J.Chem. Phys.* **92** 6492
[19] Smith D and Spanel P 1993 *Int. J. Mass Spectrom. Ion Proc.* **129** 163
[20] Le Padellec A, Sheehan C, Talbi, D and Mitchell J B A 1997 *J. Phys.* B **30** 319
[21] Geppert W D ,Thomas R, Ehlerding A, Hellberg F, Österdahl F, Millar T J, Semaniak J, Kaminska M, Källberg A, Paal A, Larsson M 2005, *Journal of Physics Conference Series* **4** 26
[22] Rowe B R, Gomet J C, Canosa A and Mitchell J B A 1992 *J. Chem. Phys.* **96** 1105
[23] Kraemer W P and Hazi A U 1998, *in Dissociative Recombination: Theory, Experiment and Applications, edited by J.B.A. Mitchell and S.L. Guberman (World Scientific, Singapore 1998)*, p. 61
[24] Bates D R 1992, *J. Phys. B* **25** 5479
[25] Larson A, Tonzani S, Santra R and Greene C H 2005 *Journal of Physics Conference Series* **4**, 148
[26] Tomashevsky M, Herbst E and Kraemer W P 1998 *Astrophysical J.* **498** 728
[27] Butler J M, Babcock L M and Adams N G 1997 *Mol. Phys.* **91** 81
[28] Wisthaler A , Hansel A, Schwarzmann M, Scheiring Ch, Lindinger W and Ferguson E E 2000 *J. Chem. Phys.* **112** 731
[29] McEwan M J 1992 , *in "Advances in Ion Chemistry" , JAI Press 1992, Vol. 1* 24
[30] Petrie S, Freeman C G , McEwan M J and Ferguson E E 1991, *Mon. Not. R. Astr. Soc.* 248 272

[31] Hansel A, Glantschnig, M, Scheiring Ch, LindingerW and Ferguson E E 1998 *J. Chem. Phys.* **109** 1743
[32] Hansel A, Glantschnig, M, Scheiring Ch, LindingerW and Ferguson E E 1998 *J. Chem. Phys.* **109** 1748
[33] Sheehan C, Le Padellec A, Lennard W N, Talbi D and Mitchell J B A 1999 *J.Phys.* B **32**, 3347
[34] Knowles P J , Werner, H-J, Hay P J and Cartwright D C 1988 *J. Chem. Phys.* **89** 7334

Seventh International Conference on Dissociative Recombination (DR2007) IOP Publishing
Journal of Physics: Conference Series **192** (2009) 012010 doi:10.1088/1742-6596/192/1/012010

Probing the low-temperature rotational population of H_2D^+

Oskar Asvany, Edouard Hugo, Serjoscha Wahed and Stephan Schlemmer

I. Physikalisches Institut, Universität zu Köln, 50937 Köln, Germany

E-mail: asvany@ph1.uni-koeln.de

Abstract. The gas phase exchange reactions of isotopologues of H_3^+ with isotopologues of H_2 are responsible for the observed deuteration in low-temperature interstellar clouds. At the prevailing cryogenic temperatures many quantum effects, as zero-point vibrational energies, large rotational level spacings and nuclear spin effects become important. In order to understand the processes on a level-to-level basis, experiments in a 22-pole ion trap are carried out accompanied by microcanonical simulations. In particular, the method of laser induced reaction (LIR) is applied to probe the four lowest rotational levels of H_2D^+.

1. Introduction

Before an interstellar cloud collapses into a proto-star, possibly surrounded by a planetary disc, it is characterized by strong depletion of heavier molecules, leaving predominantly molecular hydrogen isotopologues and helium in the gas phase, and of course electrons and ice-covered dust grains. The dominant ionic species besides H^+ and H_2^+ in such an environment is cold H_3^+ and its isotopic counterparts produced by the chain of exothermic reactions [1, 2, 3, 4]:

$$H_3^+ + HD \rightleftharpoons H_2D^+ + H_2 \tag{1}$$

$$H_2D^+ + HD \rightleftharpoons D_2H^+ + H_2 \tag{2}$$

$$D_2H^+ + HD \rightleftharpoons D_3^+ + H_2 \tag{3}$$

H_2D^+ and D_2H^+ have been detected in cold clouds [5, 2] by their low rotational transitions, and there is a wealth of singly and even multiply deuterated molecules detected up to date (see for example [6, 7, 8, 9, 10] and references therein) which are formed by secondary reactions.

In order to have a sensitive spectroscopic probe for the properties prevailing in cold interstellar clouds, it is of uttermost importance to understand the physical processes responsible for the formation and destruction of the basic ionic molecules. For H_3^+ and its isotopologues, these processes are mainly the above given ion-molecule reactions, the depletion reactions with heavier molecules (mainly CO), and the encounters with electrons. While many of the contributions to the DR2007 conference proceedings deal with the destruction of H_3^+ isotopologues by dissociative recombination with electrons, this paper gives an excerpt of the efforts to understand the exchange reactions (1-3) on a state-specific basis.

At the low-temperatures of interstellar clouds (10-100 K), many quantum mechanical effects such as zero-point vibrational energies (ZPVE), large rotational level spacings and different nuclear

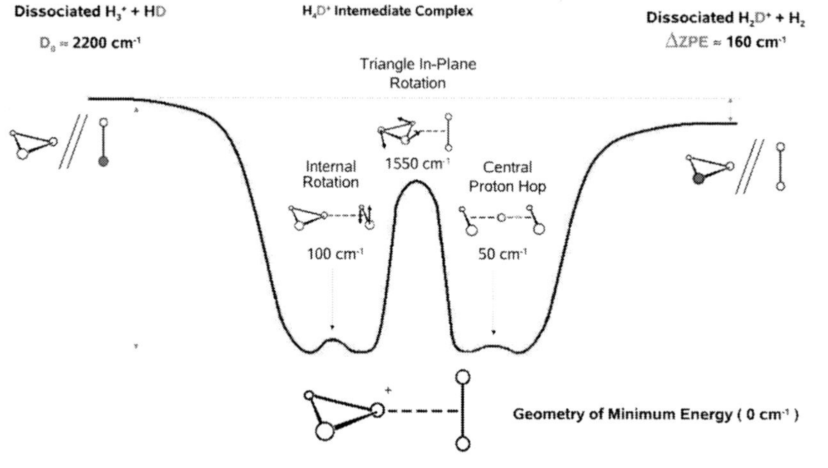

Figure 1. Schematic PES of the H_4D^+ collision system. The numbers are based on the most recent calculations on H_5^+ [12]. Internal rotation of the H_2 moiety and the hop of the central proton have a relatively low barrier, while a higher barrier has to be overcome for the in-plane rotation of the H_3^+ triangle. That this motion indeed occurs in the complex has been demonstrated by isotopic measurements [13, 14]. In the case of the H_4D^+ complex, the H_2D^+ + H_2 exit channel is lower by about 160 cm^{-1} (232 K) due to zero-point vibrational energies.

spin species determine the outcome of the deuteration reactions (1-3). For example, around 20 K, only the four lowest rotational states are substantially occupied for H_2D^+. Furthermore, it is currently discussed [11] that small amounts of o-H_2 can drive the backward direction of reaction (1), resulting in reaction rates differing from simple thermal equilibrium considerations. The influence of the ZPVE is illustrated in the schematic potential energy surface (PES) in Figure 1, where the exit channel leading to the right side of reaction (1), H_2D^+ + H_2, is exothermic by about 232 K. The details of the PES are based on the most recent *ab initio* calculations of Xie et al. [12]. As shown, the forming H_4D^+ complex can perform three types of internal motion, of which the so-called proton hop and triangle in-plane rotation lead to isotopic exchange.

To understand the underlying processes, our research group explores the exchange reactions by (i) measuring the rate coefficients as a function of temperature and different ortho/para-composition of H_2, (ii) using laser induced reactions (LIR) to probe the lowest rotational levels of the involved species and investigating their kinetics, and furthermore the measurements are complemented by (iii) microcanonical calculations to understand the results. This contribution gives some examples of our experimental and theoretical efforts.

2. Experimental Section

The experimental procedures are described in detail elsewhere [15, 16, 17, 18, 19, 20, 21]. In brief, a cryogenic 22-pole ion trap [22, 23] is used to investigate mass-selected ions produced in an external ion source. Several hundred ions are initially cooled by collision with He gas and kept in the trap to study their reactions with an admitted neutral reaction partner (usually isotopologues of H_2), and also a laser beam can be used to excite the cold ions and initiate reactions in the trap (LIR).

The determination of rate coefficients is exemplified for the reaction H_3^+ + HD. A pulse of several hundred H_3^+ ions is injected into the trap containing pure HD gas (> 98%) with known number

density. After a certain reaction period, the ionic trap content is extracted, mass analyzed and counted by a detector. The sequence of trapping, reaction and detection is sequentially repeated many times for different product masses and trapping times. By fitting numerical solutions of a suitable rate equation system to the data and knowing the HD number density, the rate coefficients of the involved reactions can be determined. Usually, the measurements are repeated at different nominal temperatures of the trap, from room temperature down to a nominal minimum of 13 K, to obtain the temperature dependence of the reaction.
With LIR, an extension of this trapping technique, a laser beam is shone into the trap to excite the stored ions. Thereby an endothermic reaction, as for example all backward directions of reactions (1-3) can be substantially promoted if the corresponding ion (H_2D^+, D_2H^+ or D_3^+) is excited. Thus ion spectroscopy can be performed by counting the product ions, those on the left side of reactions (1-3), as a function of the laser frequency, and also measurements on the reaction kinetics are possible. While this work focuses mainly on the spectroscopy and dynamics of H_2D^+ and D_2H^+ in reaction with H_2, as illustrated in Figure 2, the reaction with Ar atoms have been shown to be useful in probing the lowest rotational levels of H_3^+ isotopologues [24, 25, 26, 27].

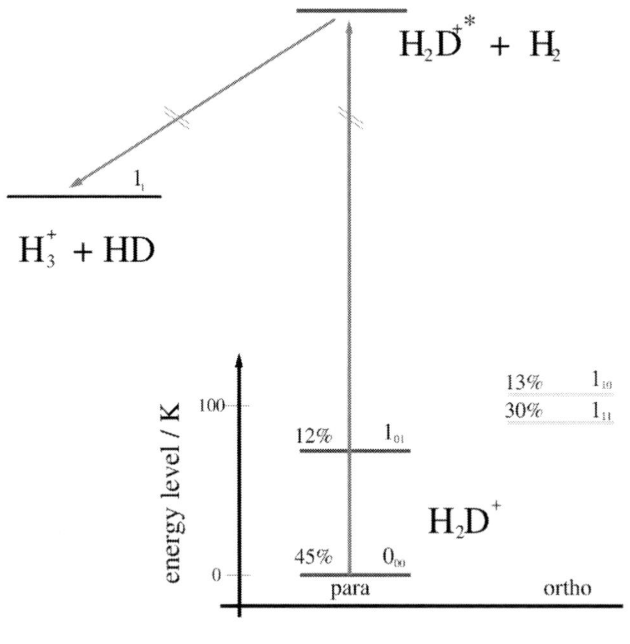

Figure 2. Explanation of LIR: H_2D^+, being in one of the four lowest rotational states at low temperature, is excited by a IR laser. Thus its reaction with H_2, no longer being endothermic by 232 K, is enhanced. Spectroscopy can be performed by measuring the number of H_3^+ reaction products as a function of the laser frequency. Furthermore, the populations of the lowest levels can be probed due to their proportionality to the H_3^+ signal. The given percental populations are explained in section 4.3.

3. Microcanonical model

A microcanonical model is developed with the aim to determine state-to-state reactive and inelastic rate coefficients consistent with experimental results. These rate coefficients are urgently needed to understand the H_3^+ isotopic fractionation phenomenon and to perform radiative transfer codes for H_2D^+ and D_2H^+.
The model describes all state-to-state thermal rate coefficients for the $H_3^+ + H_2$ collision system in all isotopic variants. The formation of the H_5^+ collision complex is described by the Langevin model. Assuming both reactants to be kinetically thermalised, the state-to-state rate coefficients are calculated by averaging the state-to-state cross sections on a Maxwellian collision energy distribution. The complex is described by three strictly conserved quantities which are (i) the total energy (ii) the total nuclear spin symmetry and (iii) the total angular momentum. The total energy is conserved and all degrees of freedom (rotational and kinetic) are assumed to mix completely their energies through a long-lived strongly bound complex. Furthermore, it

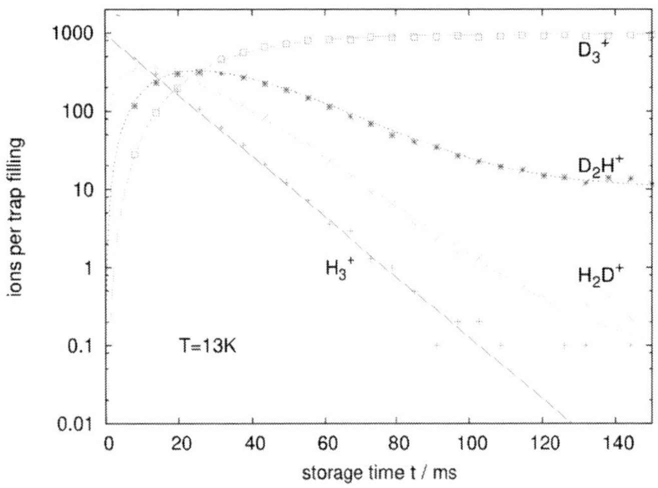

Figure 3. Measurement of the sequence of forward reactions (1-3) leading to full deuteration of H_3^+ in a bath of 13 K cold HD with a density [HD]=6.3 × 10^{10} cm^{-3}. The equilibrium reached after a storage time of 150 ms is given by endothermic hydrogenation reactions with HD. From the measured HD number density and the simulated decay rate (solid red line) for H_3^+, the rate coefficient k_1=1.3 × 10^{-9} cm^3s^{-1} is determined for reaction (1) at a nominal temperature of 13 K.

is assumed that the long lifetime of the complex leads to a full scrambling of the nuclei by the motions depicted in Figure 1. This full scrambling hypothesis enables us to derive nuclear spin statistics using the permutation group as described by Quack [28]. The rotational angular momenta statistics are derived with regular angular momentum algebra using the K (spatial) rotation group.

The mathematical details of the statistical method are described elsewhere [29]. They are quite similar to that from Park & Light [30] but it was generalised to isotopic variants. Our results are in good agreement with their results for the H_3^+ + H_2 system despite the fact that we did not account for tunnelling nor above-barrier reflection effects nor charge-quadrupole interaction terms.

The outcome of the model is a complete set of state-to-state thermal rate coefficients in the temperature range 5–50 K [29]. We did not extend the calculations to higher temperatures because (i) only low temperatures are relevant for the H_3^+-driven deuterium fractionation in interstellar clouds, (ii) breakdown of the full scrambling hypothesis is expected toward higher temperatures as more direct reactions will occur, and (iii) the calculation time increases quickly with energy because of the increasing number of accessible rotational states. The set of state-to-state rate coefficients are consistent with the detailed balance principle and the thermodynamical equilibrium constants.

4. Selected Results

4.1. Temperature dependence of rate coefficients

Already much insight can be gained into exchange processes by investigating reactions of the type (1-3), both in forward and backward direction, as a function of temperature and, if applicable, also varying the ortho/para-composition of the H_2 neutral reaction partner. As a simple example, the fundamental deuteration reaction (1), H_3^+ + HD, has been reinvestigated. Figure 3 shows an example measurement performed at the currently lowest nominal temperature of the 22-pole ion trap, T=13 K. A monoexponential decrease of the H_3^+ ions can be seen in the deuteration reaction with HD, from which the rate coefficient for reaction (1) has been deduced, k_1=1.3 × 10^{-9} cm^3s^{-1}. The rate coefficients of the following deuteration reactions leading to D_3^+ and endothermic backreactions can also be deduced from such measurements, but are not discussed here for brevity. A series of such measurements at different nominal

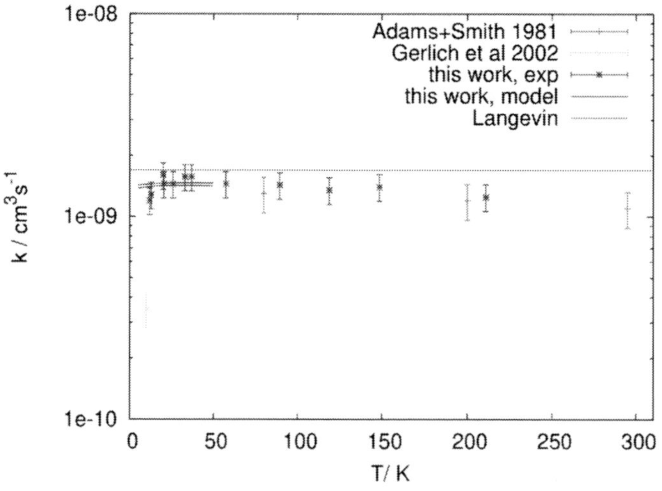

Figure 4. Temperature dependence of reaction (1). Besides the measurements of this work, with error bars in the range of 15%, also the data of Gerlich et al. [11] and Adams and Smith [1, 13] are shown, as well as the limiting Langevin rate coefficient k_L=1.7 × 10^{-9} cm^3s^{-1}. Furthermore, the results of the microcanonical model for the two lowest ortho and para rotational states of H_3^+ are shown (see text).

trap temperatures is summarized in Fig. 4 together with available literature data [11, 1]. The measured rate coefficients agree well with those of Adams and Smith [1] at higher temperature, which show an asymptotic increase towards the Langevin limit of k_L=1.7 × 10^{-9} cm^3s^{-1} when the temperature is lowered. The discrepancy to the value of Gerlich et al. [11] is disturbing as the measurements presented here were performed essentially with the same experimental setup. All calibration factors and calculations have been carefully checked. For example, the slight decrease of the rate coefficient below 30 K is most probably due to incomplete thermalization of the HD gas, leading to a temperature of the HD gas in the trap slightly higher (by an estimated 5-10 K) than the nominal wall temperature of the trap. This in turn leads to a slight overestimation of the HD density, resulting in a slightly lower value for the rate coefficient. The existence of this low-temperature artefact has been proved for an exothermic reaction which is expected to have a temperature-independent rate coefficient, as for example $N_2^+ + H_2 \rightarrow N_2H^+ + H$.
Results of the microcanonical model are also included in Fig. 4. The similarity of the curves calculated for the two lowest ortho and para rotational states shows that the rate coefficient is insensitive to the nuclear spin state of H_3^+ in this case. At the low temperatures, HD is almost exclusively in its lowest rotational state J=0. As explained in the preceeding section, the thermal rate coefficients have been calculated only for temperatures up to 50 K. The slight decrease of the rate coefficient at the low temperature end is due to endothermic reaction pathways which get energetically inaccessible. There is a good agreement with the measured data, but further measurements and calculations will have to be made to pose a more critical test upon the micocanonical model. One critical systems is D_3^++H_2 due to its endothermicity and the two possible reaction channels leading to H_2D^+ and D_2H^+.
Still, in the end, there is a discrepancy between experimental data at the lowest temperature, and further measurements have to be performed for reaction (1) and its backreaction. Gerlich et al. [11] deduced the backward rate from the forward rate with help of an equilibrium measurement using n-H_2 and p-H_2 containing the natural isotopic impurity of HD (HD/H_2= 3 × 10^{-4}). The increased value for the forward reaction rate presented here then also implies a faster backward rate of about k_{-1}=1.8 × 10^{-10} cm^3s^{-1} for n-H_2 and about k_{-1}=2.7 × 10^{-12} cm^3s^{-1} for p-H_2. To solve the discrepancies, it is planned to repeat the equilibrium measurements, as well as to determine the backward rate coefficient k_{-1} directly by injecting H_2D^+ into the ion trap containing cold n/p-H_2. As this work finds both forward and backward rates of the process (1) increased by a factor of four compared to the previous measurement [11], it is an interesting question to what extent it will influence current astrochemical deuteration models [4, 3, 31].

4.2. Overtone spectroscopy of H_2D^+ and D_2H^+

By applying the LIR detection scheme outlined in the experimental section, several rovibrational transitions originating from the four lowest levels of H_2D^+ and D_2H^+ have been measured. The detections serve not only spectroscopic purposes, but also provide a tool for further exploring low-temperature collision dynamics and non-thermal rotational populations. Similar experiments for probing the low rotational levels of H_3^+ are described in this proceedings [26]. In a first stage of the experiments, the free-electron laser FELIX [32] has been used to prove the feasibility of the detection scheme [33] shown in Fig. 2 by exciting the known fundamental vibrational bands of H_2D^+ and D_2H^+ [34, 35, 36]. Later, commercial diode laser systems operating in the range 6200-7200 cm^{-1} have been applied due to their reliability and easy handling. As the overtone and combination band spectroscopy performed with the diode lasers has been published [21], only a brief overview is given here.

Guided by high-accuracy *ab initio* calculations [37, 38, 39], 20 H_2D^+ transitions from the overtone and combination bands (0,3,0), (0,2,1) and (1,2,0) have been found, and also 7 transitions for D_2H^+ [21]. The narrow width of the lines, corresponding to a Doppler temperature in the range 24-27 K, allowed to determine the line positions with an accuracy of about 0.002 cm^{-1}. The comparison of the measured and *ab initio* predicted line positions shows differences which depend systematically on the vibrational band. In general, the differences are lower than 0.15 cm^{-1} for H_2D^+, and less than 0.05 cm^{-1} for D_2H^+, even at these high excitation energies. Furthermore, the *ab initio* predicted Einstein B coefficients have been tested for pairs of transitions starting from the same lower rotational level. No deviation between experiment and prediction could be seen within an experimental accuracy of about 10% [21]. The reliability of Einstein coefficients is not only of uttermost importance for astronomical applications, but also when is comes to the determination of level populations as shown in the next section.

4.3. Probing low rotational levels with LIR

Besides doing spectroscopy by counting the number of product ions as a function of the laser frequency, LIR can also yield additional information on the collision system. To give an example, the population of the lowest rotational states of H_2D^+ can be measured by exciting the corresponding transitions and counting the H_3^+ product ions formed by the backward direction of reaction (1). The relative population of H_2D^+ in the probed rotational state ($J_{K_aK_c}$) is then given by

$$[H_2D^+](J_{K_aK_c}) \sim \frac{[H_3^+]}{BP} \qquad (4)$$

where $[H_3^+]$ is the (background corrected) number of signal counts, B is the Einstein coefficient for absorption, and P the measured laser power. The absolute percental population of the four lowest rotational levels (see Fig. 2) can then be calculated by neglecting the population of higher-lying rotational levels with $J \geq 2$, which have an estimated population of less than 1 % at the low temperatures. First tests of this probing scheme have already been performed by exciting fundamental transitions of H_2D^+ in the 2300-3000 cm^{-1} range with the free electron laser FELIX [33], and a similar scheme using collisions with Ar atoms is used to determine the ortho/para ratio of H_3^+ ions which are used for dissociative recombination measurements at the TSR storage ring in Heidelberg [26].

Due to nuclear spin selection rules [28, 40, 29] in collisions with o/p-H_2, the low-temperature rotational population of H_2D^+ will deviate from a thermal Boltzmann distribution. As pointed out by Gerlich [11], for example, even small traces of o-H_2 can enhance the rotational population of ortho-H_2D^+, leading to an increased destruction of H_2D^+ to form H_3^+ by the backward direction of reaction (1). To probe the rotational population of H_2D^+ in a bath of n-H_2 (consisting of 75 % o-H_2 and 25 % p-H_2) at the lowest nominal temperature of 13 K, the diode laser system has been used in an automated measurement procedure, and the obtained percental population is included in Fig. 2 for the corresponding levels. As mentioned in the preceeding section, the kinetic ion temperature is measured to be about 27 K. The discrepancy to the nominal temperature has been investigated [41], and is most probably caused by small heating effects in the trap and insufficient thermalization of some parts of the trap, in particular the trapping electrodes. The measured populations yield a rotational temperature of about 27 K if one considers populations within the para and ortho species separately. On the other hand, the ortho/para ratio is enhanced. Taking into account the rotational level energies and the nuclear and rotational degeneracies, one expects only a total population of 31 % for the ortho-states at 27 K. In contrast, the measured population in Fig. 2 gives a total population of (43±4) % for the ortho-states. In summary, it is observed that the ortho-states of H_2D^+ in low-temperature collisions with n-H_2 are enhanced, and it can be speculated that this enhancement would be even more pronounced in the absence of backreactions which destroy ortho-H_2D^+ preferentially [11].

5. Future potential of low-temperature trapping and LIR

This contribution gives a brief overview of the activities to understand the exchange reactions (1-3) which are responsible for the gas-phase deuteration processes in low temperature interstellar clouds. Fig. 4 shows the temperature dependence of reaction (1), and the temperature dependence of the other isotopic exchange reactions and in particular equilibrium situations are currently under investigation as a function of reaction temperature and also for the possible different neutral collision partners: n-H_2, p-H_2, HD and n-D_2. For the production of p-H_2 a catalytic generator has been built and is now successfully used. In-situ test reactions with N^+ [14] show that a p-H_2 purity better than 99% can be obtained in the experiments.
The presented LIR-investigations applying commercial diode laser systems comprise spectroscopic determination of line positions, transition strengths and rotational populations in collision with n-H_2. Future work will extend the population measurements for different o/p-composition of the H_2 collision partner and compare the results to the microcanonical model. Furthermore, the lifetimes of highly excited states and rotationally inelastic rate coefficients of H_2D^+ ions colliding with H_2 will be measured. Such information can be obtained by considering the competition between depletion of an occupied rotational level by laser excitation and refilling by inelastic collisions with H_2. It is hoped that, together with microcanonical simulations, an insight into the state-to-state rotationally inelastic collisions can be gained.
Further interesting ions to be investigated by LIR are CH_3^+, $C_2H_2^+$ and their isotopologues. These molecules are held responsible for deuteration in high-temperature clouds due to the higher exothermicity of their exchange reactions [31, 16]. As the IR-spectroscopy of the fundamentals of these species has been well established by Oka, it is straightforward to transfer the methods presented here for H_3^+ to the carbon containing species. Unfortunately, overtone transitions for the CH_3^+ and $C_2H_2^+$ species have not yet been predicted and thus must be searched for.

Acknowledgments

This work has been financially supported by the Deutsche Forschungsgemeinschaft (DFG) via the Forschergruppe FOR 388 "Laboratory Astrophysics" and SFB494, the Nederlandse Organisatie voor Wetenschappelijk Onderzoek (NWO) via grant 614.000.415, and the European QUASAAR network. The authors thank for the excellent support by the mechanical workshop of the I. Physikalische Institut in Köln (Matthias Mondt, Jürgen Krause and their team).

References

[1] Adams N G and Smith D 1981 *Astrophys. J.* **248** 373–379
[2] Vastel C, Phillips T G and Yoshida H 2004 *Astrophys. J.* **606** L127–L130
[3] Walmsley C M, Flower D R and des Forêts G P 2004 *Astron. Astrophys.* **418** 1035–1043

[4] Roberts H, Herbst E and Millar T J 2003 *Astrophys. J* **591** L41–L44
[5] Stark R, van der Tak F F and van Dishoeck E F 1999 *Astrophys. J.* **521** L67–L70
[6] Millar T J 2003 *Space Science Reviews* **106** 73–86
[7] Lis D C, Roueff E, Gerin M, Phillips T G, Coudert L H, van der Tak F F S and Schilke P 2002 *Astrophys. J.* **571** L55–L58
[8] van der Tak F F S, Schilke P, Müller H S P, Lis D C, Phillips T G, Gerin M and Roueff E 2002 *Astron. Astrophys.* **388** L53–L56
[9] Parise B, Ceccarelli C, Tielens A G G M, Herbst E, Lefloch B, Caux E, Castets A, Mukhopadhyay I, Pagani L and Loinard L 2002 *Astron. Astrophys.* **393** L49–L53
[10] Ceccarelli C, Vastel C, Tielens A G G M, Castets A, Boogert A C A, Loinard L and Caux E 2002 *Astron. Astrophys.* **381** L17–L20
[11] Gerlich D, Herbst E and Roueff E 2002 *Planetary and Space Science* **50** 1275–1285
[12] Xie Z, Braams B J and Bowman J M 2005 *J. Chem. Phys.* **122** 224307
[13] Giles K, Adam N G and Smiths D 1992 *J. Phys. Chem.* **96** 7645–7650
[14] Gerlich D 1993 *J. Chem. Soc. Faraday Trans.* **89** 2199–2208
[15] Schlemmer S, Lescop E, v Richthofen J and Gerlich D 2002 *J. Chem. Phys.* **117** 2068–2075
[16] Asvany O, Schlemmer S and Gerlich D 2004 *Astrophys. J.* **617** 658–692
[17] Schlemmer S, Asvany O and Giesen T 2005 *Phys.Chem.Chem.Phys.* **7** 1592–1600
[18] Schlemmer S and Asvany O 2005 *J. Phys. Conf. Ser.* vol 4 pp 134–141
[19] Asvany O, Padma Kumar P, Redlich B, Hegemann I, Schlemmer S and Marx D 2005 *Science* **309** 1219–1222
[20] Asvany O, Giesen T, Redlich B and Schlemmer S 2005 *Phys. Rev. Lett.* **94** 073001
[21] Asvany O, Hugo E, Müller F, Kühnemann F, Schiller S, Tennyson J and Schlemmer S 2007 *J. Chem. Phys.* **127** 154317
[22] Gerlich D 1992 *Adv. Chem. Phys.: State-Selected and State-to-State Ion-Molecule Reaction Dynamics* vol LXXXII ed Ng C Y and Baer M (Wiley, New York) pp 1–176
[23] Gerlich D 1995 *Phys. Scr.* **T59** 256–263
[24] Mikosch J, Kreckel H, Wester R, Plašil R, Glosík J, Gerlich D, Schwalm D and Wolf A 2004 *J. Chem. Phys.* **121** 11030–11037
[25] Glosík J, Hlavenka P, Plašil R, Windisch F, Gerlich D, Wolf A and Kreckel H 2006 *Phil. Trans. R. Soc. A* **364** 2931–2942
[26] Petrignani A, Berg M, Bing D, Buhr H, Froese M, Hoffmann J, Jordon-Thaden B, Krantz C, Mendes M, Novotny O, Novotny S, Orlov D, Reinhardt S, Kreckel H and Wolf A 2007 *proceedings of DR2007*
[27] Kreckel H, Bing D, Reinhardt S, Petrignani A, Berg M and Wolf A 2008 *J. Chem. Phys.* **129** 164312
[28] Quack M 1977 *Mol. Phys.* **34** 477–504
[29] Hugo E, Asvany O and Schlemmer S 2009 *J. Chem. Phys.* **130** 164302
[30] Park K and Light J C 2007 *J. Chem. Phys.* **126** 044305
[31] Roueff E, Parise B and Herbst E 2007 *Astron. Astrophys.* **464** 245–252
[32] Oepts D, van der Meer A and van Amersfoort P 1995 *Infrared Phys. Technol.* **36** 297–308
[33] Schlemmer S, Asvany O, Hugo E and Gerlich D 2006 *IAU Symposium 231* ed Lis D, Blake G and Herbst E
[34] Foster S C, McKellar A R W, Peterkin I R, Watson J K G, Pan F S, Crofton M W, Altman R S and Oka T 1986 *J. Chem. Phys.* **84** 91–99
[35] Foster S C, McKellar A R W and Watson J K G 1986 *J. Chem. Phys.* **85** 664–670
[36] Amano T 1985 *J. Opt. Soc. Am. B* **2** 790–793
[37] Polyansky O L and Tennyson J 1999 *J. Chem. Phys.* **110** 5056–5064
[38] Ramanlal J and Tennyson J 2004 *Mon. Not. R. Astron. Soc.* **354** 161–168
[39] the full electronic lists of H_2D^+ and D_2H^+ transitions are not published, but can be obtained from Jonathan Tennyson
[40] Oka T 2004 *J. Molec. Spetrosc.* **228** 635–639
[41] Asvany O and Schlemmer S 2009 *Int. J. Mass Spectrom.* **279** 147–155

A direct, local model of dissociative recombination of HF^+

Johanna B Roos[1], **Åsa Larson**[1] **and Ann E Orel**[2]

[1] Department of Theoretical Chemistry, School of Biotechnology, Royal Institute of Technology, S-106 91 Stockholm, Sweden

[2] Department of Applied Science, University of California Davis, CA 95616, USA

E-mail: `johanna@theochem.kth.se`

Abstract. The direct mechanism of dissociative recombination of HF^+ have been studied by propagating wave packets on 30 resonant states. The relevant electronic states have been calculated *ab initio* with electron scattering calculations and multireference configuration interaction calculations. We obtain a qualitative good agreement with experiments for energies in the range from 0.04 eV to 10 eV. Some of the structures in the experimental cross section can be explained by the direct capture and dissociation along the resonant states. To fully describe the measured cross section, the electronic couplings between the neutral states cannot be neglected.

1. Introduction

Dissociative recombination (DR) is the key process in the physics of plasmas where molecular ions are present. Resonant ion-pair formation (RIP) and DR have two modes. The direct mode of DR and RIP of HF^+ can be represented by

$$\mathrm{HF}^+(v=0) + \mathrm{e}^- \longrightarrow \mathrm{HF}^{**} \quad \longrightarrow \quad \left\{ \begin{array}{lll} \mathrm{H} & + & \mathrm{F} \\ \mathrm{H}^+ & + & \mathrm{F}^- \end{array} \right. , \tag{1}$$

and the indirect mode by

$$\mathrm{HF}^+(v=0) + e^- \rightarrow \mathrm{HF}_{nv}^{Ryd} \longrightarrow \mathrm{HF}^{**} \quad \longrightarrow \quad \left\{ \begin{array}{lll} \mathrm{H} & + & \mathrm{F} \\ \mathrm{H}^+ & + & \mathrm{F}^- \end{array} \right. . \tag{2}$$

In the present model only the direct process is considered and no electronic couplings between the neutral states are included. We then obtain the total cross section that is the sum of the two reactions in (1). When the electron is captured into the resonant state it can either autoionize (re-emit the electron) or dissociate. Autoionization is possible until the neutral state has crossed the ionic ground state and after that the resonant state becomes electronically stable. This study includes wavepacket propagation on 30 resonant states and the total cross section is obtained by summing up the individual cross sections. The cross sections for DR and RIP of HF^+ have been measured in the ion-storage ring CRYRING [1]. The cross section for DR of HF^+ is smaller than for other diatomic molecules and the ratio between RIP and DR is relatively large. Both cross sections also show structures that were attributed to competition between the direct and indirect processes [1].

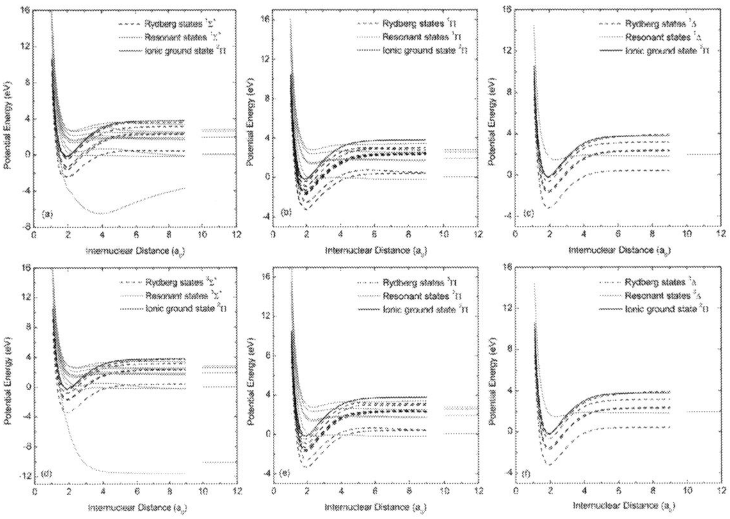

Figure 1. (a)-(e): Quasidiabatic potential energy curves of HF and HF^+

2. Potential energy curves and autoionization widths

We have determined the relevant electronic state potentials in all symmetries. The resonance energy and autoionization width are obtained from electron scattering calculations using the complex Kohn variational method [2]. In the region where the resonant states become electronically stable we have carried out a set of multireference configuration interaction (MRCI) calculations. To be able to merge the results of the different calculations we have used the same basis sets and target wave function in both the MRCI and the complex Kohn calculations.

The quasidiabatization of the resonant states relative to the continuum and the Rydberg states are performed by tracking the dominant configurations when the internuclear distance, R, is varied. Both the ionization continuum and the Rydberg states will have a configuration corresponding to the ground state of the ion $[(1\sigma)^2(2\sigma)^2(3\sigma)^2(1\pi)^3]$ plus an electron in an outer orbital. The resonant states are Rydberg states converging to electronically excited ionic cores and all of these have the (3σ) orbital singly excited. The structure calculations do not show any indications of avoided crossings between the resonant states in any of the symmetries. They are well seperated in energy so the diabatization is made by following the configuration of the excited core towards larger internuclear distances assuming no crossing between the resonant states. Also it is assumed that there is no crossing among the Rydberg states.

Outside the region where scattering calculations and structure calculations have been performed, we extrapolate autoionization widths and potentials. It is important to have a good representation of the autoionization widths in the Franck-Condon region where the overlap between the ground vibrational wave function of the HF^+ ion is nonzero out to the point where the resonant state crosses the ion potential energy curve. In this region we have performed scattering calculations and outside we extrapolate the autoionization widths smoothly towards zero. The potential energy curves are extrapolated towards their asymptotic limits which are determined from spectroscopic data for the separated atoms [3]. The Wigner-Witmer rules [4] give us the number of electronic states, of a given symmetry, that is associated with a specific asymptotic limit. The quasidiabatic potential curves are displayed in figure 1.

3. Wave packet dynamics

The electron capture induces a wave packet on the resonant state [5]

$$\Psi_i(t=0,R) = \sqrt{\frac{\Gamma_i(R)}{2\pi}}\chi_{v=0}(R), \tag{3}$$

where $\chi_{v=0}$ is the vibrational wave function of the ground state of HF^+, calculated with a finite-difference method [6]. Γ_i is the autoionization width of the resonant state i. The wave packet is then propagated by solving the time dependent Schrödinger equation with a Cranck-Nicholson propagator [7]. Autoionization is included by letting the potentials become complex above the crossing point with the ground state of HF^+

$$\tilde{V}_i(R) = V_i(R) - i\frac{\Gamma_i(R)}{2}. \tag{4}$$

This is the so called Boomerang approximation [8] that assumes that the total energy of the system is high enough to allow autoionization into a complete set of vibrational eigenstates. When this approximation breaks down, *i.e.* for the lower resonant states, a non-local operator [9; 10] for treating autoionization has to be used. This will be addressed in a future study.

4. The total cross section

When the wave packets have been propagated out in the asymptotic region they are projected onto energy normalized wave functions for the separated atoms. This gives us the transition matrix element which for the resonant state i is given by T_i

$$T_i(E) = \lim_{t\to\infty} \langle \Phi_E(R) | \Psi_i(R,t) \rangle_R . \tag{5}$$

The cross section for state i is given by[8; 11]

$$\sigma_i(E) = \frac{2\pi^3}{E} g \left| T_i(E) \right|^2 , \tag{6}$$

where g is the multiplicity ratio of the neutral resonant state to the ionic state with the free electron. The total cross section is given by the sum of the cross section of the individual states.

5. Results and conclusions

The calculated cross sections are shown in figure 2 where also the measured cross sections for DR and RIP [1] are included. Figure 2(a) and (b) show the partial cross sections for the four lowest resonant states of singlet and triplet multiplicities. In figure 2(c) we show the contributions from the resonant states with threshold energies around 1.94 eV. Finally in figure 2(d) we compare the calculated total cross section with the measured. As can be seen in figure 2(d), we have reasonably good agreement in magnitude with the experimental cross section and we find that many resonant states contribute to the total cross section. The vertical lines in figure 2(d) correspond to energies where the different resonant states become energetically open for dissociation. In agreement with our calculated cross section the experimental cross section shows a clear threshold at $E = 1.94$ eV and broad peaks in the partial cross section can be seen when the Franck-Condon overlap has a maximum for several of the resonant states. Below $E = 0.04$ eV we believe that electronic couplings need to be included in the model to reproduce the experimental cross section. Flux captured into resonant states that are energetically closed for dissociation can then leak out via the Rydberg states to states that are energetically open for dissociation. In a future study, the electronic couplings will be determined. By propagating wave packets on coupled potentials we will be able to calculate the cross section for ion-pair formation and DR branching ratios.

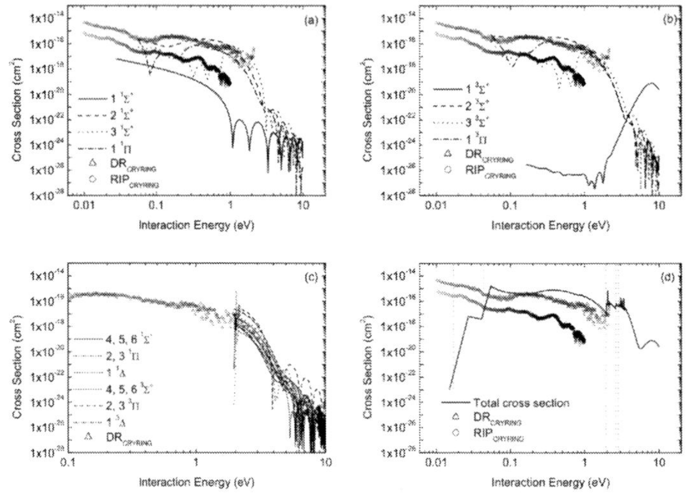

Figure 2. (a)-(c): Partial cross sections. (d): Calculated total cross section. The experimental results for DR (△) and RIP (○) measured at CRYRING are included in the figures for comparision [1]

Acknowledgments

Å.L. acknowledges support from The Swedish Research Council and A.E.O. acknowledges support from the National Science Foundation, Grant No. PHY-05-55401.

References

[1] Djurić N, Dunn G H, Al-Khalili A, Derkatch A M, Neau A, Rosén S, Shi W, Vikor L, Zong W, Larsson M, Padellec A L, Danared H and af Ugglas M 2001 *Phys. Rev.* A **64** 022713

[2] Rescigno T N, Lengsfield B H and McCurdy C W 1995 *Modern Electronic Structure Theory* vol 1 (Singapore: World Scientific) p 501

[3] Sansonetti J E, Martin W C and Young S L 2005 *Handbook of Basic Atomic Spectroscopic Data* (version 1.1.2) [online] Available: http://physics.nist.gov/Handbook [2007, April 6] national Institute of Standards and Technology, Gaithersburg, MD

[4] Herzberg G 1950 *Molecular spectra and molecular structure. Vol.1: Spectra of diatomic molecules* (New York: Van Nostrand Reinhold, 1950, 2nd ed.)

[5] McCurdy C W and Turner J L 1983 *J. Chem. Phys.* **78** 6773

[6] G D Truhlar G D 1972 *J. Comput. Phys.* **10** 123

[7] Goldberg A, Schey H M and Schwartz J L 1967 *Am. J. Phys.* **35** 177

[8] McCurdy C W and Turner J L 1983 *J. Chem. Phys.* **78** 6773

[9] Orel A E 2000 *Phys. Rev. A* **62** 020701

[10] Trevisan C S, Houfek K, Zhang Z, Orel A E, McCurdy C W and Rescigno T N 2005 *Phys. Rev. A* **71** 052714

[11] Kulander K C and Heller E J 1978 *J. Chem. Phys.* **69** 2439

Dissociative recombination as primary dissociation channel in plasma chemistry

D. C. Schram[1], R. A. B. Zijlmans[2], O. Gabriel[3], R. Engeln
Department of Applied Physics, Eindhoven University of Technology,
P.O. Box 513, 5600 MB Eindhoven, The Netherlands

E-mail: d.c.schram@tue.nl

Abstract. Molecule formation, surface modification and deposition in plasmas can in first order be described as dissociation in the plasma and association of fragments at the surface. In active plasmas ionization and dissociation by electrons is accompanied by excitation. But besides these direct electron processes also a second dissociation channel is active: that by charge transfer followed by dissociative recombination. This latter route is the dominant one in the colder recombining phase of the plasma. Atomic and molecular radicals diffuse or flow to the surface, where new molecules are formed. As a result the original molecules are, after being dissociated in the plasma, converted at the surface to new simple molecules, as H_2, CO, N_2, H_2O, O_2, NO, NH_3, HCN, C_2H_2, CH_4, to name a few in C/H/O/N containing plasmas. There is evidence that the molecular fragments resulting from dissociative recombination are ro-vibrationally (and possible electronically) excited. Also the molecules resulting from association at the surface may be ro-vibrationally or electronically excited. This may facilitate follow up processes as negative ion formation by dissociative attachment. These negative ions will be lost by mutual recombination with positive ions, giving again excited fragments. Rotational or other excitation may change considerably plasma chemistry.

1. Introduction

In many applications the use of plasma [1] relies on the first step in the process sequence: the initial dissociation of injected and produced molecules. Radicals thus formed can be transformed to other more complex radicals, but at the end they will reach the surfaces bordering the plasma. At the surface deposition [2], etching or surface modification [3] takes place. It should be realized that during these surface processes new molecules are produced. If the plasma is very active then any molecule injected or produced is dissociated at least once during the residence time. The situation is then quite different [4] from e.g. chemical vapor deposition or heterogeneous catalysis. In hot gases only very small amount of the molecules is sufficiently hot to dissociate, whereas follow up reactions are fast. Hence the surface is usually only partially occupied. In fact this is essential for catalysis to proceed. In plasmas on the contrary the radical abundance and flux to the surface is so high that surfaces become passivated and fully covered and a different "hot" chemistry can be expected.

---

1 To whom correspondence should be addressed.
2 present address: Sensor Sense, Heijendaalseweg 135, 6525 AJ Nijmegen
3 present address: PVcomB/Helmholtz-Zentrum Berlin, Schwarzschildstr. 3, 12489 Berlin, Germany

The role of radicals is evident: they form the saturated layer at the surface and they are the active ingredients in the conversion of fragments of injected molecules to new molecules (and are the precursors for deposition, etching etc.). In the formation of molecules the path through charge transfer and dissociative recombination is in many cases more important than the direct dissociation by electrons. As a nice example of this, the optical light from a nuclear fusion device, the JET Tokamak, may serve. Clearly atomic hydrogen Balmer H_α line is observed [5]. Even in this *fully atomized* atmosphere at a low pressure (< 1Pa), a significant part of this *atomic* line emission comes from dissociative recombination of *molecular* ions. These ions are a result of charge transfer of atomic ions with *molecules* coming from the surface. Hence even in this harsh atomic world the effect of the few H_2 molecules from recycling of H atoms and H^+ ions at the surface is dominantly visible in the emission. The importance of these molecular reactions, denoted by MAR, molecular assisted recombination underlines the importance of charge transfer and dissociative recombination, even in this atomic environment [6].

In earthly plasmas the situation is more molecular and more complex. It is important to realize that a plasma is out of equilibrium: it is created in the active, ionizing phase because of power input and later in time and space it is recombining when the energy sources have decayed. In the ionizing plasma the electron temperature is high to ensure its existence and thus direct ionization, dissociation (and excitation, giving line emission) is important. However, even in the ionizing phase, charge transfer from ions to molecules will give molecular ions, which then decay by dissociative recombination (giving subsequent line emission from excited atomic and/or molecular fragments) [7]. In the much longer persisting recombining phase, the direct electron ionization, excitation and dissociation are absent as the electron temperature is too low. The only reactions occurring are charge transfer and dissociative recombination. Thus for the majority of molecular plasmas dissociative recombination (preceded by charge transfer) is the dominant path for plasma fragmentation. Again (as in the atomic fusion machine) the molecules formed at the surface enter the plasma and undergo charge transfer followed by dissociative recombination (and many times line radiation). In Fig. 1 a sketch is given of a remote plasma and a photo of the actual plasma used in deposition and molecule formation studies.

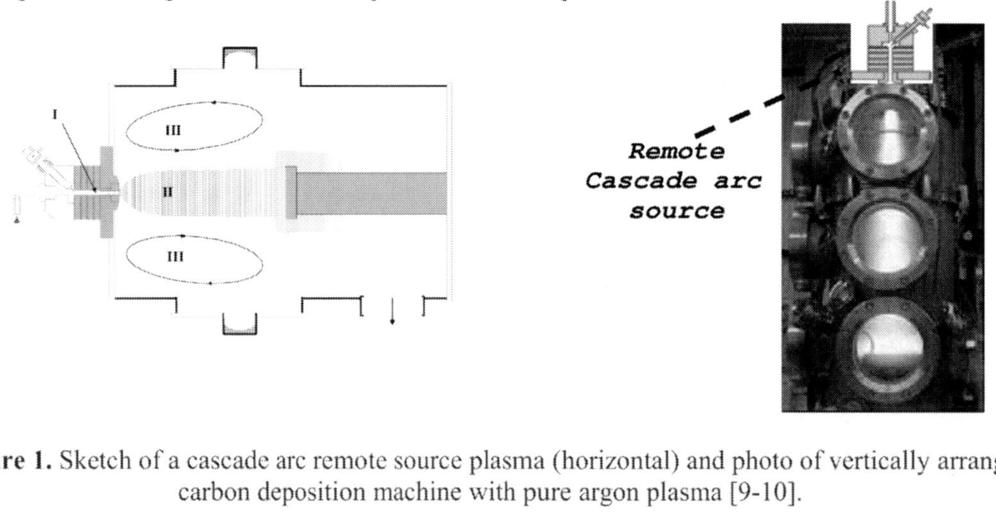

Figure 1. Sketch of a cascade arc remote source plasma (horizontal) and photo of vertically arranged carbon deposition machine with pure argon plasma [9-10].

The importance of the dissociative recombination and charge transfer was found in the search for new fast deposition methodologies [8]. Fast deposition requires a large precursor flow and this can only be reached with an efficient plasma source and an equally efficient transformation of the primary energy (Ar^+ ions for a-C:H deposition) to injected C_2H_2 molecules to radicals as C_2H one of the deposition precursors. Then a remote source approach is beneficial. In this work a high-pressure atomic source expanding in low pressure was chosen because of high source strength at moderate power. A high flux of radicals is produced just after the source by charge transfer and dissociative recombination:

$Ar^+ + C_2H_2 \rightarrow C_2H_2^+ + Ar$; $C_2H_2^+ + e \rightarrow C_2H + H$, but also $CH^* + CH$ and $C_2^* + H_2$.

The dissociative recombination is thus visible in the products and emission of CH^* and C_2^* is thus a sign of the *production* of these radicals and thus the presence of C_2H_2 and ions and electrons [9-10]. In many cases also OH^* and CN^* bands are visible, coming from ions of residual H_2O and HCN.
The process results thus in a high flux of radicals to the surface, which is capable of deposition rates up to 100 nm/s. There is however a downside in particular in the presence of H_2. If H_2 is used or generated, H radicals will generate H_2 at the surface, which will re-enter the expanding plasma, where they will cause a loss of ions, and thus chemistry, from the source.

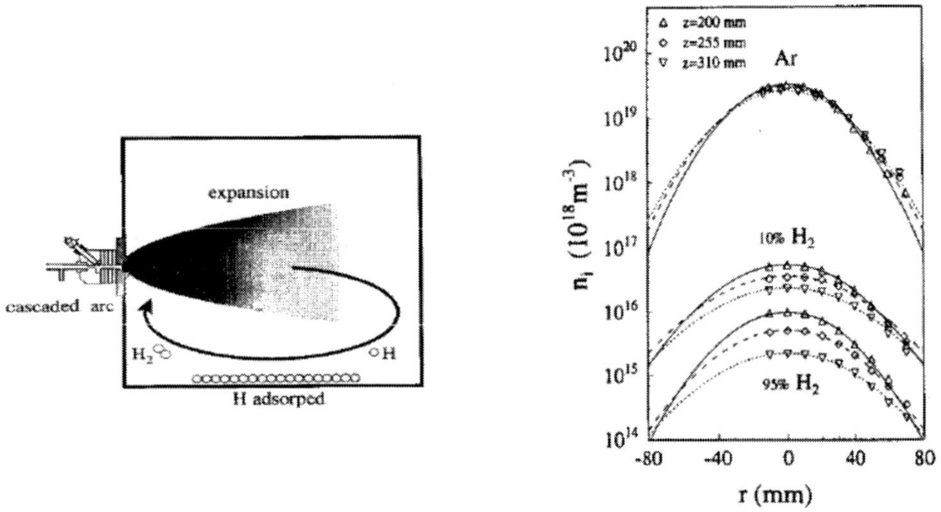

Figure 2. Hydrogen plasma (from remote cascade arc source) and electron density profiles at 40 Pa, showing the fast decrease by dissociative recombination with increasing amounts of H_2 [7].

The strong effect, which molecules have on a plasma is observed if only a small amount of hydrogen is admixed to a pure argon expanding plasma. In the atomic argon plasma the ion recombination (by 3 particle recombination) is weak and ions are preserved, whereas in the hydrogen admixture recombination of molecular ions by (two particle) dissociative recombination is strong. Even atomic ions, which normally do not recombine, are now destroyed, as they are first converted to molecular ions by charge transfer and then recombine:

$Ar^+ + H_2 \rightarrow ArH^+ + H$; $ArH^+ + e \rightarrow Ar + H^*(n \leq 4)$
$H^+ + H_2 \rightarrow H_2^+ + H$; $H_2^+ + e \rightarrow H + H^*(n \leq 3)$

Thus, if molecules are added to an atomic plasma the situation changes: charge transfer and dissociative recombination become the dominant mechanism and atomic or molecular fragment line emission from the dissociative recombination (indicated by *), forms a sign of parent molecules and the *formation* of radicals rather than their *presence*. [9,10]

2. Some ionizing and recombining plasmas

It has been argued above, that plasmas have an ionizing and a recombining phase and that the line emission shows the processes of ionization and of dissociative recombination respectively. We will illustrate this by showing some typical plasmas, varying from low pressure purely ionizing plasmas, mixed systems with ionizing and recombining parts, to purely recombining plasmas at high pressure. At the end we will return to the consequences of a high electron density within a cascade arc recombining plasma.

We will first discuss *a low pressure purely ionizing plasma*: an inductively coupled plasma (ICP) at low pressure (1 Pa). Such plasmas, used e.g. for atomic layer deposition need to be of low pressure with a large mean free path, of the order of the dimension of the system, to ensure that active particles arrive at the substrate. In such a plasma the electron temperature is high and electron impact is the dominant process for ionization and dissociation. With hydrogen the H_2^+ ions and H atoms all arrive at the surface and there new hydrogen molecules are formed by surface association. In Fig. 3 a spectrum in hydrogen is shown, taken from Heil et al [11]. In this ionizing plasma we see in excitation all neutrals present: H_2 molecules in the Fulcher bands and H atom in the H Balmer lines. As the electron and ion densities are low (10^{15} m^{-3}) the contribution of dissociative recombination to H_α emission is relatively small.

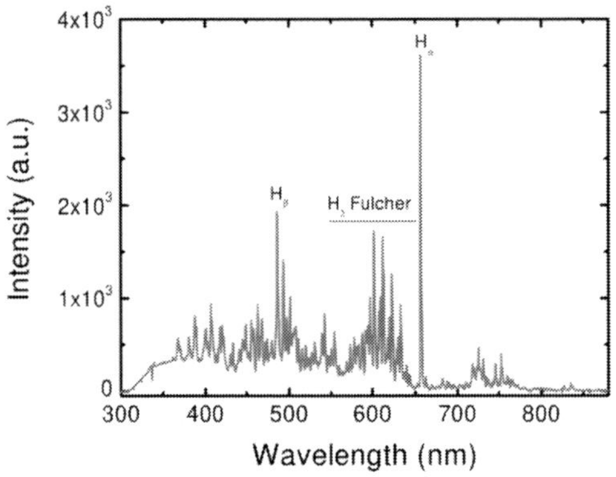

Figure 3. Light emission spectrum showing Fulcher H_2 band and H Balmer lines of a low pressure inductively coupled plasma in hydrogen [12].

If this ICP is filled with either CO or CO_2 the emission spectra are very similar: they show both the presence of CO. Apparently CO_2 is re-formed to CO (and O_2) in a sequence of dissociation and surface association. Similarly, CO* is observed in a situation that a CH_3 loaded surface is exposed to an O_2 plasma [12]. We thus see that in these purely ionizing plasmas at low pressure, with low n_e ($n_e \sim 10^{15}$ m^{-3}) and high T_e only direct excitation is visible (low n_e). Injected molecules may be totally transformed to other molecules by dissociation in the plasma and association at the surface.
The line emission associated with electron excitation can be related to ionization. For e.g. hydrogen (nuclear fusion) this factor has been calculated and (for electron density $n_e \leq 10^{19}$ m^{-3}) is typically 10^{-3} – 10^{-2}. For higher electron densities this factor decreases with n_e, because of saturation of the emission due to upward de-excitation. Thus for higher n_e the emission accompanying ionization becomes relatively small.

In recombining molecular plasmas the situation is different. Now each ion can, by charge transfer, be transformed into a molecular ion. If the electron density is large enough (> 10^{17} m^{-3} for an electron life time, τ_e, of 10^{-4} s), each ion recombines through this channel. Hence for higher n_e in ionizing plasmas and always in recombining plasmas, the light emission associated with dissociative recombination of a molecular ion (formed by charge transfer with a molecule) dominates. It is for this reason that the spectra get empty and contain only very persistent bands and lines: H^* from H_2^+, CH^* and C_2^* from $C_2H_2^+$, CN^* from HCN^+ etc. and thus spectra get very similar for different plasmas.
Dissociative recombination induced radiation becomes thus dominant if the electron density is high enough in terms of the electron life time: $n_e\ \tau_e > 10^{13}$ s/m^3. This value is met for most high density plasmas with $n_e > 10^{17}$ m-3 and life times of 10^{-4} s. But even the very low density plasmas in dark astrophysical clouds with $n_e \sim 10^2$ m^{-3} and electron life time $\tau_e \leq 10^{11}$ s will emit mostly emission lines from dissociative recombination processes. Hence a molecular plasma is easily in this domain and

dissociative recombination is (preceded by charge transfer) *the* reaction determining the chemistry and emission of the plasma.

A nice example of mixed ionizing/recombining plasma is displayed in Fig. 5: the planar microwave discharge of INP Greifswald [13]. A detail shows the existence of a small ionizing part close to the window and a larger recombining part. Part of the spectra of these low $n_e \sim 10^{17}\ m^{-3}$ plasmas, is from ionizing origin and part from the recombining one. This plasma has been used to investigate the generation of molecules in plasmas, a subject, which met renewed interest in the framework of deposition and molecule conversion by plasmas. The role of the plasma is primarily to dissociate injected molecules, whereas the formation of new molecule takes place by association of molecular fragments at the surface. The latter is saturated with radicals from the plasma. These processes should be rather similar in totally differing plasmas as dissociation needs not to be specific and association takes place at passivated surfaces. It proved that one of the easiest methods to address the issue of molecule conversion is to measure the abundances of molecules in situations that the dissociation of molecules is effective. This was done at the planar microwave reactor in INP in Greifswald [13] (and with the earlier mentioned recombining plasma of the ETP method) [14]. The Greifswald experiment (aluminum with quartz windows) is equipped with advanced infrared absorption, with which the formation of new molecules can be measured [15].

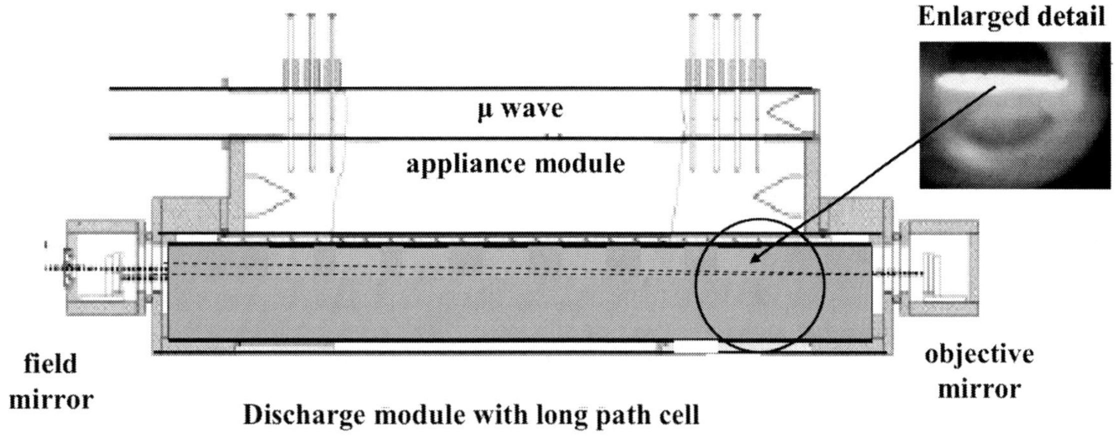

Figure 4. Sketch of planar microwave discharge at INP Greifswald; detail shows clearly ionizing and recombining regions [13, 15].

A general result is that in mixtures some dominant molecules are formed from fragments of injected molecules. In e.g. N_2 and O_2 mixtures (but also in a NO gas) the main product molecules are N_2 and O_2, but with a significant 2-5 % NO. In mixtures with H/N NH_3 is formed besides N_2 and H_2. With H/N/O mixtures H_2O appears and with carbon, CO is the dominantly produced molecule.
As an example we show in Fig. 5 the decay of formation of NH_3 and the increase of H_2O in hydrogen/nitrogen plasma to which oxygen is added. It proves also that the formation of NO is not much changed, if H_2 is added to a N_2/O_2 gas feed [14]. The explanation is that dissociated molecular fragments do stay a finite time at the surface. There they form together molecules, which subsequently desorb from the surface.
In pure gases e.g. hydrogen $H^s + H^s \rightarrow H_2$ hydrogen atoms (H^s is surface hydrogen) are re-cycled that way. In mixtures it depends on the radical fragments residing on the surface. In the absence of oxygen N and H radicals can build up to NH_2, which then, with an H atom, can form NH_3. With oxygen the OH radicals become dominant and H_2O is produced.

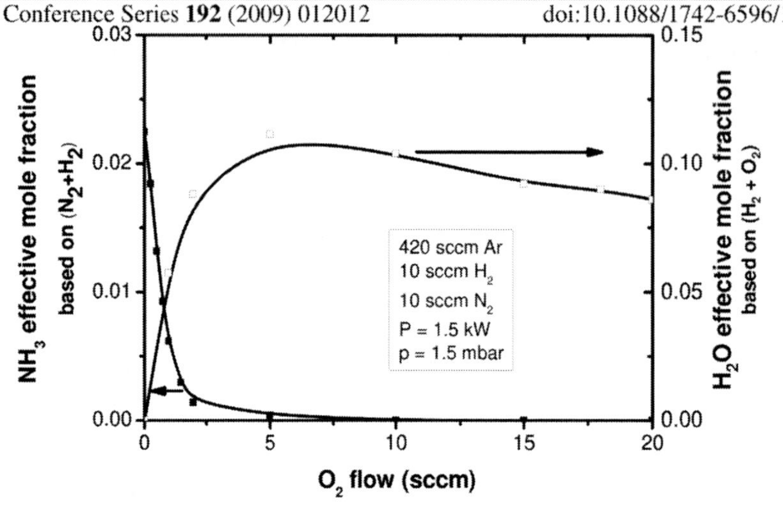

Figure 5 Decrease of NH_3 effective mol fraction and increase of H_2O with O_2 flow [14]

In the totally different Eindhoven experiment, expanding thermal plasma with stainless steel walls, very similar results have been obtained [16]. One way to explain this is that in plasma environment the surfaces are passivated and that thus the chemical specificity is decreased.

In a third very different experiment, a high-pressure plasma in a water bubble in water by P. Bruggeman [17], the light emission is primarily a result from the recombining phase. The ionization duration is short, whereas the recombination phase takes a long time. Only some bands/lines are visible in the spectrum from 200-800 nm: OH* band, H lines and a Na (or K) line in 'salt' water. Thus only two molecular ions make themselves visible in the dissociative recombination path: OH^* and H^* from H_2O^+ (or H_3O^+) and Na^* probably from NaH^+. Seeing the complexity of this experiment, the spectrum is thus remarkably simple and this leads us to a general rule: One observes *the recombination of dominant molecular ions* (in this case H_2O^+). Another fact, which becomes clear from the UV part of the spectrum in fig 6 is that a large overpopulation is visible in the OH band. In this figure also a simulated spectrum is shown with Trot ~ 1900 K necessary to cover the spectrum to 315 nm and a vibrational temperature of 7000 K to represent the peak at 315 nm. The part of the spectrum above 315 nm is due to higher J transitions and indicates a second temperature of ~ 8000 K.

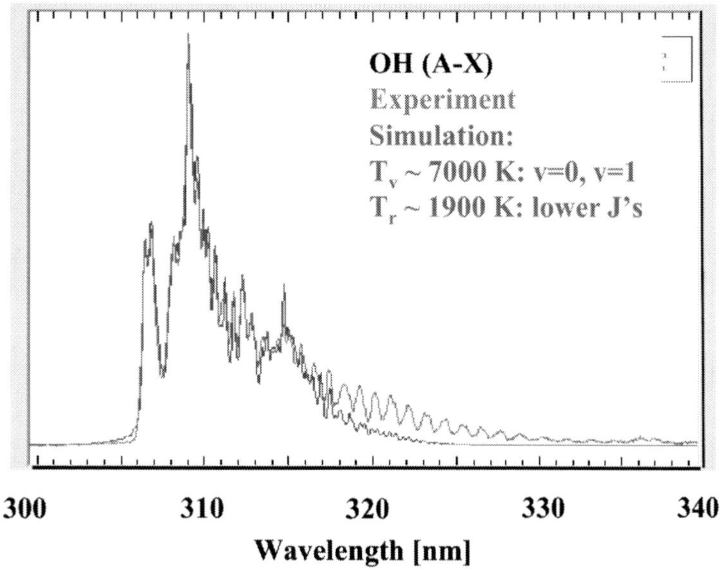

Figure 6 OH band emission spectrum from a high pressure plasma in a water bubble recorded by Bruggeman [17]. The emission at $\lambda = 315$- 340 nm points to a strong non-thermal overpopulation of high rotational levels (J>14).

This experiment is at atmospheric pressure and thus this non-equilibrium excitation points to direct observation of the dissociative recombination process $H_2O^+ + e \rightarrow OH^* + H$.
Apparently dissociative recombination of molecular ions leads to a strong overpopulation in rotation. The low levels are collisionally coupled, but the higher levels with energy differences above kT are heavily overpopulated [18]. This is an interesting fact, as this overpopulation can be rather persistent and offers a secondary energy reservoir for chemical reactions. Dissociative attachment is highly facilitated and thus the production of negative ions.

4. Recombining low T_e plasmas in remote plasma processing

We will now return to the expanding purely recombining plasmas with a remote source used for fast plasma deposition and molecule conversion developed at the Technical University of Eindhoven. In this low temperature ($T_e \leq 0.2$ eV) plasma, electron excitation is absent and charge transfer and dissociative recombination are the most important reactions. Recombination is very strong and was called initially anomalous [7]; later in the fusion community it is referred to as MAR processes [6]. We described above that this recombination forms also the main source of line emission. This sequence of charge transfer from atomic ions to molecules, forming molecular ions, which dissociatively recombine, is very present in hydrogen plasmas. The spectrum shows a large series of atomic hydrogen Balmer lines, as is shown in Fig. 7, even though the main constituents are hydrogen molecules, associate at the surface [7].
In Fig. 8 it is shown that not only red emission (H_α) but also blue emission from the higher n Balmer lines is observed. This emission is likely caused by mutual recombination of H^- and H_n^+ ions [8]. Blue emission is considered a signal of negative ion formation.

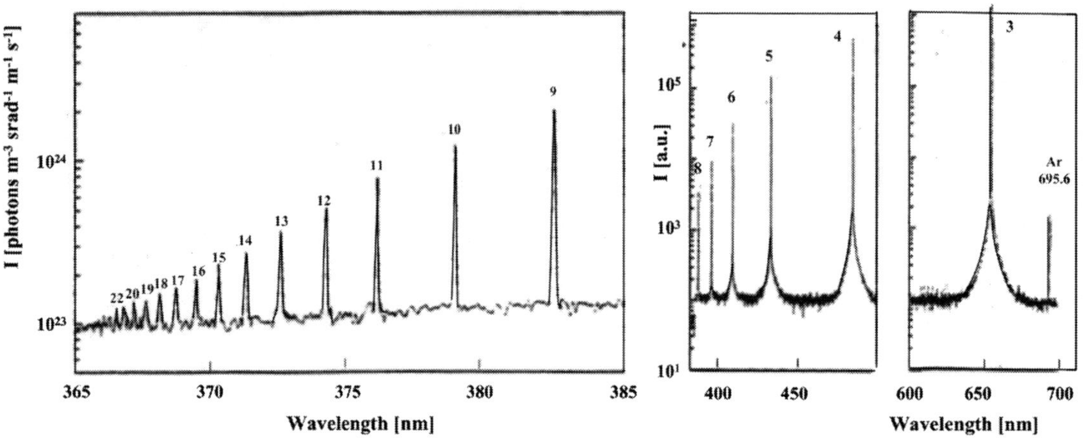

Figure 7. Measurement of Balmer series in a hydrogen expanding plasma [7]. The principal quantum number of the emitting state is indicated.

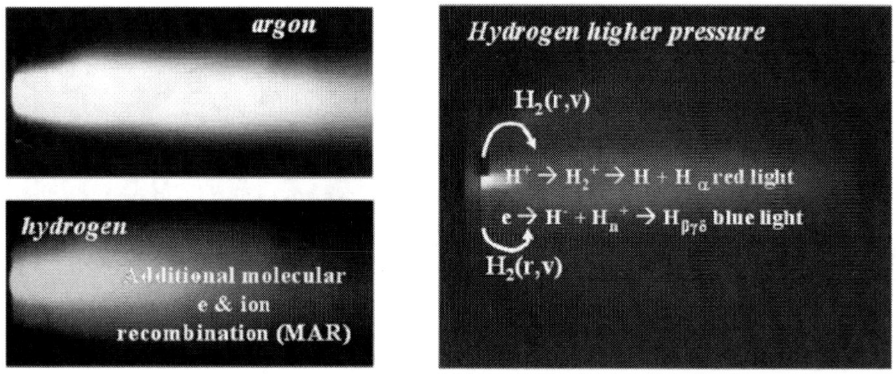

Figure 8. *Left*: Expanding plasmas in Ar and in Ar/H_2 plasmas at pressure of 40 Pa. *Right*: expanding plasma in H_2 at higher pressure, showing the blue periphery around the red central expansion.

A necessary ingredient for the process of dissociative attachment is the presence of ro-vibrationally excited molecules $H_2(r,v)$. To investigate their presence we have used VUV laser induced fluorescence, calibrated using earlier measurements with CARS. Results shown in Fig. 9 give clearly evidence for an appreciable excitation in $H_2(r,v)$ [18]. In this figure only rotational levels for v= 0, 1, 2, 4 and 6 are shown; in ref [19] more detail and also rotational levels of higher vibrational levels are given.

The main chemical route in these recombining plasmas is thus: Dissociation by plasma, transport (and possibly modification) of radicals to surfaces and then association at the surface to molecules, which can be excited. In this process the high radical flux plays also another role: it leads to a passivated and thus modified surface with possibly different chemistry. One suggestion is the existence of mobile,

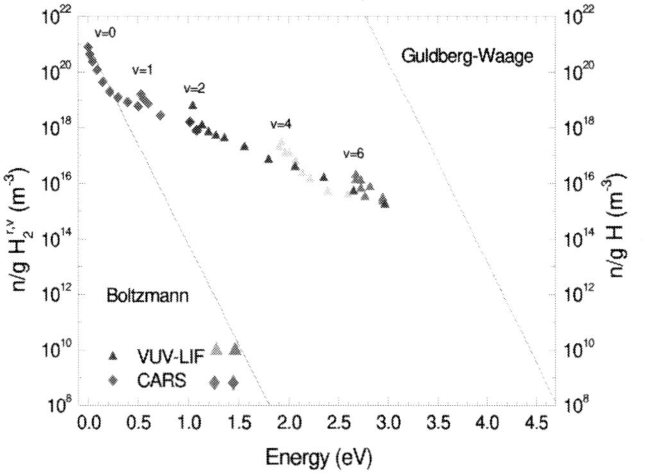

Figure 9. Measurements of $H_2(r,v)$ with CARS and VUV-LIF in hydrogen expanding plasma.

weakly bound, H atoms, which by "hot-LH" reactions produce the excited hydrogen molecules. These excited $H_2(r,v)$ molecules give then the possibility to form e.g. negative ions. Additionally, the presence of $H_2(r,v)$ may facilitate the conversion of H^+ to H_2^+ , which is then converted to H_3^+, even in very cold atmospheres. The presence of $H_2(r,v)$ may thus indirectly promote further chemistry.

More evidence of excited molecules produced at the surface can be found in cold recombining plasmas in nitrogen, where substantial radiation is observed (N* lines, first positive system of N_2(B→A) and the first negative system of N_2^+(B→X)). The N* emission is assigned to the weak three particle recombination of N^+ ions. The molecular bands cannot be explained by volume processes. One possibility, suggested by extra emission close to a substrate, is that N atoms create *excited* molecules at the surface, possibly in the B or a' state [20]. By B→A radiation these decay to metastable N_2(A) molecules. We note that also here high rotational excitation is observed. Charge transfer from N_2(A) molecules to N^+ ions can then lead to excited molecular ions N_2^+(B), which decay through the first negative system N_2^+(B→X). Hence also this observation of *excited* molecules, being desorbed from the surface could indicate the existence of weakly absorbed "hot" fragments at the surface. Another example is the appearance of the orange shuttle glow close to the surface in N_2/O_2 downstream plasmas, which is attributed to NO_2* [21]. Also this emission can only be explained, if weakly bound pre-cursors at the surface cause the generation of excited molecules.

6. Conclusions

In cold strongly recombining plasmas, as present in downstream plasmas used in remote source plasma processing, dissociative recombination in combination with charge transfer is the main channel for dissociation of molecules injected or produced. Dissociative recombination is thus a dominant process in chemical plasmas. It is there the most important source of visible and UV emission and we observe thus the dissociation process. Many times the molecular products prove to be highly excited (in particular rotationally). High rotational excitation is possibly rather stable and can facilitate further chemistry. Dissociative recombination can also be used to register the formation of new molecules by the footprint they give in emission, as CN* for $HCN^{(+)}$, OH* for $H_2O^{(+)}$, H* for $H_2^{(+)}$ etc..

In low-pressure plasmas with a high radical flux, molecule formation proceeds primarily by association of molecular fragments at saturated surfaces. In this process substantial (rotational) excitation has been observed and the example of H^- formation indicates the importance of this excitation for further chemistry. Likewise rotational excitation can facilitate the process of dissociative recombination. Hence a study of rotational excitation influence and consequence would be an important subject for plasma chemistry.

Acknowledgement
The authors acknowledge the partial support from the Dutch society for research NWO for support through a program with the Russian federation, a grant from FOM, and Euratom in fusion related studies. We thank S. Heil and other members of the group PMP at TU/e, J. Röpcke of the Leibniz Institute in Greifswald and P. Bruggeman from Gent University for clarifying discussions and for sharing observations.

References

[1] Capitelli M, Ferreira C M, Gordiets B F, Osipov A I 2000 *Plasma kinetics in atmospheric gases*. Springer-Verlag, Berlin,
[2] Kessels W M M, Severens R J, Smets A H M, Korevaar B A, Adriaenssens G J, Schram D C, van de Sanden M C M 2001 *J. Appl. Phys.* **89** 2404-2413
[3] Brussaard G J H, Letourneur K G Y, Schaepkens M, van de Sanden M C M, Schram D C 2003 *J. Vac. Sci. Technol. B* **21** 61-66
[4] Schram D C 2002 *Pure Appl. Chem.* **74** 369
[5] http://www.jet.efda.org/pages/multimedia/gallery/insidevessel/CP05j-438-01.html
[6] Ohno N, Ezumi N, Takamura S, Krasheninnikov S I, Pigarov A Yu 1998 *Phys. Rev. Lett.* **81** 818 & *comment* Meulenbroeks R F G, van de Sanden M C M, Schram D C *Phys. Rev. Lett.* **82** 2215 & *reply Phys. Rev. Lett.* **82** 2216
[7] de Graaf M J F, Severens R, Dahiya R P, van de Sanden M C M, Schram D C 1993 *Phys. Rev.* **E 48** 2098
[8] Schram D C, Otorbaev D K, Meulenbroeks R F G, van de Sanden M C M, Eerden M J J, van der Mullen J A M 1995 *Proceedings of the 1995 Workshop on Dissociative Recombination: Theory, Experiment and Applications*: Ein Gedi, Israel 29
[9] Beulens J J, Gastineau C, Gbuerrasimov N, Koulidiatti J, Schram D C 1994 *Plasma Chem. Plasma Process.* **14** 15
[10] de Graaf A, van Hest M F A M, van de Sanden M C M, Letourneur K G Y, Schram D C 1999 *Appl. Phys. Lett.* **74** 2927-2929
[11] Heil S B S, van Hemmen J L van de Sanden M C M, Kessels W M M 2008 *J. Appl. Phys.* **103** 103302.
[12] Heil S B S, Kudlacek P, Langereis E, Engeln R, van de Sanden, M C M, Kessels W M M 2006 *Appl. Phys. Lett.* **89** 131505).
[13] Zijlmans R A B, Gabriel O, Welzel S, Hempel F, Röpcke J, Engeln R, Schram D C 2006 *Plasma Sources Science & Technology* **15** 564
[14] Zijlmans R A B 2008 *Molecule conversion in recombining plasmas, PhD thesis Technische Universiteit Eindhoven (www.tue.nl)*
[15] Roepcke J, Mechold L, Kaening M, Anders J, Wienhold G, Nelson D, Zahniser M 2000 *Rev. Sci. Instrum.* **71** 3706
[16] van Helden J H 2006 *The generation of molecules through plasma surface interactions,* PhD thesis Technische Universiteit Eindhoven (www.tue.nl)
[17] Bruggeman P, Degroote J, Leys C and Vierendeels J, *J. Phys. D: Appl. Phys.* 41 (2008) 194007 Bruggeman P, Schram D C, Gonzalez M A, Rego R, Kong M G and Leys C, *Plasma Sources Sci. Technol.* 18 (2009) 025017
[18] Vankan P, Schram D C, Engeln R 2005 *Plasma Sources Science & Technology* **14** 744
[19] Gabriel O, Schram D C, Engeln R 2008 Phys Rev **E 78** 016407
[20] Brussaard G J H, Aldea E, van de Sanden M C M, Dinescu G, Schram D C 1998 *Chemical Physics letters* **290** 379
[21] van Helden J H, Zijlmans R A B, Engeln R, Schram D C 2005 *IEEE Trans. Plasma Sci.* **33** 390

Electron collisions with $^4He_2^+$ at the TSR

H Buhr[1,2], H B Pedersen[1], S Altevogt[1], V M Andrianarijaona[1], H Kreckel[1], L Lammich[1], S Novotny[1], D Strasser[2], J Hoffmann[1], M Lange[1], M Lestinsky[1], M B Mendes[1], M Motsch[1], O Novotný[1], D Schwalm[1,2], X Urbain[3], D Zajfman[2,1] and A Wolf[1]

[1] Max-Planck-Institut für Kernphysik, Heidelberg, Germany
[2] Department of Particle Physics, Weizmann Institute of Science, Rehovot, Israel
[3] Département de Physique, Université catholique de Louvain, Louvain-la-Neuve, Belgium

E-mail: henrik.buhr@mpi-hd.mpg.de

Abstract. The dissociative recombination (DR) of $^4He_2^+$ has been investigated at the heavy-ion storage ring TSR in Heidelberg. Rate coefficients were measured up to collision energies of 40 eV. Vibrational level populations were monitored using the Coulomb explosion imaging technique showing relaxation to the $v = 0$ level (>95%) through collisions with cold electrons within 50 s. Low-energy DR rate coefficients are derived for $v = 0$, 1 and ≥ 2 which show a strong v-dependence. A low-energy super-elastic collision (SEC) rate coefficient of $\alpha_{SEC}^{v=1\to 0}(E_d = 0) \approx 1.8\times 10^{-7}\,\mathrm{cm}^3\,\mathrm{s}^{-1}$ was found.

1. Introduction

The dissociative recombination (DR) of He_2^+ has raised the interest of experimental and theoretical physicists over a long time. The molecular ion potential curve has a curve crossing with a neutral dissociative state only at relatively large internuclear distances favourable for the DR of vibrational levels v=3 and higher. This should result in a very low DR rate coefficient for ground state He_2^+ at low electron energies. This was predicted by theory for both $^3He\,^4He^+$ and $^4He_2^+$ [1] and already confirmed by experiment for $^3He\,^4He^+$ [2, 3]. An energy dependent DR rate coefficient for $^4He_2^+$ was calculated in the low [1] and high (up to 15 eV) energy range [4] but only measured for vibrationally hot $^4He_2^+$ [3]. Here we report on experimental results obtained for vibrationally hot and cold $^4He_2^+$ at the TSR heavy-ion storage ring in Heidelberg [5]. The absence of a dipole moment eliminates radiation as a process that changes the vibrational distribution of $^4He_2^+$, so that in contrast to the case of $^3He\,^4He^+$, where vibrational relaxation to the ground state (>98.9 %) occurs in ~3 s [2], the effects of collisions with low-energy electrons and residual gas molecules can be studied undisturbed and over considerably longer times. Time and energy dependent DR rate coefficients under collisions with cold electrons and residual gas are presented and the changes of vibrational level populations are discussed. In addition, vibrational level dependent DR rate coefficients are derived for $v = 0$, 1 and ≥ 2. Further details are reported in a full publication [6].

2. Experiment

The experiment was performed at the TSR at the MPI für Kernphysik in Heidelberg. He_2^+ was produced in a standard duoplasmatron source, accelerated to E_i=3.83 or 8.32 MeV by an rf accelerator and injected into the storage ring, where the ion beam was stored for periods of up to 100 s. The electron cooler [7] provided velocity matched electrons for phase space cooling during the first 5 s and electrons as collision partners at variable detuning velocities v_d compared to the ion beam for the collision experiments. The detuning energy is given by $E_d = m_e v_d^2/2$ (m_e is the electron mass) and describes well the average collision energy for $E_d > kT_\perp = 11.5\,\text{meV}$, where $T_\perp$ is the transversal electron beam temperature. During DR rate coefficient measurements [8], E_d was changed periodically following a cycle of cooling (E_d^c=0), measurement (E_d^m) and reference for normalization (E_d^r) with dwell times of 25–125 ms. For E_i=3.83 MeV the electron beam density at cooling energy was $n_e(E_d^c)=5.5\times10^6\text{cm}^{-3}$ and the reference energy E_d^r=7.3 eV (Type A), for E_i=8.32 MeV it amounted to $n_e(E_d^c)=1.1\times10^7\text{cm}^{-3}$ and E_d^r=23.5 eV (Type B). Absolute DR rate coefficients were measured similarly to the way described in [2], finding $\alpha_{DR}^{v=0}(E_d= 7.4\,\text{eV})=2.0(4)\times10^{-8}\text{cm}^3\,\text{s}^{-1}$. Using the Coulomb explosion imaging (CEI) technique [9] the vibrational population distribution was measured for several periods of electron cooling between 5 and 50 s.

3. Results

3.1. Time dependence of the DR rate coefficient

The time dependence of the DR rate coefficient $\alpha_{DR}(E_d^c)$ was measured for both Type A and B parameter sets (figure 1(a)). The rate coefficient decreases strongly especially in the beginning.

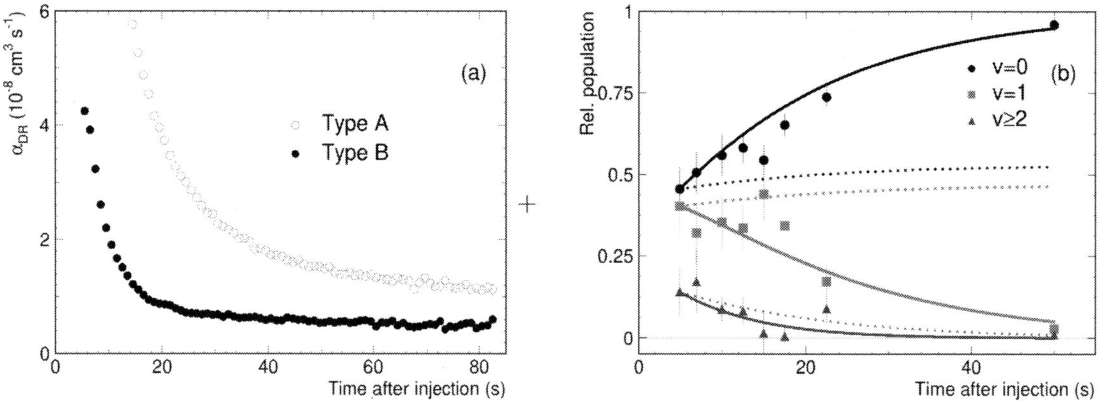

Figure 1. Measured DR rate coefficient and population distribution of vibrational levels in $^4He_2^+$ as a function of time. (a) Measured DR rate coefffficient at E_d^c as function of time for the two parameter sets and a dwell time ratio of $t_c : t_m : t_r = 5 : 0 : 1$. (b) Measured population distribution of $^4He_2^+$ with permanent electron cooling (markers, Type B) as function of time compared to model calculations using the extracted values $\alpha_{DR}^v(E_d^c)$ including (full line) and neglecting (dotted) super elastic collisions.

This effect is much stronger for the high electron density (Type B) and after 80 s a rate coefficient lower by a factor of 2 than in the Type A experiment is reached. This indicates a strong change in the vibrational population due to electron-$^4He_2^+$collisions, since the observed DR rate coefficient $\alpha_{DR}(E_d^c)$ is given by the weighted average of the vibrational-level specific DR rate coefficients $\alpha_{DR}^v(E_d^c)$

$$\alpha_{DR}(E_d^c) = \sum_v p_v \alpha_{DR}^v(E_d^c) \tag{1}$$

where p_v is the relative population of the vibrational level v, assuming the influence of rotational excitations on the α_{DR}^v to be negligible. A direct measurement of the relative populations using

Table 1. Vibrational-level specific low-energy DR rate coefficients $\alpha^v_{DR}(E^c_d)$ derived for v=0, 1 and ≥2.

vibrational level v	0	1	≥2
$\alpha^v_{DR}(E^c_d)$ (10^{-9}cm^3 s^{-1})	3.4(2.0)	$3.2^{+17.8}_{-3.2}$	202^{+153}_{-62}

the CEI technique at different times between 5 and 50 s and constant electron cooling shows that in the beginning there are 40–45 % of the $^4He_2{}^+$ ions in both the v=0 as well as in the v=1 state and ~15 % in all other states (figure 1(b)) while ~95 % are in the ground state after 50 s. A least-squares-fit combining the results for $p_v(t_{CEI})$ obtained at 8 different times t_{CEI} and $\alpha_{DR}(E^c_d)(t_{DR})$ in equation 1 with $\alpha^v_{DR}(E^c_d)$ as fitting parameters for v=0, 1 and ≥2 yields the results given in table 1. The times t_{DR} were chosen in a way that the total time spent at E^c_d matched with t_{CEI}, i. e., the time spent at electron cooling was equal in CEI and DR. Based on these values for $\alpha^v_{DR}(E^c_d)$ a simple model for the evolution of the relative vibrational populations p_v in the stored $^4He_2{}^+$ beam was applied, starting from the distribution measured after 5 s and taking into account only ion loss due to the DR reaction while the losses due to collisions with residual gas are assumed to be v-independent. The p_v resulting from this model are plotted as a function of time in figure 1(b) for the best values given in table 1 (dotted); they clearly do not agree with the measured time dependence. By including a super-elastic collision process assumed to change the vibrational level v from ≥2 to 1 and from 1 to 0 with a rate coefficient $\alpha^v_{SEC}(E^c_d)=1.8\times10^{-7}$cm^3 s^{-1}, the experimental results can be nicely modelled (full lines).

It was also tested, whether collisions with residual gas might explain the observed time dependence. For this, after 22.5 s of electron cooling, the electron beam was switched off for a certain time and the $^4He_2{}^+$ was only subject to collisions with the residual gas. Examining these ions with CEI revealed that only marginal changes in the vibrational level population occured towards higher excitation, while DR rate measurements after increasing non-cooling times showed a strongly increasing DR rate coefficient, indicating that very small fractions of the stored ions were excited to higher vibrational levels with very high $\alpha^v_{DR}(E^c_d)$. So the vibrational relaxation to v=0 as displayed in figure 1(b) is the result of super-elastic electron collisions at very low energies.

No similar time dependence of $\alpha_{DR}(E_d)$ was observed in an earlier experiment on $^4He_2{}^+$ [3], probably because SEC was suppressed due to lower electron density and higher electron beam temperature as well as due to higher residual gas density leading to vibrational excitation of the ions by collisions.

3.2. Energy dependence of DR rate coefficient as function of time

The strong time dependence of the vibrational populations p_v under electron cooling and hence of $\alpha_{DR}(E^c_d)$ implies also a strong effect on the energy dependent DR rate coefficient. This can be seen in figure 2 which compares $\alpha_{DR}(E_d)$ for different vibrational population distributions reached by different times and intensities of electron cooling. One finds that for E_d<7 eV $\alpha_{DR}(E_d)$ is decreasing with increasing electron cooling, where the effect is strongest for small E_d. At higher energies $\alpha_{DR}(E_d)$ does not change. Comparison with the measured p_v (figure 1(b)) suggests that the rate coefficient for Type B (≥58.5 s) is dominated by the vibrational ground state (p_0>95 %).

This vibrationally cold DR rate coefficient shows a broad structure between 2 and 12 eV with the peak at ~7.5 eV, which is also present, but less prominent in the measurements of Type A. Between 20 and 35 eV another structure with 3 peaks is found. Similar structures were observed in the DR of $^3He^4He^+$ [2]. The structure around 7.5 eV is also present in the calculated DR rate coefficient for the vibrational ground state [4] with the peak at 7.1 eV. However, the calculated height is considerably smaller than the one measured here, namely 0.91×10^{-8}cm^3 s^{-1}

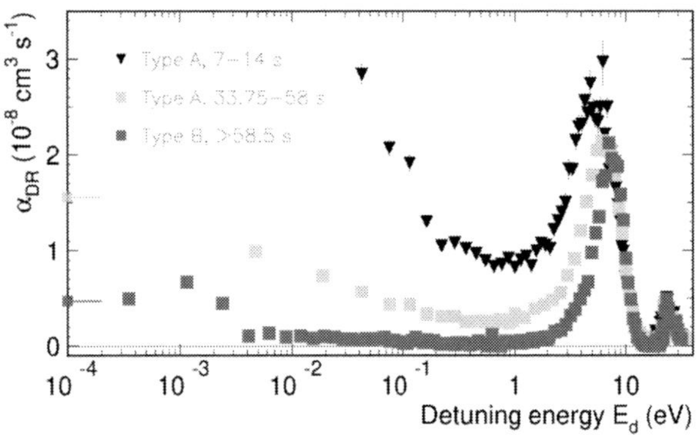

Figure 2. Measured DR rate coefficient of $^4He_2{}^+$ as function of detuning energy in the range of E_d=10^{-4}–40 eV for different total electron cooling: Type A parameters and storage times of 7–14 s and 33.75–58 s and Type B parameters and storage times>58.5 s. The electron beam temperature was $kT_\perp = 11.5$ meV.

vs. $2.0(4)\times10^{-8}$cm^3 s^{-1}. Support for both values comes from previous measurements of the absolute DR rate coefficient of $^3He\,^4He^+$ [2, 3]. The values of the resonant dissociative excitation rate coefficient from calculation and the present experiment agree well [4].

4. Conclusions

Using results of CEI and DR rate coefficient measurements, we deduced vibrational level dependent DR rate coefficients, which increase by 2 orders of magnitude when going from v=0 ($3.4(2.0)\times10^{-9}$cm^3 s^{-1}) to $v\geq2$ ($202^{+153}_{-62}\times10^{-9}$cm^3 s^{-1}). Similar increases were found in theroretical [1] and previous experimental [2] works on $^3He\,^4He^+$. The strong decrease of $\alpha_{DR}(E_d^c)$ as a function of time reflects the vibrational cooling of the stored ions due to SEC with a rate coefficient of $\alpha_{SEC}(E_d)$=1.8×10^{-7}cm^3 s^{-1}.

Acknowledgments

This work has been funded in part by the German-Israeli Foundation for Scientific Research (GIF) under contract I-707-55.7/2001. H. B. P. acknowledges support from the European Community program IHP under contract No. HPMF-CT-2002-01833. X. U. is Senior Research Associate of the F.R.S.-FNRS.

References

[1] Carata L, Orel A E and Suzor-Weiner A 1999 *Phys. Rev.* A **59** 2804
[2] Pedersen H B *et al* 2005 *Phys. Rev.* A **72** 012712
[3] Urbain X *et al* 2005 *J. Phys.* B **38** 43
[4] Royal J and Orel A E 2007 *Phys. Rev.* A **75** 052706
[5] Habs D *et al* 1989 *Nucl. Instrum. Methods Phys. Res.* B **43** 390
[6] Buhr H *et al* 2008 *Phys. Rev.* A **77** 032719
[7] Kilgus G, Habs D, Schwalm D, Wolf A, Badnell N R and Müller A 1992 *Phys. Rev.* **A** 5730
[8] Al-Khalili A *et al* 2003 *Phys. Rev.* A **68** 042702
[9] Wester R *et al* 1998 *Nucl. Instrum. Methods Phys. Res.* A **413** 379

A Comparative Study of the DR Reactions of c-$C_3H_3^+$ and l-$C_3H_3^+$: Preliminary Theoretical Studies

D. Talbi[1], **A. P. Hickman**[2], **D. Kashinski**[2], **R. F. Malenda**[2], **and P. Redondo**[3]

[1] Université Montpellier II - GRAAL, CNRS - UMR 5024, place Eugène Bataillon, 34095 Montpellier, France

[2] Department of Physics, Lehigh University,16 Memorial Dr. E, Bethlehem, PA, 18015, USA

[3] Departamento de Quimica Fisica, Facultad de Ciencias, Universidad de Valladolid, 47005 Valladolid, Spain

E-mail: Dahbia.Talbi@graal.univ-montp2.fr, aph2@lehigh.edu

Abstract. Preliminary calculations related to the dissociative recombination (DR) of electrons with $C_3H_3^+$ have been carried out. Both the linear and cyclic isomers of this ion exist in the interstellar medium, and accurate DR rate constants for both isomers are needed for astrophysical models. The electronic structure calculations reported here yield quasi-diabatic potential energy curves that can be used to assess the efficiency of dissociation of a CH bond. The calculations confirm a favorable position of a dissociative state for the cyclic isomer and suggest that dissociation of the linear isomer is less probable. More detailed dynamical studies are planned, and a normal mode analysis of the vibrational modes of $C_3H_3^+$ is reported as the first step in that direction.

1. Introduction

The three-membered ring molecule cyclopropenylidene (c-C_3H_2), discovered in the interstellar medium (ISM) in 1985 [1], has gained attention very rapidly because of its ubiquity in space. It is not a common molecule on Earth, and it is not used in the laboratory for any synthesis. However, it is a stable interstellar species surviving even in UV exposed regions. Its linear counterpart l-C_3H_2 was detected in 1991 [2] and appears to coexist in the ISM with c-C_3H_2. It is considered in the astrochemical models that both are formed from the dissociative recombination of the corresponding ions:

$$\text{c-}C_3H_3^+ + e^- \rightarrow \text{c-}C_3H_2^+ + H \quad (1)$$

$$\text{l-}C_3H_3^+ + e^- \rightarrow \text{l-}C_3H_2^+ + H \quad (2)$$

In 1993, Abouelaziz *et al* [3] attempted an experimental determination of the rate coefficient for the DR of $C_3H_3^+$. At 300K, they measured this coefficient to be $(7 \pm 2) \times 10^{-7}$ cm^3 s^{-1}. This value, indicating an efficient DR process, is in agreement with the 1×10^{-7} cm^3 s^{-1} determined earlier at 2000K [4]. However, there are two uncertainties related to the Abouelaziz *et al* experiment. These are the state of excitation and the isomeric form of the recombining ion including whether or not $C_3H_3^+$ was in its cyclic form.

In the ISM, the c-C_3H_2/l-C_3H_2 abundance ratio, observed to be varying from the dense molecular clouds to the diffuse ones [5, 6], has been proposed to serve as a tool to probe the electronic abundances in the ISM [7]. However, in order to deduce accurate values of the fractional ionization of astrophysical objects from the c-C_3H_2 and the l-C_3H_2 abundances, the branching ratios and kinetic rates used in the related chemical networks must be more accurately known. Among the uncertainties concerning the interstellar chemistry of C_3H_2: the efficiency of the DR of the c-$C_3H_3^+$ and l-$C_3H_3^+$ ions. For the cyclic isomer, the above experimental rate constant [3] determined at 300K is used in the astrochemical models. For the linear isomer, the rate constant of the cyclic one is assumed. Since it is difficult to get experimentally a rate constant for each of these isomers at the low temperature of the ISM, a quantum chemical study of the two DR processes appears to be a good alternative. Indeed, quantum chemical calculations have proven to be extremely useful for understanding these recombining processes. Such processes may occur by direct and/or indirect mechanisms. While the former involves a direct transition from the ionic state to the final neutral dissociative state, the second one is a two-step process involving the formation of a rovibrationally excited state of the neutral molecule (mainly a Rydberg state). For this last case, either autoionization takes place or the state formed after binding the electron predissociates. Understanding the DR of an ion requires knowledge of the potential energy surfaces of the ionic, Rydberg and dissociative states of the corresponding neutral molecule. This is not a trivial task because, very often, the nature of the states changes along the reaction coordinate. Carefully designed wave functions are needed to accurately follow those transformations. But the difficulty is also in the even handed treatment required for all the states involved in the mechanism. Moreover in the present case, where we are interested by the relative efficiency of the formation of c-C_3H_2 and l-C_3H_2 from the DR of the corresponding ionic isomers, potential energy surfaces involved in both DR of c-$C_3H_3^+$ and l-$C_3H_3^+$ has to be treated at the same theoretical level.

To ensure an even-handed treatment of both the ionic, Rydberg and dissociative states of c-C_3H_3 as well as of l-$C_3H_3^+$, the correlation energy must be treated through the same CI (Configuration Interaction) procedure so as not to favor one state over another. In order to fulfill this requirement, we have adopted a methodology, already defined[8, 9], that is based on the use of orbitals of strong localized character to build the CI wave functions. It is then possible to separate, valence bonding from antibonding orbitals and antibonding from Rydberg orbitals, each of them having a well-defined spatial expansion and a clear physicochemical meaning. The advantage of such a method is that the CI n-particle space is defined in terms of orbitals whose chemical significance is clearly identified. Configurations can be selected according to chemical evidence and the type of arbitrariness associated with energy related selection processes can be avoided. More important, the CI expansion is directly obtained in a configuration space that contains the approximate diabatic representation of the problem [10]. It is therefore possible to extract an approximate diabatic representation of the states (hereafter referred to as quasi-diabatic) according to the weights of their relevant configurations in the CI wave function.

The methodology used to study the DR of c-$C_3H_3^+$ is presented elsewhere [11] and therefore will not be repeated again here. The same one has been applied to build the potential energy surfaces involved in the DR of l-$C_3H_3^+$, including the size and nature of the atomic basis set, the procedure to get the localized orbitals and the CI partitioning. The dissociation path has been chosen so as to preserve the C_{2v} symmetry of the molecule. The Alchemy program [12] has been used for the calculations.

2. The Potential Energy Curves

Thanks to the use of orbitals with a strong localized character to build the electronic states configurations for the c-C_3H_3 and l-C_3H_3 systems, it has been possible to extract quasi-diabatic potential energy curves from their CI adiabatic wave functions according to the weights of the

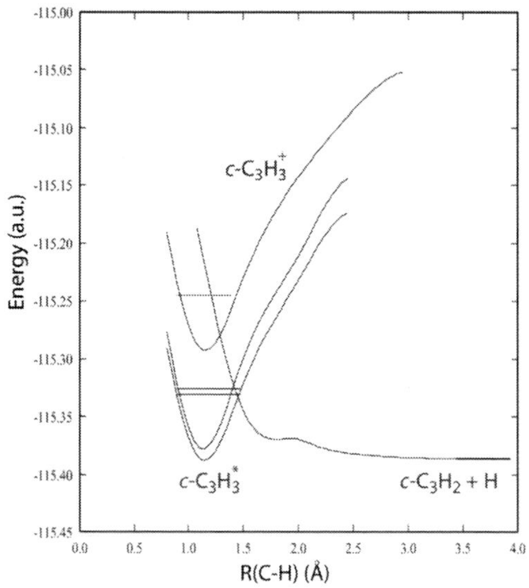

Figure 1. Quasi-diabatic potential energy surfaces for the lowest Rydberg and lowest repulsive states of c-C_3H_3 and for the lowest state of the ion. The $v = 0$ levels are shown.

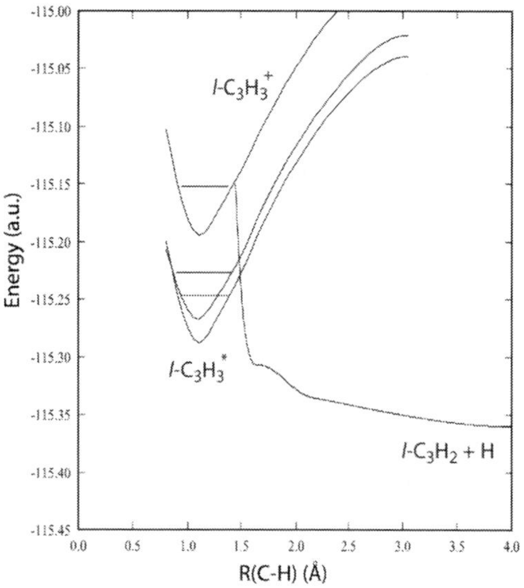

Figure 2. Quasi-diabatic potential energy surfaces for the lowest Rydberg and lowest repulsive states of l-C_3H_3 and for the lowest state of the ion. The $v = 0$ levels are shown.

relevant configurations in these CI wave function. The quasi-diabatic potential energy surfaces for the cyclic and linear isomers of C_3H_3 are reported in Figures 1 and 2, respectively. For more clearness, we have only reported the curves of interest for the present DR mechanisms (the ion, the two lowest bound Rydberg and the lowest repulsive states) but the reader has to keep in mind that there exist plenty of valence excited states of c-C_3H_3 and l-C_3H_3 not of interest for the present mechanism showing up at the energies of the figures and that have been omitted and that there exist a series of Rydberg states converging to the ion that has not been calculated.

2.1. c-$C_3H_3^+ + e \rightarrow$ c-C_3H_2 + H

The quasi-diabatic potential energy surfaces of the lowest repulsive and of the two lowest bound Rydberg states of c-C_3H_3, all of $^2A_1$ symmetry, are reported as a function of the CH linear dissociation coordinate in Figure 1 together with the potential energy curve of the corresponding ion. The next higher repulsive state of c-C_3H_3 is of $^2A_2$ symmetry. It crosses the ion at very high energies and has therefore not be reported here since it is not of interest for the present study. As can be seen the lowest $^2A_1$ repulsive state of c-C_3H_3 dissociating to ground state fragments c-C_3H_2 and H, intersects the $^1A_1$ curve of the ion at its minimum energy below the $v = 0$ energy level. The position of the crossing argues for an efficient DR mechanism for c-$C_3H_3^+(v = 0)$ to dissociate through the direct mechanism to c-C_3H_2 and H.

2.2. l-$C_3H_3^+ + e \rightarrow$ l-C_3H_2 + H

The quasi-diabatic potential energy surfaces for the lowest $^2A_1$ repulsive state of l-C_3H_3 as well as those of the two lowest $^2A_1$ Rydberg states and of the $^1A_1$ corresponding ionic state are reported Figure 2 as a function of the CH linear dissociation coordinate. A second repulsive state (of $^2B_1$ symmetry) is much higher in energy and is not shown. Figure 2 shows that the crossing between the lowest repulsive state of l-C_3H_3 (which dissociates to ground state l-C_3H_2) and the l-$C_3H_3^+$ ion is located above the $v = 0$ ground vibrational energy level of the ion. Contrary to

Journal of Physics: Conference Series **192** (2009) 012014 doi:10.1088/1742-6596/192/1/012014

the cyclic isomers, there does not exist for l-$C_3H_3^+$($v = 0$) an efficient curve crossing with the repulsive state leading to ground state l-C_3H_2.

Even if preliminary, these theoretical studies show that the formation of c-C_3H_2 from the DR of c-$C_3H_3^+$ should be more efficient that the formation of l-C_3H_2 from the DR of l-$C_3H_3^+$. This theoretical conclusion corroborates the DR experiments of Adams *et al* [13], who determined using afterglow techniques that the cyclic ion recombines more rapidly than the linear isomer at 300K. This agreement should encourage further studies on these two DR processes, including calculations of the corresponding rate coefficients and the investigation of other dissociation paths such as the C-C bond breaking. The experimental results of Mitchell *et al* [14] provide relevant branching ratios for the c-$C_3H_3^+ + e^-$ reaction.

Our discussion so far has considered dissociation of c-$C_3H_3^+$ into C_3H_2 along one of the C-H bonds; we implicitly assume that the initial CH vibration in c-$C_3H_3^+$ is governed by a single ionic potential curve. We can improve the analysis by including the energy redistribution among the three equivalent CH stretching modes (and other modes as well) by considering normal coordinates. This analysis is described next.

3. Vibrational Analysis of c-$C_3H_3^+$

As a preliminary step in the treatment of the dynamics, we have calculated the normal modes of c-$C_3H_3^+$. The normal mode analysis will enable us to formulate a realistic initial vibrational wave function for the molecular ion.

3.1. Structure, Force Constants, and Classical Dynamics

As shown in Figure 3, $C_3H_3^+$ is a planar molecule whose symmetry group is D_{3h}. An analysis using group theory shows that the molecule has 12 nontrivial vibrational modes: four are degenerate and eight come from four degenerate pairs. We set up a secular equation in terms of spring constants $k_{\rm CH}$, $k_{\rm CC}$, $k_{\rm p}$ and $k_{\rm z}$ (defined in Figure 3) and factored it into two identical 3×3 blocks (E'), one 2×2 block (A'_1) and several uncoupled (1×1) roots.

The spring constants were determined using the GAMESS code [15]. Figure 3 shows the results of our preliminary CI calculations, which were carried out at the level of single and double excitations (CISD). Table 1 compares these results with others available in the literature.

The key feature in the dynamics of $C_3H_3^+$ that may influence the dissociative recombination is the flow of energy into and out of the vibration of a single CH bond. Even if we neglect anharmonic effects and assume the energy in each normal mode is constant, then the amplitude of the CH bond vibration will not be a constant of the motion. Instead, the amplitude will fluctuate. The quasi-diabatic potential curves shown in Figure 1 show that the dissociating

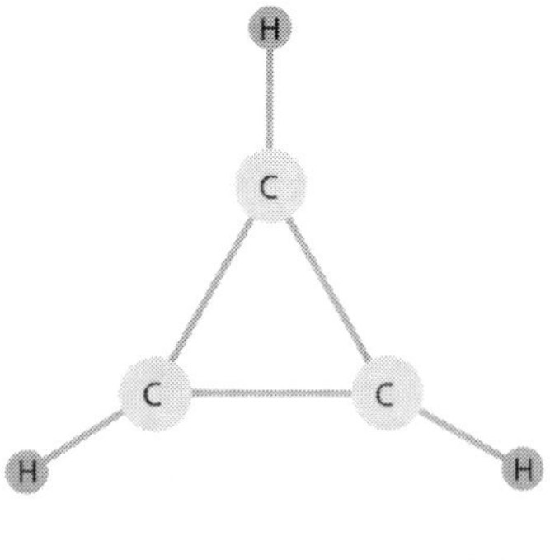

force constant	definition	value ($E_{\rm h}/{\rm Å}^2$)
$k_{\rm CH}$	CH stretch	1.439
$k_{\rm CC}$	CC stretch	1.816
$k_{\rm p}$	H wagging (in plane)	0.109
$k_{\rm z}$	H flapping (out of plane)	0.078

Figure 3. Structure and force constants of $C_3H_3^+$. We set up a secular equation in terms of the spring constants $k_{\rm CH}$, $k_{\rm CC}$, $k_{\rm p}$ and $k_{\rm z}$, and the values of these constants were calculated.

Table 1. Fundamental vibrational frequencies of $C_3H_3^+$ in cm^{-1}.

		present calculation	observed (Craig *et al*)[16]	MP2 calculation (Forney *et al*)[17]
A_1'	(breathing)	3519	3183	3333
		1810	1626	1646
A_2'	(pinwheel)	1160	(1030)	1037
E'	(in plane)	3476	3138	3283
		1460	1290	1328
		884	927	948
A_2''	(drumhead)	842	758	768
E''	(out of plane)	980	(990)	1009

curve for the cyclic isomer does cross the ionic curve. The details of the overlap between the initial state and the dissociating curve would therefore determine the magnitude of the DR rate. Speaking claassically, one would expect that at those times when the energy in a particular CH bond is large, the probability of disssociation along that bond would be large, and that at other times the probability might be weaker. This analysis supports the idea that a careful determination of the distribution of CH bond lengths should be incorporated into the theory.

3.2. Quantum Mechanical Analysis and Comparison with Classical Model

The ground state harmonic oscillator wave function for a mass-weighted normal coordinate is

$$\psi(Q) = \left(\frac{\omega}{\pi\hbar}\right)^{1/4} \exp\left(-\frac{\hbar}{2\omega}Q^2\right).$$

where Q is a normal coordinate and ω is the corresponding frequency. If a variable z can be expressed as a linear combination of normal coordinates,

$$z = \sum_{i=1}^{N} c_i Q_i,$$

one can calculate analytically the distribution of values of z:

$$P(z) = \int dQ_1\, \psi_1(Q_1)^2 \cdots \int dQ_N\, \psi_N(Q_N)^2\, \delta\left(z - \sum_{i=1}^{N} c_i Q_i\right).$$

We can interpret $P(z)$ as the quantum mechanical probability density for the distribution of values of z. By letting z be the CH internuclear distance, we can thereby obtain the range of CH distances that may occur in ground state $C_3H_3^+$. In this case, we can regard $P(z)$ as the square of the quantum mechanical wave function of the variable $z = R_{\rm CH}$. There is no closed form expression for $P(z)$, but we wrote a simple recursive algorithm to evaluate it.

In order to explore the classical motion of the $C_3H_3^+$ we performed classical simulations using the normal coordinate analysis. We randomly selected initial conditions for which the energy in each normal mode was the quantum mechanical zero point energy and then monitored the evolution of the CH bond length over a long period of time. The results are shown in Figure 4.

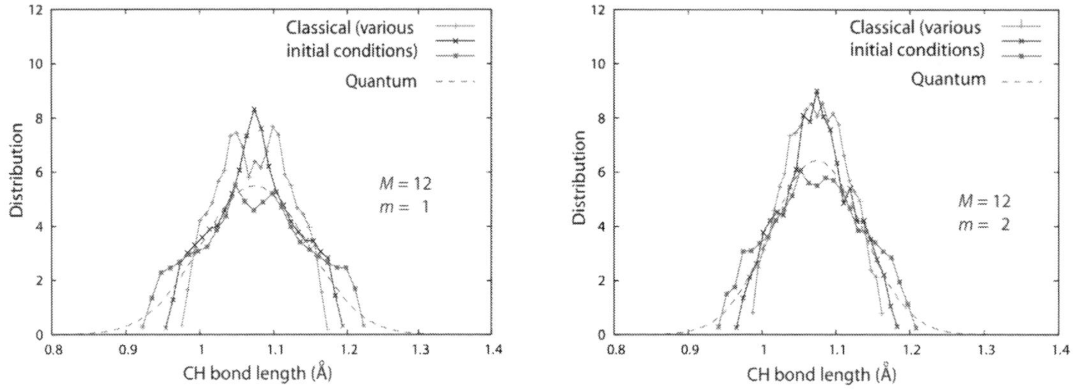

Figure 4. Distribution of CH bond lengths using the quantized normal mode formula (dashed line) and a classical simulation based on putting the zero point energy into each normal mode. M and m denote the masses of C and H in amu, respectively.

4. Concluding Remarks

The *ab initio* calculations reported here confirm the existance of a direct mechanism for the DR of electrons with c-$C_3H_3^+$ and show how a normal mode analysis of the ion can provide an appropriate initial vibrational wave function. Further work will involve the application of the block diagonalization method [18] to obtain more quantitative diabatic potential curves and coupling terms. Detailed dynamical calculation are also planned.

Acknowledgments

All electronic calculations have been performed at the French Supercomputer centers CINES and IDRIS, and the project was financially supported by the French National Program PCMI. All are gratefully acknowledged. APH and DK were supported by NSF grant PHY-0652938, and RFM was supported by REU site grant PHY-0353620 to Lehigh University.

References

[1] Thaddeus P, Vrtilek J M and Gottlieb C A 1985 *Ap. J.* **299** L63–L66
[2] Cernicharo J, Gottlieb C A, Guélin M, Killian T C, Paubert G, Thaddeus P and Vrtilek J M 1991 *Ap. J.* **368** L39
[3] Abouelaziz H, Gomet J C, Pasquerault D, Rowe B R and Mitchell J B A 1993 *J. Chem. Phys.* **98** 1
[4] Graham S M and Goodings J M 1984 *J. Mass Spectrom. Ion Proc.* **56** 1205
[5] Cernicharo J, Cox P, Fossé D and Güsten R 1999 *Astron. and Astrophys.* **351** 341
[6] Turner B E, Herbst E and Terzieva R 2000 *Ap. J. Suppl.* **126** 427
[7] Fossé D, Cernicharo J, Gerin M and Cox P 2001 *Ap. J.* **552** 168–174
[8] Talbi D, Pauzat F and Ellinger Y 1988 *Chem. Phys.* **126** 291–300
[9] Talbi D and Ellinger Y 1998 *Chem. Phys. Lett.* **288** 155–164
[10] Bacchus-Montabonnel M C and Vermeulin P 1983 *Comput. Phys. Commun.* **163** 30
[11] Talbi D 2003 *Proceedings of the ACS Meeting on Dissociative Recombination of Molecules with Electrons* ed Guberman S L (Kluwer Academic/Plenum) p 203
[12] MacLean A, Yoshimini M, Lengsfield B H, Bagus P S and Liu B ALCHEMY II in MOTEC-9
[13] Adams N and Babcock L 2005 *J. Phys. Conf. Ser.* **4** 38
[14] Mitchell J B A, Angelova G, Rebrion-Rowe C, Novotny O, Le Garrec J L, Bluhme H, Seiersen K, Svendsen A and Andersen L H 2005 *J. Physics Conf. Series* **4** 198–204
[15] Schmidt M W, Baldridge K K, Boatz J A, Elbert S T, Gordon M S, Jensen J H, Koseki S, Matsunaga N, Nguyen K A, Su S, Windus T L, Dupuis M and Montgomery Jr J A 1993 *J. Comp. Chem.* **14** 1347–1363
[16] Craig N C, Pranata J, Reinganum S J, Sprague J R and Stevens P S 1986 *J. Am. Chem. Soc.* **108** 4378–4386
[17] Forney D, Jacox M E, Lugez C L and Thompson W E 2001 *J. Chem. Phys.* **115** 8418–8430
[18] Hickman A P, Miles R D, Hayden C and Talbi D 2005 *Astron. and Astrophys.* **438** 31–37

A quantum chemical study of the $N_2H^+ + e^- \rightarrow NH + N$ reaction

D Talbi[1]
Université Montpellier II - GRAAL, CNRS - UMR 5024
place Eugène Bataillon, 34095 Montpellier, France

E-mail: Dahbia.talbi@graal.univ-montp2.fr

Abstract. A theoretical investigation of the DR of N_2H^+ to give N + NH has been undertaken because it is of interest for astrochemistry but also because the DR of N_2H^+ has been determined experimentally leading to controversial results. Using the same level of theory used to investigate the DR of N_2H^+ along the N_2 + H reaction path, the present study shows that the lowest repulsive state of N_2H leading to the NH and N fragments in their ground electronic states is likely going to cross the N_2H^+ ionic curve far above its v=0 vibrational energy level. This study suggests that the DR of N_2H^+ to give N + NH should not be efficient in the low temperature of the interstellar clouds.

1. Introduction

N_2H^+ is detected in a large number of astronomical objects [1-6]. In astrochemical models, it is considered that N_2 is almost completely recycled through the electronic dissociative recombination (hereafter DR) of N_2H^+

$$N_2H^+ + e- \rightarrow N_2 + H$$

In 2001, Geppert et al. [7] measured the DR of N_2H^+ using the CRYRING ion storage apparatus throwing doubt on the validity of the astrophysical hypothesis. The unexpected experimental results posing the problem of the mechanism involved by the DR of N_2H^+ as well as of the abundance of N_2 in space has motivated an extensive theoretical study [8] of the DR of N_2H^+. The DR of N_2H^+ along the N_2 + H channel has been first undertaken and is published elsewhere [8]. In this study and using state of the art methods of quantum chemistry it is shown that the lowest repulsive Σ state of N_2H is very low in energy. Its curve passes below the $^1\Sigma$ state of N_2H^+ and below the lowest bound $^2\Sigma$ state of linear N_2H (a Rydberg state). However it is also shown that there exist a higher repulsive $^2\Sigma$ state of N_2H (the second repulsive state) crossing the ion curve for its v=0 vibrational energy level. Moreover, even if this crossing is shown to occur at the turning point of the vibrational v=0 wave function of the ion and of the nuclear wave function of the second $^2\Sigma$ repulsive state it is also shown that for slightly bend geometries of the ion (5-10°) this second repulsive $^2\Sigma$ state of N_2H is lowered toward the minimum of the potential energy curve of the ion (see Figure 1) leading to a very efficient curve crossing because of a good overlap between the vibrational wave functions of N_2H^+ (v=0) and of the second $^2\Sigma$ repulsive state of N_2H. It is therefore expected that the DR of N_2H^+ will be efficient for slightly bent geometries of the ion, geometries that can be expected at the low temperature of the cold interstellar clouds for this loose ion.

[1] To whom any correspondence should be addressed.

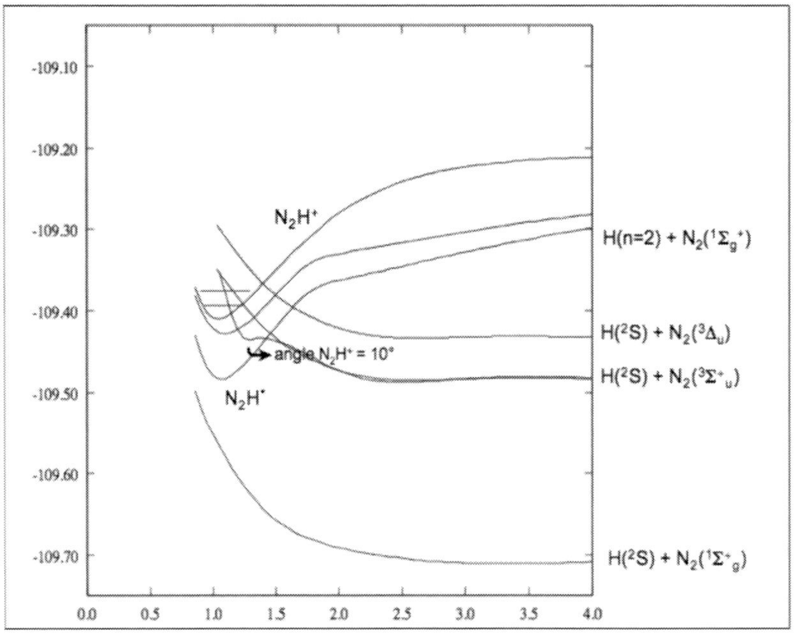

Figure 1 : Quasi-diabatic potential energy surfaces for the N_2H^+ + e- → N_2 + H reaction. Energies (in a.u.) are reported as a function of the N_2-H distances (Å). The top bound curve is the ion. The lower bound curves are the Rydberg states. While bending the NNH angle by 10° does not affect the shape of the potential energy surfaces of the ion nor of the Rydberg states, it do lower the energy of the repulsive states. This lowering is reported only for the second repulsive state, the one of interest for us in the present case.

The same theoretical study has been extended to the N + NH channel. It is presented below.

2. The theoretical approach

The theoretical method used to study the DR of N_2H^+ for the N_2 + H channel has been applied here. Therefore only the philosophy of this approach is recalled and the reader is sent to the paper of D. Talbi [8] for more details. For a quantitative study of the DR of N_2H^+ along the N + NH channel, the even handed treatment of both the ionic, the excited (mainly Rydberg) and the dissociative states of N_2H has to be insured which implies that the correlation energy for all these states has to be treated through the same CI procedure (so as not to favour one state over another). To fulfil this requirement, molecular orbitals with a strong localized character are used to build the CI wave functions. The advantage of such a method is that the CI n-particle space is defined in terms of orbitals whose chemical significance is clearly identified. Configurations can be selected according to chemical evidence and the type of arbitrariness associated with the energy related selection processes can be avoided. Moreover the CI expansion is directly obtained in a configuration space that contains the approximate diabatic representation of the problem (hereafter quasi-diabatic). Indeed, thanks to the use of orbitals with a strong localized character to construct the configuration space, it is possible to identify most configuration state functions with a quasi-diabatic state (repulsive, excited valence, Rydberg) according to the weights of the relevant configurations in the CI wave functions. The orbitals of localized character have been obtained with the MCSCF n-particle space of table 1 given in Reference [8] but with a different number of electrons in set1 and set2. In the present case where we are interested by the dissociation along the NN reaction coordinate, two electrons have been distributed in set1 (NH and NH*) and three in set2 (NN and NN*). In the same way the CI n-particle space designed to evaluate the correlation energy and to calculate the potential energy surfaces is the one reported in table 2, in Reference [8]. The 6-311G(d,p) and 4s3p1d atomic basis extended by diffuse functions, used respectively for nitrogen and hydrogen, are also the ones used in Reference [8]

. Let us mention here that thanks to the design of the same CI n-particle space to built the potential energy surfaces along the two dissociation channels i.e N_2 + H and N + NH, the DR of N_2H^+ along these two channels is treated at the same correlation level and the relative position of respective potential energy surfaces can be compared.

All the calculations have been performed in C_{2V} symmetry using ALCHEMY-CI II program [9]

3. The adiabatic and quasi-diabatic potential energy surfaces

The NH bond being almost identical in N_2H^+ and NH, the study of the [N_2H^+ + e-] complex along the NN dissociative reaction coordinate, has been carried for fixed NH bond optimized in the ion at the present MRCI level i.e. NH=1.033Å. The expected reliability of the present CI calculations has already been assessed in the previous paper [8] from the energetics deduced from the N_2H ($^2\Sigma$) and N_2H^+ ($^1\Sigma$) energies calculated at the dissociation limits within the super molecule approach. It will not be discussed here.

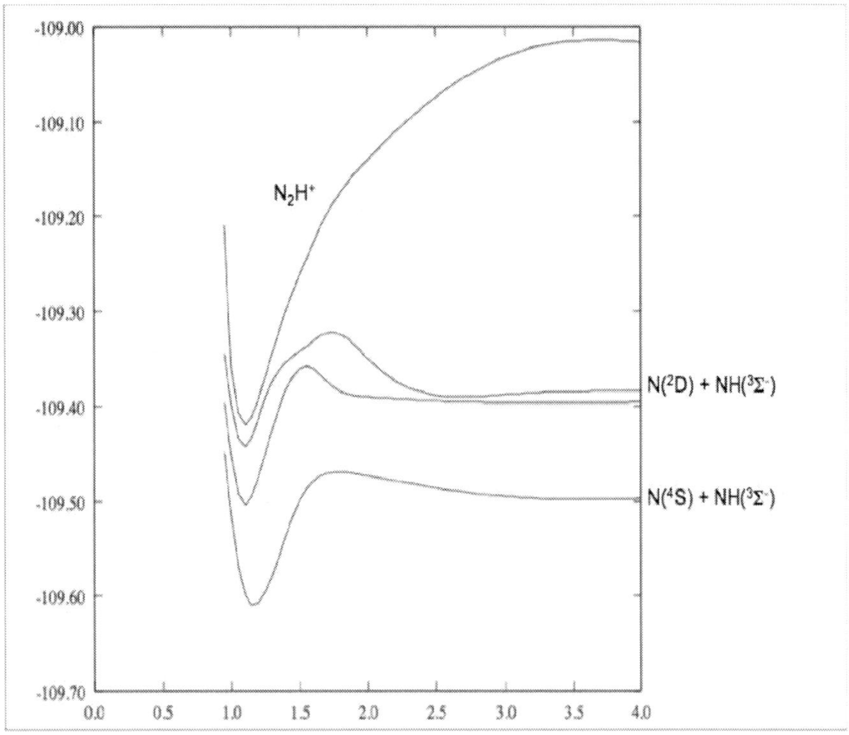

Figure 2 : $^2A_1$ adiabatic potential energy surfaces for the N_2H^+ + e- → N + NH reaction. Energies (a.u.) are reported as a function of the N-NH distances (Å). The top bound curve is the ion.

The adiabatic potential energy surfaces for the three lowest $^2A_1$ state of N_2H (states of Σ and Δ symmetry in $C_{\infty v}$ notation) as well as that of the ion N_2H^+ (X $^1\Sigma_g^+$) are presented in Figure 2. They are calculated along the NN reaction coordinate keeping the NH bond frozen at 1.033Å. This figure shows curves with strong avoided crossings indicating a change in their character. The shape of these adiabatic surfaces as well as the analysis of the corresponding MRCI wave functions argue for the existence of curve crossing in a quasi-diabatic representation, between N_2H bound excited states and repulsive ones.

The quasi-diabatic potential energy surfaces (extracted from the analysis of the MRCI wave functions) for the lowest Rydberg state and for the two lowest repulsive states of linear N_2H are reported in

Figure 3 together with the potential energy curve of the ion. Even if only the lowest Rydberg state of N_2H is shown in the figure, the reader has to keep in mind that above it exists a series of parallel Rydberg curves converging towards the ionic state. Figure 3 shows that the two lowest repulsive states of linear N_2H correlating to N (^2D) + NH (X $^3\Sigma^-$) are high in energy and are not expected to cross the curve of the ion close to its minimum of energy i.e. below v=0.

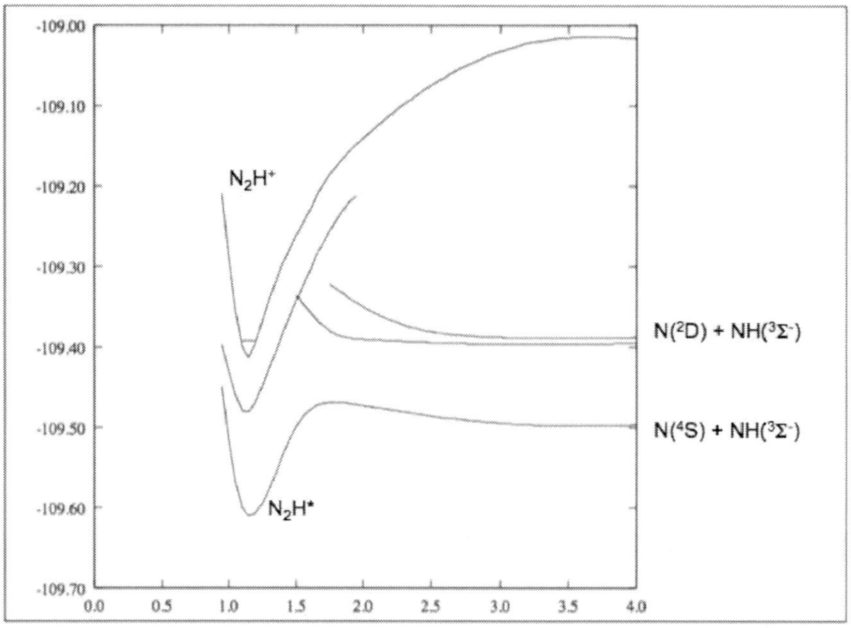

Figure 3 : Quasi-diabatic potential energy surfaces for the N_2H^+ + e- → N + NH reaction. Energies (a.u.) are reported as a function of the N-NH distances (Å). The top bound curve is the ion. The one below is the lowest Rydberg state.

4. Conclusion

The present study of the DR of N_2H^+ to give N + NH, completes the study undertaken on the DR of N_2H^+ to give N_2 + H. It appears from this study that the lowest repulsive state of N_2H along the N + NH dissociation path is much too high in energy to lead to an efficient curve crossing with the corresponding ion. Bent geometries has not yet been considered along this dissociation channel, but regarding the position of the lowest repulsive state of N_2H, we do not expect to improve the position of the crossing to a point which will lead to an efficient DR of N_2H^+ to N + NH. At the low temperature of the interstellar medium the DR of N_2H^+ should lead essentially to N_2. This conclusion is in total agreement with the experimental results of N. Adams [10].

Acknowledgments: All electronic calculations have been performed at the French Super Computer centers CINES and IDRIS and the project financially supported by the French National Program PCMI. All are gratefully acknowledged. Prof. P. Hickman is also gratefully acknowledged for constructive discussions.

References

[1] Turner B E 1974 *Astrophys. J.* **193** L83
[2] Fuente A, Martin-Pintado J, Cernicharo J and Bachiller R 1993 *Astron. Astrophys.* **276** 473
[3] Turner B E 1995 *Astrophys. J.* **449**, 635
[4] Krasnopolsky V A, Geophys J. Res. 107 (E12), DOI:10.1029/2001JE001809
[5] Caselly P, Walmsley C M, Zucconi A, Tafalla M, Doré L and Myers P C 2002 *Astrophys. J.* **565**, 473
[6] Tatematsu K, 2005 *J. of the Korean Astron. Society* **38**, 279
[7] Geppert W D *et al* 2004 *Astrophys. J.* **609**, 459
[8] Talbi D, 2007 *Chem. Phys.* **332**, 298
[9] MacLean A D, Yoshimini M, Lengsfield B H, Bagus P S and Liu B ALCHEMYII in MOTEC-90
[10] C.D. Molek, J.L. McLain, V. Poterya and N.G. Adams, 2007 *J. Phys. Chem. A*, **111**, 6760

Journal of Physics: Conference Series **192** (2009) 012016 doi:10.1088/1742-6596/192/1/012016

Rotational excitation of interstellar molecular ions by electrons

A Faure[1], J Tennyson[2], V Kokoouline[3] and Chris H Greene [4]

[1] Laboratoire d'Astrophysique, UMR 5571 CNRS, Université Joseph-Fourier, B.P. 53, 38041 Grenoble cedex 09, France

[2] Department of Physics and Astronomy, University College London, Gower Street, London WC1E 6BT, UK

[3] Department of Physics, University of Central Florida, Orlando, Florida 32816, USA

[4] Department of Physics and JILA, University of Colorado, Boulder, Colorado 80309-0440, USA

E-mail: afaure@obs.ujf-grenoble.fr

Abstract.
Electrons are known to be efficient in rotationally exciting molecular ions in cold ionized media. Rotational effects have also been shown to affect the dissociative recombination (DR) process. Electron collisions are thus expected to play a significant role in the thermalization and dissociation dynamics of molecular ions, both in the laboratory and in space. Using the molecular **R**-matrix method combined with the Adiabatic-Nuclei-Rotation (ANR) approximation corrected for threshold and closed-channel effects, we have computed new rate coefficients for the rotational excitation of H_3^+ and HCO^+ by electrons at temperatures from 10 to 1 000K. At temperatures above rotational thresholds, rotational rates are found to compete or even dominate those of dissociative recombination, suggesting that electron collisions provide a possible source of rotational (de)excitation in DR measurements.

1. Introduction

Since the discovery of CH^+ in space 70 years ago [1], more than a dozen of positive molecular ions have been detected in the interstellar medium. Among these, H_3^+ and HCO^+ and their deuterated isotopologues are particularly important in the context of interstellar gas-phase chemistry [2]. Along with CH^+, H_3^+ and HCO^+ have also been detected in diffuse clouds where the fractional ionization is high ($n(e)/n(H) \sim 10^{-4}$). In such environments, electrons play a very important role in the destruction (through DR) and in the thermalization (through inelastic collisions) of positively charged species. Moreover, these two processes are closely related because rotational effects can significantly affect the DR process, as shown for example in the case of H_2^+ [3, 4] or H_3^+ [5, 6]. These experimental and theoretical studies indeed suggest that rotational effects can increase or decrease the DR rate coefficients by factors up to ~ 3. Thus, the most recent calculations on H_3^+ [6] have shown that an improved agreement with the experimental DR measurements can be obtained by assuming rotational temperatures significantly larger than those derived in the experiments, suggesting the possibility that the ions get rotationally 'hot' before the DR rate is measured. Interestingly, in the case of diatomic ions, a theoretical study has shown that rotation plays a role mainly for the indirect DR process and for light molecules [7]. It is therefore not clear whether rotational effects are significant in the case of heavy polyatomic ions like HCO^+, even if the indirect mechanism is dominant [8].

In the astrophysical context, the sensitivity of the DR process to the rotational temperature of the ions is particularly important because the rotational populations of interstellar molecules are generally not at thermal equilibrium. The accurate determination of the ion abundances therefore requires rotationally resolved rate coefficients for both inelastic collisions and DR. This is particularly important for H_3^+ whose simple chemistry has been used to infer the primary cosmic-ray ionization rate in the diffuse interstellar medium [9]. It is to be noted, in this context, that the rotational dependence of the H_3^+ DR rate is ignored in current astrochemistry studies (e.g. [10, 11]). In the present paper, we discuss the competitive role of rotational and dissociation dynamics at low energy for the two astrophysically relevant ions H_3^+ and HCO^+. The objective is to illustrate the relative rate of rotational (de)excitation and DR for a non-polar and a highly polar ion. Section 2 briefly describes the theoretical approach employed to compute the rotational rate coefficients. Illustrative results are presented in Section 3 while conclusions are drawn in Section 4.

2. Theory of rotational excitation

The reference method for obtaining electron-impact excitation rates has been the Coulomb-Born (CB) approximation [12, 13]. This approach assumes that the collisional excitation rates are determined by long-range interactions. A standard further approximation is to consider only the dominant long-range term. Within this model, the CB theory predicts that only single jumps in rotational quanta ($\Delta J = 1$) are allowed for polar species. Recent **R**-matrix calculations combined with the adiabatic-nuclei-rotation (ANR) approximation have been applied to several molecular ions and have shown that the inclusion of short-range interactions actually lead to significant rates for transitions with $\Delta J > 1$, and particularly $\Delta J = 2$ [14]. On the other hand, the collisional propensity rules were generally found to be consistent with the CB theory. As a result, dipole-allowed transitions are always preferred for highly polar ions like HCO^+ and the K quantum number is essentially conserved for symmetric-top molecules like H_3^+ [15].

These **R**-matrix studies were, however, hampered by the use of the ANR approximation which is expected to become invalid close to a rotational threshold [16]. Thus, the previous **R**-matrix calculations applied the "kinematic correction" factor k'/k, where k (k') is the initial (final) momentum of the electron, designed for neutral targets (see [17] and references therein). Very recently, we have shown in the case of H_3^+ that this correction is in fact incorrect for molecular ions [18]. Thus, by comparison with rovibrational quantum defect theory (QDT) calculations based on the treatment described in [19], pure ANR excitation cross sections have been shown to be accurate for kinetic energies above the resonance regime caused by rotational closed-channels. These resonances were found to occur for transitions with $\Delta J = 1$ and $\Delta K = 0$ and for these, an analytical formula for averaging transition probabilities over the resonance structure was formulated. In the case of transitions with $\Delta J > 1$, the ANR theory was shown to be accurate down to threshold, provided a simple "Heaviside correction" is applied to the excitation cross sections. Full details can be found in [18]. The corrected H_3^+ rotational cross sections were thus shown to be large and finite at threshold, in accord with the Wigner law for an attractive Coulomb field, with a significant but moderate contribution from closed channels. The range of validity of the adiabatic theory is therefore much wider than the usual classically derived condition that the impacting electron energy be large compared to the threshold energy.

These results contrast with the case of neutral non-polar targets such as H_2 for which rotational cross sections fall steeply near threshold (e.g. [17]). This fondamental difference between ions and neutrals, which was actually predicted sometime ago (although the role of closed-channels was not considered) [20], reflects the influence of the strong Coulomb field which ensures that the time scale for electron motion is always rapid compared to the nuclear motion.

At the same time, the Coulomb potential provides the energy to access closed channels, thus creating substantial resonance effects. As such, H_3^+ is actually quite an unfavorable system for the ANR theory owing to its large rotational spacings that make threshold and closed-channel effects important up to large kinetic energies (> 10 meV). In the case of HCO^+, however, resonances due to the transitions $\Delta J > 1$ are not expected to contribute significantly to the dipolar rate coefficients owing to the large HCO^+ dipole (3.9 D) and its small rotational spacings. As a result, the Heaviside correction has been applied to the ANR excitation cross sections of HCO^+ but closed-channel effects were ignored. These latter will be investigated in detail in a future work.

Finally, it should be noted that the rotational rate coefficients presented below are different from those previously published [14, 21] since the threshold behaviour of the cross sections has been reconsidered and corrected. The differences are however generally modest except at low temperatures ($T < 300$ K). These new rates will be published elsewhere soon.

3. Results

Examples of rotational (de)excitation rate coefficients for H_3^+ are presented in Fig. 1 along with the experimental DR rate [22]. The rotational temperature of H_3^+ in the CRYRING experiment was estimated to be $\sim$ 20–60 K, indicating that the only levels with significant population are the para (1, 1) and ortho (1, 0) levels. It can be noticed in Fig. 1 that the rotational excitation rates peak at relatively high temperatures, well above 100 K, as a consequence of the large rotational excitation energies of H_3^+ (e.g. 250 K for (1, 1)$\rightarrow$ (2, 1)). Thus, rotational excitation rates are orders of magnitude lower than the DR rates below 100 K. On the other hand, rotational deexcitation and DR rates have comparable magnitude down to 10 K. This suggests that rotational cooling by electrons is an important mechanism in low temperature DR experiments while rotational heating is expected to occur only at larger temperature. Such effects have indeed been observed in recent long-time storage-ring experiments at the TSR facility in Heidelberg [5, 23]. A proper model of these measurements, i.e. the time dependence of the DR rate as a function of cooling time, clearly requires the inclusion of accurate state-to-state rotational rates.

Fig. 1 shows also the rotational (de)excitation rate coefficients for the same transitions, (1, 1) $\leftrightarrow$ (2,1) calculated using the scattering matrix obtained in the QDT approach. Details of the approach can be found elsewhere (see for example, Refs. [18, 19]). In this treatment, only p-wave states of the incident electron are accounted for. The s-wave does not contribute into the final rate of rotational (de)excitation, because J is conserved for the s-wave scattering. Waves with $l > 1$ do not contribute significantly because H_3^+ is a relatively compact ion: the scattering phase shifts in $l > 1$ waves are much smaller than in the p-wave. These considerations are in agreement with the results obtained with ANR method: since p-wave $H_3^+ + e^-$ collisions have the largest scattering amplitude, the rate coefficients of rotational (de)excitation are the largest for $\Delta J = 1, 2$. Examples of rotational transitions with high rate coefficients are (1,1) $\leftrightarrow$ (2,1) $\leftrightarrow$ (3,1); (1,1) $\leftrightarrow$ (3,1); (1,0) $\leftrightarrow$ (3,0). The energy-dependent cross-sections for this type of transitions have many Rydberg resonances, which belong to excited rovibrational states of the ion (see for example, Fig. 2 of Ref. [18]). The Rydberg states associated with the rotational state, which is closed for ionization, can play a significant role in rotational (de)excitation. Thermal rates do not have any resonant structure because the Rydberg resonances are smeared out by the Maxwell-Boltzmann distribution. However, the presence of the resonances in the rotational (de)excitation spectrum increases the average thermal rate (see Fig. 3 of Ref. [18]). The difference between the two theoretical results (ANR and QDT) in Fig. 1 is mainly due to these Rydberg states, which are accurately included in the QDT approach but only approximatively in ANR (through an analytical averaging [18]). Nevertheless, the agreement between the two approaches is good.

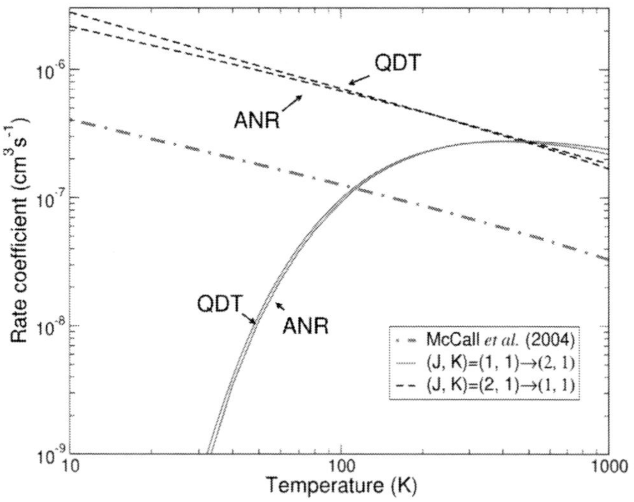

Figure 1. Rate coefficients as a function of temperature for electron$-H_3^+$ collisions. The solid and dashed lines denote typical rotational transitions for para-H_3^+, calculated using two different methods: ANR and QDT. The dash-dotted line gives the power law fit of the experimental DR rate, obtained from CRYRING ion storage-ring measurements [22].

Examples of rotational (de)excitation rate coefficients for HCO^+ are presented in Fig. 2 along with the experimental DR rate [24]. It is striking to note that the DR rate is comparable to that of H_3^+ (i.e. about 10^{-7} cm^{3}s^{-1} at 100 K) while the rotational rates are much larger, by typically two orders of magnitude. This reflects the large electron-dipole interaction which, for dipoles larger than about 2 D, completely dominate the dipolar transitions [14]. Furthermore, owing to the small HCO^+ rotational spacings (e.g. 4 K for $0 \to 1$), excitation rates for the low lying transitions do not decrease at the lowest temperatures investigated here (e.g. the $0 \to 1$ rate coefficient is still above 10^{-5} cm^{3}s^{-1} at 10 K). This suggests that both a significant rotational cooling and rotational heating by electron collisions is expected in low temperature DR measurements in HCO^+. It is, however, not clear yet whether these rotational effects will significantly affect the dissociation dynamics.

4. Conclusions

The competitive role of electron-impact rotational (de)excitation and dissociative recombination at low energy has been discussed and illustrated in the case of two astrophysically relevant ions: H_3^+ and HCO^+. Electron collisions have been shown to provide a possible source of rotational (de)excitation in DR measurements, in particular for polar and heavy ions like HCO^+. In low density astrophysical media, these collisions can also contribute to the thermalization of the ions. In this context, it should be noted that the TSR storage-ring experiments have shown that H_3^+ gets trapped in long-lived rotational states [25], an effect that is beginning to be observed astrophysically [26]. Under these circumstances electron collisions could be particularly important for determining actual (nonthermal) rotational distributions [27]. We finally note the recent suggestion to use rotational emission of HCO^+ caused by electron-impact to probe electron densities in interstellar C-type shocks [28].

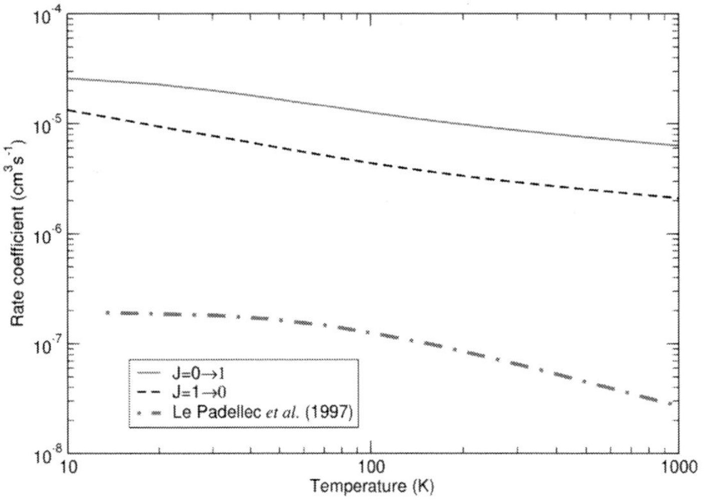

Figure 2. Rate coefficients as a function of temperature for electron–HCO^+ collisions. The solid and dashed lines denote typical rotational transitions calculated using the ANR theory. The dash-dotted line gives the experimental DR rate obtained from the experimental cross sections of Le Padellec *et al.* (1997) [24].

Acknowledgments
This work has been supported by the National Science Foundation (NSF) under PHY- 0427460 and PHY-0427376, by an allocation of NERSC supercomputing resources, and by an allocation of NCSA supercomputing resources (project No. PHY-040022). CHG also thanks the scientists at the Max-Planck-Institute for Nuclear Physics for their extraordinary hospitality, and also the Alexander von Humboldt Foundation for its partial support of this work.

References

[1] Dunham T and Adams W S 1937 *Publ. Astron. Soc. Proc.* **49** 26
[2] Herbst E 2005 *J. Phys.: Conf. Series* **4** 17
[3] Takagi H 1993 *J. Phys. B: At. Mol. Opt. Phys.* **26** 4815
[4] Zhaunerchyk V, Al-Khalili A, Thomas R D, Geppert W D, Bednarska V, Petrignani A, Ehlerding A, Hamberg M, Larsson M, Rosen S and van der Zande W J 2007 *Phys. Rev. Lett.* **99** 013201
[5] Wolf A, Kreckel H, Lammich L and et al 2006 *Roy. Soc. Lond. Phil. Trans. Series A* **364** 2981
[6] Dos Santos S F, Kokoouline V and Greene C H 2007 *J. Chem. Phys.* **127** 124309
[7] Vâlcu B, Schneider I F, Raoult M, Strömholm C, Larsson M and Suzor-Weiner A 1998 *European Physical Journal D* **1** 71
[8] Mikhailov I A, Kokoouline V, Larson Å, Tonzani S and Greene C H 2006 *Phys. Rev. A* **74** 032707
[9] McCall B J, Huneycutt A J, Saykally R J, Geballe T R, Djuric N, Dunn G H, Semaniak J, Novotny O, Al-Khalili A, Ehlerding A, Hellberg F, Kalhori S, Neau A, Thomas R, Österdahl F and Larsson M 2003 *Nature* **422** 500
[10] Le Petit F and Roueff E 2006 *Roy. Soc. Lond. Phil. Trans. Series A* **364** 3043
[11] Indriolo N, Geballe T R, Oka T and McCall B J 2007 *Astrophys. J.* **671** 1736–1747 (*Preprint* 0709.1114)
[12] Chu S I and Dalgarno A 1974 *Phys. Rev. A* **10** 788
[13] Chu S I 1975 *Phys. Rev. A* **12** 396
[14] Faure A and Tennyson J 2001 *Mon. Not. Roy. Astron. Soc.* **325** 443
[15] Faure A and Tennyson J 2002 *J. Phys. B: At. Mol. Opt. Phys.* **35** 3945
[16] Lane N F 1980 *Reviews of Modern Physics* **52** 29
[17] Morrison M A 1988 *AdV. At. Mol. Phys.* **24** 51
[18] Faure A, Kokoouline V, Greene C H and Tennyson J 2006 *J. Phys. B: At. Mol. Opt. Phys.* **39** 4261
[19] Kokoouline V and Greene C H 2003 *Phys. Rev. A* **68** 012703

[20] Chang E S and Temkin A 1970 *J. Phys. Soc. Jap.* **29** 172
[21] Faure A and Tennyson J 2003 *Mon. Not. Roy. Astron. Soc.* **340** 468–472
[22] McCall B J, Huneycutt A J, Saykally R J, Djuric N, Dunn G H, Semaniak J, Novotny O, Al-Khalili A, Ehlerding A, Hellberg F, Kalhori S, Neau A, Thomas R D, Paal A, Österdahl F and Larsson M 2004 *Phys. Rev. A* **70** 052716
[23] Shafir D, Novotny S, Buhr H, Altevogt S, Faure A, Grieser M, Harvey A G, Heber O, Hoffmann J, Kreckel H, Lammich L, Nevo I, Pedersen H B, Rubinstein H, Schneider I F, Schwalm D, Tennyson J, Wolf A and Zajfman D 2009 *Physical Review Letters* **102** 223202
[24] Le Padellec A, Sheehan C, Talbi D and Mitchell J B A 1997 *J. Phys. B: At. Mol. Opt. Phys.* **30** 319–327
[25] Kreckel H, Tennyson J, Schwalm D, Zajfman D and Wolf A 2004 *New Journal of Physics* **6** 151
[26] Oka T, Geballe T R, Goto M, Usuda T and McCall B J 2005 *Astrophys J.* **632** 882–893 (*Preprint* arXiv:astro-ph/0507463)
[27] Faure A, Wiesenfeld L, Valiron P and Tennyson J 2006 *Roy. Soc. Lond. Phil. Trans. Series A* **364** 3113
[28] Jiménez-Serra I, Martín-Pintado J, Viti S, Martín S, Rodríguez-Franco A, Faure A and Tennyson J 2006 *Astrophys. J.* **650** L135–L138 (*Preprint* arXiv:astro-ph/0609538)

Non-adiabatic effects in dissociative recombination of molecular ions

Viatcheslav Kokoouline[†], **Roman Čurík**[*], **and Chris H. Greene**[‡]

[†]Department of Physics, University of Central Florida, Orlando, Florida 32816, USA;
[*]Department of Theoretical Chemistry, J.Heyrovsky Institute of Physical Chemistry, Prague, Czech Republic;
[‡]Department of Physics and JILA, University of Colorado, Boulder, Colorado 80309-0440, USA

E-mail: `slavako@mail.ucf.edu`

Abstract. The article discusses the role of non-Born-Oppenheimer couplings in dissociative recombination (DR) of triatomic and diatomic molecular ions. For highly symmetric molecular ions the couplings play an important role, which results in a high rate for the *indirect* DR process. The non-Born-Oppenheimer couplings are particularly strong in highly-symmetric ions owing to the degeneracy of electronic and vibrational states at the equilibrium configuration of the nuclei. This is a generic property of polyatomic molecular ions, frequently associated with the Jahn-Teller or Renner-Teller effect, although analogous situations should occur in other species having conical intersections. In this respect, DR in triatomic ions differs significantly from DR in diatomic systems, where indirect DR is usually slow, although one exception to this rule is reviewed here. We concentrate here on the dissociative recombination of two highly symmetric triatomic ions: H_3^+ and HCO^+. In both ions, non-adiabatic couplings between doubly-degenerate vibrational and doubly-degenerate electronic states play an important role and produce a fast indirect DR. We also compare DR in these triatomic ions with the exceptional diatomic ion LiH^+, where non-adiabatic effects are strong due to l-mixing between p and d waves and this contributes to a high indirect DR rate. For all the systems considered, good agreement with experiments has been achieved, although some discrepancies remain to be resolved in the future.

1. Introduction

The theory of dissociative recombination for diatomic ions, e.g.

$$AB^+ + e^- \rightarrow AB^* \rightarrow A + B, \tag{1}$$

is relatively well-developed and has now successfully interpreted experimental DR rate coefficients for several such ions, including H_2^+ and its isotopologues, HeH^+, LiH^+, O_2^+, and NO^+. Historically, it was first suggested that only the direct DR mechanism could lead to a high DR rate[1]. In this mechanism, the dissociative surface crosses the ionic potential surface in the Franck-Condon region of the nuclear configuration space. In this case, the DR process can occur without proceeding through any other intermediate state AB^*. The importance of the indirect mechanism, which proceeds through intermediate Rydberg states (Eq. 1) that exhibit no such crossing between the ionic and the dissociative neutral potential curves, was recognized later [2] and found to be important in such diatomic systems as HeH^+ [3] and LiH^+ [4].

The quantitative theory of polyatomic DR, e.g.

$$\mathrm{ABC}^+ + \mathrm{e}^- \rightarrow \mathrm{ABC}^* \rightarrow \text{dissociation}, \tag{2}$$

arrived much later [5, 6], although earlier qualitative and semi-quantitative discussions exist in the literature [7]. For example, the high DR rate for the simplest triatomic ion H_3^+ proved difficult to interpret theoretically at electron energies below 2 eV. Although the experimental DR rate coefficient measured in storage rings was large ($\sim 10^{-7}\mathrm{cm}^3/\mathrm{s}$ at 300 K), it was recognized [5] that direct DR is not possible in that energy range: there is no favorable crossing between the ground ionic potential surface and the dissociative state of the neutral molecule. This property seems to be the generic feature for closed shell polyatomic target ions: the dissociative state(s) of the neutral molecule usually lies (lie) far below the energy of the closed-shell target ion, which makes the Franck-Condon factor for direct DR quite small at low incident electron energies. The high value of the experimental DR rate coefficient, despite the absence of a favorable Franck-Condon factor, indicates that the non-adiabatic couplings driving the predissociation of ABC* in Eq. (2) should play an important role. Nevertheless, the first attempts to interpret experimental data in H_3^+ at energies below 2 eV were not very successful [6]. The situation with another closed-shell molecular target ion, HCO^+, is similar: The experimental thermal DR rate is about $\sim 10^{-7}\mathrm{cm}^3/\mathrm{s}$ at 300 K, but the dissociative potential does not cross the ionic potential [8] in the Franck-Condon region.

Recently, we have developed a theoretical method [4, 9, 10, 11] to study indirect dissociative recombination in diatomic (Eq. 1) and highly-symmetric triatomic (Eq. 2) molecular ions, where strong non-adiabatic effects lead to relatively high DR cross-sections. In this article we review the theoretical approaches developed and discuss the manner in which these non-adiabatic couplings control the DR reaction in small molecular ions, in particular, in highly symmetric triatomic as well as diatomic ions.

This article is organized as follows. The next section is devoted to a qualitative discussion of non-adiabatic effects in highly symmetric triatomic ions. Section 3 discusses the formalism describing the way that non-adiabatic couplings are included in our theoretical model. In sections 4 and 5 we briefly review the way we calculate vibrational wave functions of the molecular ions and the final DR rate coefficients. Section 6 presents the calculated DR rate coefficients. Section 7 contains concluding remarks.

2. The role of non-adiabatic effects in dissociative recombination of triatomic and diatomic ions

Experiment has shown that DR cross-sections for H_3^+ and HCO^+ are quite large at energies below 2 eV. In the absence of any direct DR mechanism in this energy range for these molecules, it was suggested [5] that non-adiabatic couplings should be responsible for the fast DR rate. Theory that included only the direct mechanism gives a very low DR rate coefficient [6], orders of magnitude lower than storage ring measurements. The first attempt to include non-adiabatic couplings via intermediate Rydberg state potential surfaces in the theoretical description of H_3^+ DR was carried out by Ref. [6]. That calculation obtained a greatly increased cross-section, far above that obtained by the initial exponentially small estimates that had ignored such non-adiabatic Rydberg state couplings; but the result of this improved calculation was still orders of magnitude smaller than the storage-ring experiments. A similar situation arose with the HCO^+ ion. Larson *et al.* [8] have considered DR in HCO^+ and included non-adiabatic coupling among the $ns\sigma$, $np\sigma$, and $nd\sigma$ states. The resulting theoretical DR rate coefficient was still much lower than the available experiments [12, 13]. It appeared, therefore, that the theoretical approach which had successfully described the indirect DR mechanism for diatomics might not be adequate for these triatomic ions.

In 2001 [9], it was suggested that the Jahn-Teller mechanism, which is specific to an electron in the field of the H_3^+ ion (or other symmetric ions), can play a role in the H_3^+ DR reaction. From the point of view of Born-Oppenheimer potential surfaces, Jahn-Teller coupling has essentially the same meaning and implications as non-adiabatic coupling in diatomic ions. However, a major difference from diatomics comes from the fact that some triatomic (or more generally polyatomic) systems possess degenerate vibrational modes. If the state of the incident electron is also nontrivially degenerate in the equilibrium nuclear geometry, as is the case for $p\pi$ states of the incident electron in the field of either the HCO^+ or H_3^+ ion, the symmetric configuration of the system exhibits a phase instability (owing to the Jahn-Teller theorem, [14]). This instability causes efficient electron capture by the ion, followed by rapid dissociation of the HCO or H_3 states that are formed. From the perspective of the Born-Oppenheimer surfaces of the neutral system (i.e., of the ion plus incident electron), the presence of degenerate electronic states of the incident electron means that the bound Rydberg state potential surfaces of the neutral system exhibit the topology of a conical intersection. Moreover, this property of the bound state surfaces carries over into the fixed-nuclei electron-ion reaction- or scattering-matrix, as one expects on the basis of quantum defect theory. The position of the conical intersection corresponds to the nuclear configuration of the highest symmetry. For all other configurations, the symmetry is lower and therefore the electronic states are not degenerate. Since the non-adiabatic couplings are inverse proportional to the splitting between electronic states, they become very strong near the conical intersection. In reality, the coupling is dynamic because the system vibrates going through the conical intersection back and forth (dynamical Jahn-Teller and Renner-Teller effects). This mechanism does not exist in diatomic molecules because they cannot have degenerate vibrational modes. It is also usually not effective for molecular ions whose symmetry group does not have doubly or triply-degenerate irreducible representations. (In certain polyatomic molecules that do not have degenerate irreducible representations, similar phenomena may also be present provided there is a conical intersection of electronic states in the Franck-Condon region. However, it is not a general situation in such non-symmetric molecules.) Therefore, in highly symmetric triatomic (and polyatomic) ions the electron-vibration coupling is much stronger than in diatomic ions, and this is likely to be the case in other species possessing low energy conical intersections.

3. How to include Jahn-Teller and Renner-Teller coupling in DR treatment of triatomic ions, and non-adiabatic effects into the DR of diatomic ions

The Renner-Teller and Jahn-Teller couplings are generic non-adiabatic couplings between electronic and vibrational degrees of freedom when a $p-$wave electron moves in the field of a linear (e.g., HCO^+) or equilateral (e.g., H_3^+) polyatomic molecules. The couplings involve electronic states with non-zero projection on the molecular axis of symmetry. Such states are doubly-degenerate when the nuclear configuration is symmetric: linear configuration for HCO and equilateral configuration for H_3. The strongest Renner-Teller and Jahn-Teller couplings involve $np\pi$ states in HCO^+ and H_3^+ correspondingly. In particular, np-wave electronic states of the continuum are responsible for the large DR cross-section in both of these ions. In the following, we limit our considerations to np electronic states only.

In the present treatment the couplings are taken into account using the reaction or scattering matrix, which describes the collision of the molecular ion (HCO^+ or H_3^+) with the electron. The starting point for the construction of the scattering matrix is of course the fixed-nuclei Hamiltonian containing the non-adiabatic coupling physics in either a diagonalized (adiabatic) or off-diagonal (diabatic) representation.

For the closed-shell linear ion HCO^+, its interaction Hamiltonian H_e with a bound np-electron depends on the configuration $\mathcal{Q}$ of the three nuclei and has the following form in the basis of the $np\pi^{-1}$, $np\sigma$ and $np\pi^{+1}$ states [15]

$$H_e(Q) = \begin{pmatrix} E_\pi & \delta e^{i\varphi} & \gamma e^{2i\varphi} \\ \delta e^{-i\varphi} & E_\sigma & \delta e^{i\varphi} \\ \gamma e^{-2i\varphi} & \delta e^{-i\varphi} & E_\pi \end{pmatrix}, \tag{3}$$

where E_σ and E_π are the electronic energies of the $np\sigma$ and $np\pi^{\pm 1}$ states at the linear ionic configuration; δ and γ are the real, non-Born-Oppenheimer coupling elements [16]. The couplings δ and γ depend on the three internuclear coordinates, the distance $R_{\rm CH}$ between C and H, the distance $R_{\rm CO}$ between C and O, and the bending angle θ of the ion. The non-diagonal elements in the above Hamiltonian have additional phase factors $e^{\pm i\varphi}$ and $e^{\pm 2i\varphi}$, where φ is the polar angle of the bending orientation.

The non-diagonal coupling parameters δ and γ vanish at the linear configuration ($\theta = 0$). Thus, at the linear geometry two eigenvalues of the Hamiltonian are degenerate. For small non-zero bending angle $\theta > 0$, the coupling δ is proportional to θ. The coupling $\gamma \sim \theta^2$ is much smaller than δ in HCO.

The diagonalization of $H_e(Q)$ gives three eigenvalues $V_{\pi''}, V_\sigma$, and $V_{\pi'}$ and is accomplished by the unitary transformation matrix U [15]:

$$U^\dagger[H_e]U = \mathrm{diag}\{V_{\pi''}, V_\sigma, V_{\pi'}\}, \tag{4}$$

with eigenvalues

$$\begin{aligned} V_{\pi''} &= E_\pi - \gamma, \\ V_\pi &= \frac{1}{2}\left(E_\pi + E_\sigma + \gamma + \sqrt{(E_\pi - E_\sigma + \gamma)^2 + 8\delta^2}\right), \\ V_\sigma &= \frac{1}{2}\left(E_\pi + E_\sigma + \gamma - \sqrt{(E_\pi - E_\sigma + \gamma)^2 + 8\delta^2}\right). \end{aligned} \tag{5}$$

The eigenvalues $V_{\pi''}, V_\sigma$, and $V_{\pi'}$ depend on three internuclear coordinates and represent the adiabatic potential energy surfaces. In this approach, they are obtained from *ab initio* calculations [8]. The matrices U in Eq. (4) and H_e in Eq. (3) can be reconstructed from the adiabatic potential surfaces $V_{\pi',\pi'',\sigma}(Q)$. For example, we obtain from Eq. (5)

$$\begin{aligned} E_\sigma &= V_\sigma(\theta = 0), \\ E_\pi &= V_{\pi'}(\theta = 0), \\ \delta &= \frac{1}{\sqrt{8}}\sqrt{(V_\pi - V_\sigma)^2 - (E_\pi - E_\sigma + \gamma)^2}, \\ \gamma &= E_\pi - V_{\pi''}. \end{aligned} \tag{6}$$

The next step is the calculation of quantum defects $\mu_{\pi'}(Q)$, $\mu_{\pi''}(Q)$, and $\mu_\sigma(Q)$. To obtain them, in addition to $V_{\pi',\pi'',\sigma}(Q)$, one also needs the potential surface $V^+(Q)$ of the ground state of the ion:

$$\mu_{\pi',\pi'',\sigma} = n - \frac{1}{\sqrt{2(V^+ - V_{\pi',\pi'',\sigma})}}, \tag{7}$$

where n is the principal quantum number of the corresponding electronic state. The three values $\mu_{\pi'}, \mu_{\pi''}, \mu_\sigma$ make the diagonal matrix of quantum defects $\hat{\mu}(Q)$. The scattering $\hat{S}(Q)$ and reaction $\hat{K}(Q)$ matrices in the basis of the $np\pi^{-1}$, $np\sigma$ and $np\pi^{+1}$ electronic states are then obtained as

$$\begin{aligned} \hat{S}(Q) &= U \exp\left[2\pi i\hat{\mu}(Q)\right] U^\dagger, \text{and} \\ \hat{K}(Q) &= U \tan\left[\pi\hat{\mu}(Q)\right] U^\dagger. \end{aligned} \tag{8}$$

The matrices depend on the three internuclear distances and, in addition, on the polar angle φ of the bending orientation.

The structure of the K-matrix is similar to the structure of the Hamiltonian:

$$\hat{K}(\mathcal{Q}) = \begin{pmatrix} K_\pi & K_{\sigma\pi}e^{i\varphi} & K_{\pi\pi}e^{2i\varphi} \\ K_{\sigma\pi}e^{-i\varphi} & K_\sigma & K_{\sigma\pi}e^{i\varphi} \\ K_{\pi\pi}e^{-2i\varphi} & K_{\sigma\pi}e^{-i\varphi} & K_\pi \end{pmatrix} . \quad (9)$$

For the small angle $\theta \ll 1$, the diagonal elements K_π and K_σ are almost independent on θ; the non-diagonal elements $K_{\sigma\pi}$ and $K_{\pi\pi}$, representing the Renner-Teller coupling, are proportional to θ and θ^2 correspondingly, similar to the coupling coefficients δ and γ in Eq. 3.

For H_3^+, the electron-ion scattering matrix is constructed in a similar way. The ground electronic state of the ion is the closed-shell state. The ion has the equilateral configuration at the equilibrium position. The interaction between the ion and the np electron is described by the following Hamiltonian in the basis of $np\pi$ states:

$$H_e(\mathcal{Q}) = \begin{pmatrix} E_\pi & \delta e^{i\varphi} \\ \delta e^{-i\varphi} & E_\pi , \end{pmatrix} , \quad (10)$$

where the non-diagonal elements represent the Jahn-Teller coupling between the $np\pi$ states. In contrast to the Renner-Teller coupling, the $np\sigma$ states are not involved in the Jahn-Teller physics. The real coupling parameters δ are zero at the equilateral configuration and linear with respect to the asymmetric stretch normal coordinate ρ [11, 17] (assuming that the symmetric stretch coordinate is fixed). The angle φ is the polar angle of asymmetric stretch mode [11, 17]. The diagonal elements E_π depend weakly on the asymmetric stretch ρ for small values of ρ. The two eigenvalues of the Hamiltonian are

$$V_{1,2} = E_\pi \pm \delta . \quad (11)$$

Using a procedure similar to the one applied to matrix of Eq. (3), parameters $E_\pi(\mathcal{Q})$ and $\delta(\mathcal{Q})$ as well as the matrix U diagonalizing the Hamiltonian $H_e(\mathcal{Q})$ can be reconstructed from *ab initio* calculation. Then, the body-frame K-matrix is obtained applying Eqs. (7) and (9). It has the structure similar to Eq. (10):

$$K_e(\mathcal{Q}) = \begin{pmatrix} K_\pi & K_{\pi\pi}e^{i\varphi} \\ K_{\pi\pi}e^{-i\varphi} & K_\pi , \end{pmatrix} , \quad (12)$$

where the non-diagonal elements $K_{\pi\pi}$ are linear with ρ, the diagonal elements depend weakly on ρ.

4. Vibrational states in theory of DR in triatomic ions

Since degenerate vibrational modes have to be taken into consideration, a reduced dimensionality treatment should only be applied with caution. The included vibrational degrees of freedom should be able to represent the required degenerate modes. It means that at least two vibrational coordinates have to be explicitly taken into account. In addition, some dissociation coordinate has to be included into consideration. This makes at least three vibrational coordinates (or three vibrational modes) that must be treated explicitly. Two of the three vibrational modes can probably be approximated by normal harmonic modes for simplicity.

The complete set $\mathcal{Q}$ of vibrational coordinates for the HCO^+ ion includes the R_{CH} and R_{CO} distances, the bending angle θ of the molecule, and the polar angle φ of the bending. To represent the vibrational motion of H_3^+ we use the hyper-spherical coordinates: the hyper-radius and two hyperangles [9, 10, 11]. In the HCO^+ case, we consider the R_{CH} distance as the dissociation coordinate, because experiment shows that the main DR channel for HCO^+ is H + CO. (It would be possible to utilize the hyperradius for the dissociation coordinate in the HCO calculation also, but if so, it will be essential to not adopt an adiabatic hyperradial approximation as was done in the initial calculations for H_3^+). For H_3^+, the dissociative coordinate is the hyper-radius. The hyper-radius represents uniformly both sets of H_3^+ DR channels: H_2 + H and H + H + H. To discuss the DR in HCO^+ and H_3^+ in the same way, we use symbol R for the dissociative coordinate, and Ω for the remaining vibrational coordinates $\mathcal{Q}$.

The two different approaches to calculate the DR rate coefficients discussed in the following account for the vibrational dynamics in slightly different ways. The "simplified" [18] approach utilizes vibrational states for which the dissociative coordinate R is not quantized, but all other vibrational coordinates must be quantized. We call such partially-quantized vibrational states as adiabatic states, meaning that the dissociative coordinate R is treated as a slow-changing coordinate. The adiabatic vibrational states φ_a are obtained by diagonalizing the fixed-R vibrational Hamiltonian in the space of the Ω coordinates:

$$H^{ad}\varphi_a(R;\Omega) = U_a(R)\varphi_a(R;\Omega)\,. \tag{13}$$

$U_a(R)$ are adiabatic energies depending parametrically on the dissociation coordinate R.

In the second approach, the exact vibrational eigenstates $\Phi_{vib}(R,\Omega)$ of the vibrational Hamiltonian of the ion are needed. These states are represented as an expansion in the basis of the functions $\varphi_a(R;\Omega)$. The coefficients of expansion $\psi_a(R)$ depend on the coordinate R

$$\Phi_{vib}(R,\Omega) = \sum_a \psi_a(R)\varphi_a(R;\Omega) \tag{14}$$

and are obtained by solving the multi-channel one-dimensional Schrödinger equation using the slow-variable discretization technique [19] (see also Refs. [20, 21]).

Note that we do not need only the bound states of the multi-channel Schrödinger equation for the ion. To allow this approach to obtain outgoing continuum states also, an complex absorbing potential is placed at a large value of the the dissociative coordinate; thereby one also obtains states with complex energies whose real part lies well above the dissociation threshold of the ion. Such states of the vibrational ionic continuum are of course very highly excited with respect to the ground vibrational level of the ion. These ionic states cannot be populated in the DR process at low energies of the incident electron. But the lower Rydberg states of the neutral molecule that are "attached to" these states can be populated and, in fact, they represent the only possible dissociation pathways during the electron-ion scattering. The inclusion of these states in the electron-ion scattering matrix makes the matrix subunitary. The "defect" of unitarity is then interpreted as the DR probability [10, 11, 22].

5. Calculation of rate coefficients

Once the scattering (or reaction) matrices of Eqs. (9) and (12) are available, the rate coefficients for HCO^+ and H_3^+ have been calculated in the following way.

In the first theoretical approach discussed here, we take into account only the vibrational motion of the molecular ions. Rotation is neglected. The body-frame scattering S (or reaction K) matrix (Eqs. (9) or (12)) is then transformed into the representation of adiabatic states

$$S_{j,j'}(R) = \langle\Phi_i(R;\Omega)|S_{\Lambda,\Lambda'}|\Phi_{i'}(R;\Omega)\rangle_{\Omega}\,. \tag{15}$$

The indices Λ and Λ' numerate the electronic states included into the treatment. The indices j, j' correspond to the basis of that representation and include the quantum numbers describing the electronic state (Λ or Λ') and the vibrational quantum number i or i' of the adiabatic state.

Formally, the amplitude $S_{j,j'}(R)$ represents the scattering of the electron from one vibronic state j' to another j, while the dissociative coordinate R is unchanged.

The scattering (or reaction) matrix $S_{j,j'}(R)$ is then used to obtain the potential curves $U_d(R)$ of the neutral molecule HCO^+ or H_3^+ as discussed in Refs. [9, 18, 16]. These potential curves have usually a non-zero autoionization width $\Gamma_d(R)$. The autoionization width is proportional to the probability to capture the electron into the corresponding state of the neutral molecule. If the electron is captured, the neutral molecule can autoionize or undergo the pre-dissociation process. Assuming that the autoionization is negligibly slow compared to pre-dissociation, from the curves $U_d(R)$ and widths $\Gamma_d(R)$ we calculate DR cross-sections and rate coefficients. We have estimated that for H_3^+, the pre-dissociation is fast relative to the autoionization [9] for the states important to DR. For HCO^+, only a partial analysis has been done so far, and it suggests that the two processes are competitive.

For H_3^+, we also used a second approach for calculation of the DR rate coefficients, which accounts for the competition between autoionization and pre-dissociation. The DR cross-section and rate coefficient are obtained directly from the actual physical scattering matrix S^{phys}, which is calculated from the body frame matrices of Eqs. (9) and (12) by the techniques of rovibrational frame transformation [11, 23, 24] and closed-channel elimination [25]. This approach is discussed in detail in Refs. [11, 26]. An important advantage of this approach is that one obtains the DR cross-section (and the rate coefficient) directly from the scattering matrix: There is no need for the resonance search as in the first approach. Another advantage is that the competition between autoionization and predissociative is accounted for implicitly: There is no need to estimate the effect of the autoionization on the DR cross-section that is obtained. The disadvantage of the approach is that the size of the scattering matrix, after implementation of the rovibrational frame transformation, could be very large (in the thousands, typically). In addition, many rotational states have to be included in the calculation to obtain the DR rate coefficient for high temperatures. This requires extensive numerical calculations, and the high density of narrow resonances means that a dense energy grid is required as well.

6. Results

6.1. H_3^+

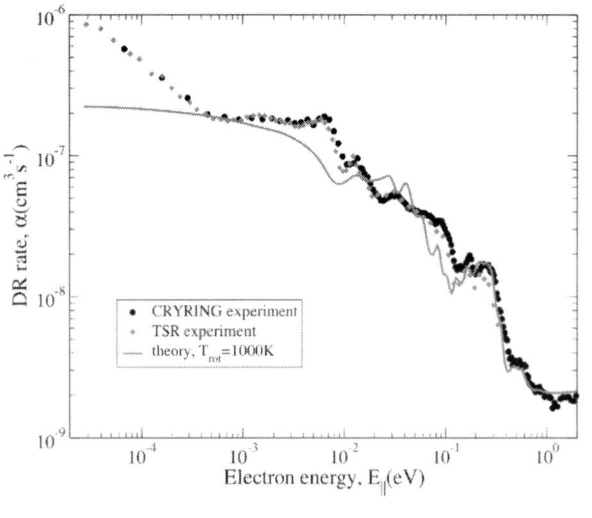

Figure 1. Theoretical (line) and experimental (symbols) H_3^+ DR rate coefficients given as functions of the parallel component of electron energy (for details, see Ref. [26]). Circles correspond to the CRYRING experiment [27], diamonds correspond to the TSR experiment [28]. In the experiments, the rotational temperature of the ions is estimated to be 40 K in CRYRING and 13 K in TSR. The theoretical curve is calculated for a higher temperature, 1000 K.

Figure 1 compares the theoretical H_3^+ DR rate coefficient with experimental data from two storage ring experiments [27, 28]. The theoretical curve is calculated using the second

Journal of Physics: Conference Series **192** (2009) 012017 doi:10.1088/1742-6596/192/1/012017

"improved" approach. The H_3^+ rate coefficient calculated within the "simplified" approach is presented in Refs. [9, 18]. Agreement between theory and experiment is generally quite good. However, the temperature of rotational excitation is 1000 K for the theoretical curve, and only 40 K and 13 K in the experiments. The theoretical curve calculated for a lower rotational temperature, < 100 K, has much more pronounced resonance structure than is observed experimentally (see the detailed discussion in Ref. [26]). Another important difference between the theoretical and experimental curves is their behavior at energies below 0.2 meV. The experimental curves raise steeply when the energy approaches to zero in contrast to the theoretical curve, which is almost flat at low energies. It should be mentioned that in the experiments the electrons have a finite-width energy distribution (2 meV at CRYRING and 0.5 meV at TSR). Therefore, a change 0.1 meV in parallel component of the electron energy should not have a large effect on the rate coefficient. This is what the theoretical rate demonstrates, but the experimental rate coefficients behave approximately as $E_{\|}^{-1/2}$ at even smaller parallel energies. The reasons for the two disagreements, (1) the sharper resonance structure in the theoretical rate and (2) the $E_{\|}^{-1/2}$ behavior in the experimental data at low energies, remain unresolved.

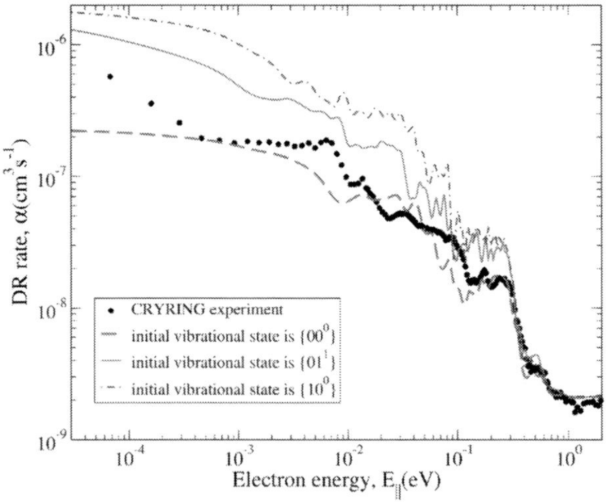

Figure 2. Theoretical H_3^+ DR rate coefficients obtained for three different initial vibrational states of the ion. The theoretical curves are all calculated for 1000 K rotational temperature. The experimental curve is from Ref. [27] and obtained for H_3^+ initially prepared in the ground vibrational level $\{00^0\}$.

The results presented in Fig. 1 correspond to the H_3^+ ion initially prepared in the ground vibrational level. Figure 2 shows the calculated DR rate coefficients for two different excited vibrational states of the ion. The DR rate coefficients for excited vibrational states are significantly larger than for ion initially in the ground vibrational state.

6.2. HCO^+

The HCO^+ DR rate coefficients have been calculated so far only in the "simplified" approach. An additional simplification was made in the calculation: the R_{CO} distance was fixed at the equilibrium position R_{CO} =2.0877 a.u. This additional simplification makes the system more "rigid" in the calculation, which seems likely to unphysically lower the calculated probability to capture the incident electron. The results (thermal rate coefficients) of the calculation are demonstrated in Fig. 3 and compared with two different experiments [12, 13]. The theoretical results are calculated and presented for both the ground and vibrationally exited states of the ion.

As was the case for H_3^+, and not surprisingly, the DR rate coefficient is larger for the vibrationally excited state. Agreement with experiment is poorer than was the case for H_3^+. The

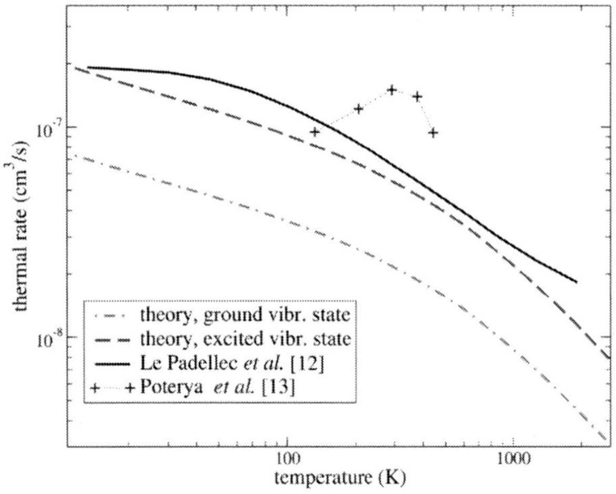

Figure 3. Theoretical thermal HCO^+ DR rate coefficients compared with the experimental data. The dash-dotted line represents the theoretical DR rate for ions prepared initially in the ground vibrational level, whereas the dashed line corresponds to the excited vibrational state of the ion, where one quantum of the bending mode is excited. The solid line is the experimental thermal rate from Ref. [12], the experimental data shown as crosses are taken from Ref. [13].

reason could be due to the fixed R_{CO} distance that was adopted in the present calculation. If the vibration motion along R_{CO} is allowed, this should increase the density of available Rydberg states where the incident electron can be captured. On the other hand autoionization following the initial electron capture has been neglected in this approach. If it is accounted for, the DR rate becomes smaller. Currently, a better analysis of DR in HCO^+, including the vibration along R_{CO}, is under development. At the same time, we point out that the experimental DR rate coefficients are quite different in different experiments, so it is possible that some systematic experimental difficulties remain to be resolved as well. In any case, it is safe to say that the problem of HCO^+ DR is far from being solved.

6.3. Indirect DR for the diatomic ion LiH^+

Some diatomic molecular ions exist for which there are no discernable low-energy pathways that can produce a dissociative recombination reaction by the conventional direct mechanism of Bates. One famous example that has received considerable attention is the HeH^+ ion. This has been measured experimentally at both the CRYRING and TARN II storage rings, [29, 30] and tackled theoretically by Takagi and others [8, 31, 32]. The study by Takagi successfully extracted the total DR rate coefficient at low energies, and obtained good agreement with experiment, and in addition it clarified the relevant pathways and suggested that for HeH^+, both rotational coupling and the nonadiabatic coupling between different electronic partial waves can be important. Another example of a molecular ion with a comparatively fast low-energy DR rate is the LiH^+ ion. The 139K rate coefficient had been extracted from experimental measurements in 2001 at TSR [33, 34, 35] to be $(6 \pm 2) \times 10^{-7}$ cm^3/s, whereas the first theoretical attempt to describe this system gave a much lower rate coefficient more than an order of magnitude smaller. That discrepancy triggered our own studies[4, 36], with the two goals of *(i)* obtaining an independent theoretical calculation to see whether theory is in fact consistent with the lone existing measurement of the total DR rate, and *(ii)* pinpointing the DR mechanism and pathways for this simple molecular ion in order to gain insight into the dynamical indirect DR process as a whole.

To understand the problem, i.e. the reason why it is not obvious that the low energy DR process should be so efficient, consider Fig. 4. It shows the Born-Oppenheimer potential curves for the ionic ground state, for several molecular Rydberg states, and the lower electronically autoionizing potential curves that lie above the ionic ground state potential curve. One sees immediately in this figure that the singlet and triplet potential curves of excited LiH molecules do not cross the ionic ground state potential curve, unless it is at small internuclear distances

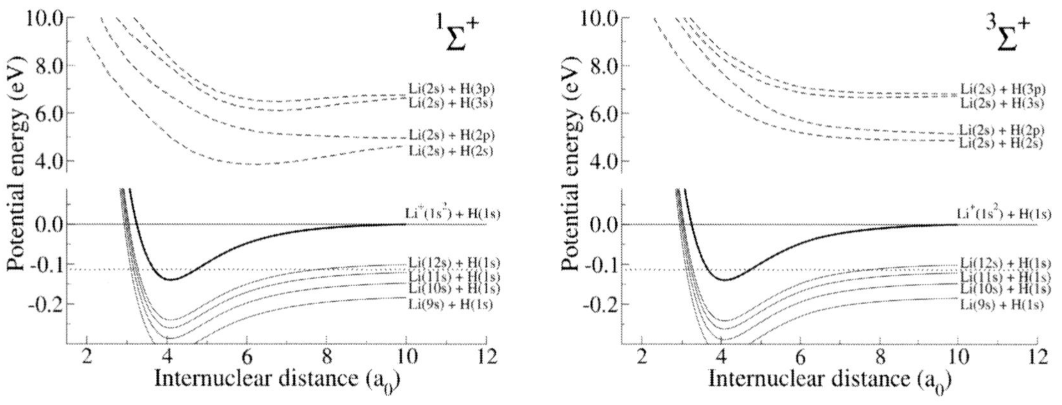

Figure 4. Several singlet and triplet Σ Born-Oppenheimer potential energy curves (thin, solid) for LiH are shown along with the ground state of LiH^+ (thick, solid). The potentials that lie above the ionic ground state are electronically autoionizing states (blue, dashed), whose width is not explicitly indicated in the figure. Because these exhibit no curve crossing between any neutral LiH potential curve and the ground state of the ion, direct DR is expected to occur rarely if at all. Two other important energies are shown as horizontal lines, the $v = 0$ vibrational level of LiH^+ (dotted) and the dissociation threshold of LiH^+ (thin, dark, solid). (From Ref.[36])

$R < 2$ a.u., at energies too high to be relevant to the low-energy DR process under consideration here. For many molecules without a curve-crossing in the Franck-Condon region of the ion, e.g. H_3^+, there are often one or more that become relevant at higher energies, but for the LiH system, no crossing occurs at radii beyond the ionic potential curve minimum, even in that higher energy range.

In studies to date of DR carried out using multichannel quantum defect theory in conjunction with the rovibrational frame transformation, it has usually been possible to identify a single dominant partial wave of the incident electron that plays a controlling role. In our LiH^+ DR calculations, however, this was not found to be the case, as coupling between different incident orbital momenta has vital importance for determining the correct rate. This has been adequately documented elsewhere, e.g. in Fig.4 of Ref.[36], where a strong avoided crossing occurs between the ns and nd partial waves, precisely at the internuclear distance where the ionic potential curve has a minimum, $R \approx 4$ a.u. While both the $^1\Sigma$ and $^3\Sigma$ symmetries show such a crossing, our calculations suggest that the sharper avoided crossing for the $^3\Sigma$ symmetry is the one which dominates low energy DR for this system.

The DR rate has been calculated using elements from the theoretical toolkit discussed above, including MQDT and the rovibrational frame transformation based on vibrational Siegert pseudostates calculated within a finite radius box in R. The resulting plot of the rate versus the parallel energy component, convolved with the experimental parallel and perpendicular energy resolutions as described above, is shown in Fig. 5. This shows that the new calculations reported in Refs. [4, 36] have in fact reproduced the observed experimental results from TSR. What we would like to focus on here, however, is more what we have learned from our analysis of the DR pathways, rather than dwelling on this question of agreement or disagreement between theory and experiment. The main point is that the agreement gives reason to take seriously the conclusions we have been able to draw about the DR mechanism in LiH^+, which presumably still has relevance to other indirect DR species.

The following is our physical picture for how we now view the DR process in LiH^+. First, the incident electron can be written as a partial wave expansion, and typically only one or just a few partial waves dominate the capture processes that ultimately result in DR. In these

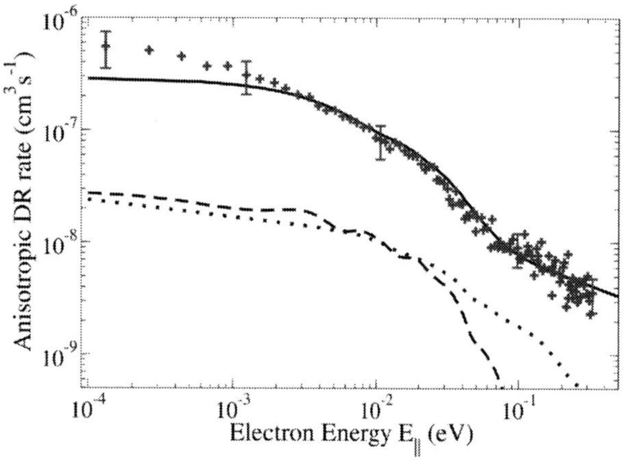

Figure 5. DR rate: The solid curve is our calculated, anisotropically-averaged rate for $\Delta E_{\|}=$ 0.1 meV and $\Delta E_{\perp}$ = 12 meV. The broken curve shows the calculation of [37]. The dotted curve is our truncated result obtained by neglecting the off-diagonal couplings in the quantum-defect matrix. The crosses denote the experimental data [33, 34], with a few characteristic error bars shown. (Taken from Ref.[4])

LiH^+ calculations, we have included up to $l = 2$, and in fact it is the $d\sigma$ partial wave that is the strongest, although couplings off-diagonal in l are also important. The d-wave electron is captured into a Rydberg state, predominantly one that is attached to the closed $v^+ = 1$ vibrational threshold. In other words, there is an energy exchange between the electronic and vibrational energy, which in the frame transformation is largely mediated by the R-dependence of the quantum defect or body-frame scattering matrix. As is typical of an MQDT calculation, an "unphysical" scattering matrix is utilized which has closed channel as well as energetically-open channel indices. The initial capture can be viewed as a simple scattering excitation process in which the incident electron strikes the molecule in a $|v^+ = 0\rangle$ state and then leaves it in a $|v^+ = 1\rangle$ state, with the important difference from ordinary scattering theory that this excitation happens when the electron is at small-r, and eventually as the electron recedes towards infinity, the fact that this channel is closed will ultimately turn the electron back around to recollide with the molecular ion again. What has just been described is all familiar to those who have followed the Rydberg capture picture of indirect DR for the past decade or two or three.

However, a new aspect of this problem was unraveled in the course of the study presented in Refs.[4, 36], namely the role of the so-called "complex resonances" of multichannel spectroscopy, which have an intermingled nature with a (low-v, high-n) Rydberg channel that is quasi-degenerate with a (higher-v, lower-n) Rydberg channel. We were able to see that such a quasi-degeneracy tends to correspond to regions of enhanced indirect DR, by factors ranging from 3-20, and in this sense, the complex resonances of multichannel spectroscopy can be viewed as providing the dominant "doorway" for indirect DR reactions.

Another insight gleaned from this new study of LiH^+ is the fact that the R-dependent angular coupling among the different partial waves is also decisively important. A test calculation conducted in Ref.[4] was to artificially turn off that coupling; once it was omitted, the resulting calculation gave a DR rate for LiH^+ that was very close to the earlier theoretical results obtained by Florescu *et al.* [37], i.e., which was more than an order of magnitude smaller than the experimental rate (and our full LiH^+ DR calculation) at low energies. This shows that an avoided crossing which happens to fall right in the middle of the Franck-Condon region can be extremely important in its affect on the low-energy DR rate; this is quite likely a general result that is applicable to other systems in addition to this particular molecular ion.

7. Concluding remarks

We have reviewed two different theoretical approaches for calculation of DR cross-sections in triatomic ions, discussed the results obtained for two different triatomic ions, H_3^+ and HCO^+, and compared them with the diatomic system of LiH^+ where the non-adiabatic effects are also strong.

The two approaches presented are generic and can be applied to theoretical DR studies for other small polyatomic molecular ions, where only a limited number (3-4) of vibrational modes can realistically be taken into account. The "doorway" mechanism of complex multichannel resonances, identified through the prototype system of an electron that collides with LiH^+, appears to show promise for understanding this frequently surprising indirect DR process.

In general, the DR mechanism in highly symmetric triatomic molecular ions is different from DR in diatomic ions owing to the presence of degenerate vibrational modes. The incident electron arriving in a degenerate electronic state such as $\pi^{\pm}$ (or $d\delta^{\pm}$) is strongly coupled to the degenerate ionic vibrational modes because of the Jahn-Teller theorem. The strong non-adiabatic coupling causes rapid energy transfer from the electronic to the vibrational degrees of freedom. In HCO^+ and H_3^+ the coupling produces recombination that is comparatively faster than will normally be the case for systems where the direct DR process is impossible.

The theoretical DR rate coefficients for both H_3^+ and LiH^+ are in good agreement with existing experiments. For HCO^+ the theoretical DR rate coefficient is approximately a factor of 2-3 smaller than the experimental values [12, 13]. However, at the present time, different experiments disagree on the value of the HCO^+ rate coefficient, so it remains unclear how seriously one should view the current discrepancy between theory and experiment.

Acknowledgments

This work has been supported by the National Science Foundation (NSF) under PHY-0427460 and PHY-0427376, by an allocation of NERSC supercomputing resources, and by an allocation of NCSA supercomputing resources (project No. PHY-040022). Support has also been appreciated from NSF grants CHE-0446688, OISE-0532040, the project Electron Induced Processing at the Molecular Level (EIPAM) sponsored by the European Science Foundation (grant No. PESC7-20), and from the Academy of Sciences of the Czech Republic (grants No. A100400501 and 1ET400400413). CHG also thanks the scientists at the Max-Planck-Institute for Nuclear Physics for their extraordinary hospitality, and also the Alexander von Humboldt Foundation for its partial support of this work.

References

[1] Bates D R 1950 *Phys. Rev.* **78** 492
[2] Bardsley J N 1968 *J. Phys. B: At. Mol. Phys.* **1** 365
[3] Guberman S L 1994 *Phys. Rev. A* **49** R4277
[4] Curik R and Greene C H 2007 *Phys. Rev. Lett.* **98** 173201
[5] Orel A E and Kulander K C 1993 *Phys. Rev. Lett.* **71** 4315
[6] Schneider I F, Orel A E and Suzor-Weiner A 2000 *Phys. Rev. Lett.* **85** 3785
[7] Bates D R 1991 *J. Phys. B: At. Mol. Phys.* **24** 3267
[8] Larson Å, Tonzani S, Santra R and Greene C H 2005 *J. Phys.: Conf. Ser.* **4** 148
[9] Kokoouline V, Greene C H and Esry B D 2001 *Nature* **412** 891
[10] Kokoouline V and Greene C H 2003 *Phys. Rev. Lett.* **90** 133201
[11] Kokoouline V and Greene C H 2003 *Phys. Rev. A* **68** 012703
[12] LePadellec A, Sheehan C, Talbi D and Mitchell J B A 1997 *J. Phys. B: At. Mol. Opt. Phys.* **30** 319
[13] Poterya V, McLain J L, Adams N G and Babcock L M 2005 *J. Phys. Chem. A* **109** 7181
[14] Landau L and Lifshitz E 2003 *Quantum Mechanics: Non-relativistic Theory* (Burlington MA: Butterworth Heinemann)
[15] Koppel H, Domcke W and Cederbaum L S 1981 *J. Chem. Phys.* **74** 2945
[16] Mikhailov I A, Kokoouline V, Larson Å, Tonzani S and Greene C H 2006 *Phys. Rev. A* **74** 032707
[17] Stephens J A and Greene C H 1995 *J. Chem. Phys.* **102** 1579
[18] Greene C H and Kokoouline V 2006 *Phil. Trans. Royal Soc. A* **364** 2965
[19] Tolstikhin O I, Watanabe S and Matsuzawa M 1996 *J. Phys. B: At. Mol. Opt. Phys.* **29** L389
[20] Kokoouline V and Greene C H 2005 *Phys. Rev. A* **72** 022712
[21] Blandon J, Kokoouline V and Masnou-Seeuws F 2007 *Phys. Rev. A* **75** 042508
[22] Hamilton E L and Greene C H 2002 *Phys. Rev. Lett.* **89** 263003
[23] Chang E S and Fano U 1972 *Phys. Rev. A* **6** 173

[24] Greene C H and ChJungen 1985 *Adv. At. Mol. Phys.* **21** 51
[25] Aymar M, Greene C H and Luc-Koenig E 1996 *Rev. Mod. Phys.* **68** 1015
[26] Santos dos Fonseca S, Kokoouline V and Greene C H 2007 *J. Chem. Phys.* **127** 124309
[27] McCall B J, Huneycutt A J, Saykally R J, Geballe T R, Djuric N, Dunn G H, Semaniak J, Novotny O, Al-Khalili A, Ehlerding A, Hellberg F, Kalhori S, Neau A, Thomas R, Osterdahl F and Larsson M 2003 *Nature* **422** 500
[28] Kreckel H, Motsch M, Mikosch J, Glosik J, Plasil R, Altevogt S, Andrianarijaona V, Buhr H, Hoffmann J, Lammich L, Lestinsky M, Nevo I, Novotny S, Orlov D A, Pedersen H B, Sprenger F, Terekhov A S, Toker J, Wester R, Gerlich D, Schwalm D, Wolf A and Zajfman D 2005 *Phys. Rev. Lett.* **95** 263201
[29] Strömholm C, Semaniak J, Rosén S, Danared H, Datz S, van der Zande W and Larsson M 1996 *Phys. Rev. A* **54** 3086
[30] Tanabe T, Katayama I, Inoue N, Chida K, Arakaki Y, Watanabe T, Yoshizawa M, Saito M, Haruyama Y, Hosono K, Honma T, Noda K, Ohtani S and Takagi H 1995 *Phys. Rev. A* **49** R1531
[31] Takagi H 2004 *Phys. Rev. A* **70** 022709
[32] Guberman S L 1994 *Phys. Rev. A* **49** R4277
[33] Krohn S, Lange M, Grieser M, Knoll L, Kreckel H, Levin J, Repnow R, Schwalm D, Wester R, Witte P, Wolf A and Zajfman D 2001 *Phys. Rev. Lett.* **86** 4005
[34] Krohn S 2001 *PhD thesis* (University of Heidelberg)
[35] Zajfman D, Schwalm D and Wolf A 2003 *Hyperfine Interact.* **146/147** 265
[36] Čurík R and Greene C H 2007 *Molec. Phys.* **105** 1565
[37] Florescu A, Suzor-Weiner A, Leininger T and Gadéa F X 2004 *Phys. Scrip.* **T110** 172

Recombination of KrD^+ and KrH^+ ions in afterglow plasma

I Korolov, T Kotrik, R Plasil, M Hejduk, J Varju, P Dohnal, J Glosik

Department of Surface and Plasma Science, Faculty of Mathematics and Physics, Charles University in Prague, Czech Republic

E-mail: juraj.glosik@mff.cuni.cz

Abstract. Reported is flowing afterglow (FALP) study of recombination of KrH^+ and KrD^+ ions with electrons at 250 K in mixtures of He/Kr/H_2 and He/Kr/D_2, respectively. The influence of fast recombining cluster ions formation on apparent effective recombination rate coefficients (α_{eff}) was measured and used in data analysis. The obtained binary rate coefficients for recombination of KrH^+ and KrD^+ are $\alpha_{KrH+} = 2\times10^{-8}$ cm^{3}s^{-1} and $\alpha_{KrD+} = 1\times10^{-8}$ cm^{3}s^{-1}.

1. Introduction

The recombination of protonated rare gas atoms is important from theoretical [1], experimental and technological aspects. The ions HeH^+, NeH^+, ArH^+ and KrH^+ do not have curve crossings between the ion ground state and a repulsive neutral state that could lead to a fast recombination process [2,3]. The recombination of HeH^+, NeH^+ and ArH^+ ions was also studied in storage ring experiments [4,5,6]. These ions recombine at thermal energies very slowly (with rate coefficients $\alpha(300\ K) < 1\times10^{-8}$ cm^{3}s^{-1}) and the study is not possible in flowing afterglow experiments limited to rate coefficients $\geq10^{-8}$ cm^{3}s^{-1}. Ions KrH^+ and XeH^+ are too heavy for existing storage ring experiments and they are difficult to handle in plasma experiments. The recombination rate of these ions is difficult to predict, because already light hydride ions recombine in a very nonstandard way. That is why several theoretical and experimental studies of KrH^+ and XeH^+ ions and their recombination with electrons were performed [7,8]. There are two FALP recombination studies of KrH^+ and XeH^+ ions; nevertheless they give just the upper limits for the values of recombination rate coefficients [8,9]. The interesting question is how deuterated analogues of these ions will recombine. Kr and Xe hydride ions recombine slowly but KrD^+ and XeD^+ ions recombine even slower. Recently we adapted our Flowing Afterglow Langmuir Probe (FALP) [10] for measurements of recombination rate coefficients as small as 5×10^{-9} cm^{3}s^{-1}. This FALP is used in present study to measure the rate coefficients of KrH^+ and KrD^+ ions at 250 K. In FALP experiment the recombining ions are formed in the decaying afterglow plasma in helium, so in principle Kr and H_2 or D_2 have to be added to the carrier gas to form KrH^+ and KrD^+. In such plasmas the formation of faster recombining molecular ions (Kr_2^+, H_3^+ or D_3^+) with recombination rate coefficients $\geq10^{-7}$ cm^{3}s^{-1} or cluster ions (e.g. KrD_3^+, Kr_2D^+, H_5^+ or D_5^+) with even larger recombination rate coefficients ($\geq10^{-6}$ cm^{3}s^{-1}) cannot be excluded from data analyse. The influence of these ions can be accounted for if dependencies on partial pressures of the reactants are measured and kinetics of processes in the flowing afterglow is considered.

2. Experiment

The FALP apparatus used in the present experiment was designed to study slowly recombining ions. It is constructed so that plasma decay can be monitored up to 95 ms. To obtain such long decay times helium pressures up to 2000 Pa have to be used in the flow tube. The schematic of the FALP is shown in Fig. 1. The plasma is generated by a microwave discharge in the upstream part of the flow tube and then it is driven by He along the flow tube. The reactant gases can be added via ports P1 and P2, i.e. 3 and 35 ms after the discharge, respectively. The axially movable Langmuir probe [11] is used to measure the electron density along the flow tube. From the evolution of the electron density along the flow tube recombination rate coefficients are calculated [12].

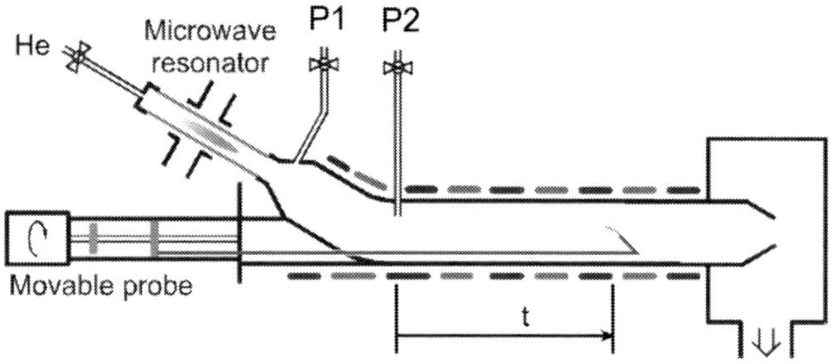

Figure 1. Principle of the FALP method. Flowing helium carries the afterglow plasma from the discharge region along the flow tube. The decay time is measured from position of port P2, i.e. $t_{P2} = 0$. Upstream from P2 $t < 0$ and downstream from P2 $t > 0$.

In the discharge He^+ ions, electrons and helium metastables are formed. In the sequence of ion molecule reactions after addition of krypton via port P1, Kr^+ dominated plasma is formed. Via the port P2 hydrogen (deuterium) is then added to an already relaxed plasma (see also accompanying paper by Plasil *et al.* in present volume [13]). In following ion molecule reactions KrH^+/H_3^+ plasma is formed (or KrD^+/D_3^+). The rate coefficients of ion molecule reactions are known [14] and the process of formation can be predicted using a numerical model. The calculated evolution indicating the formation of KrH^+/H_3^+ plasma is plotted in Fig. 2. Similar results can be obtained for the formation of KrD^+/D_3^+ dominated plasma.

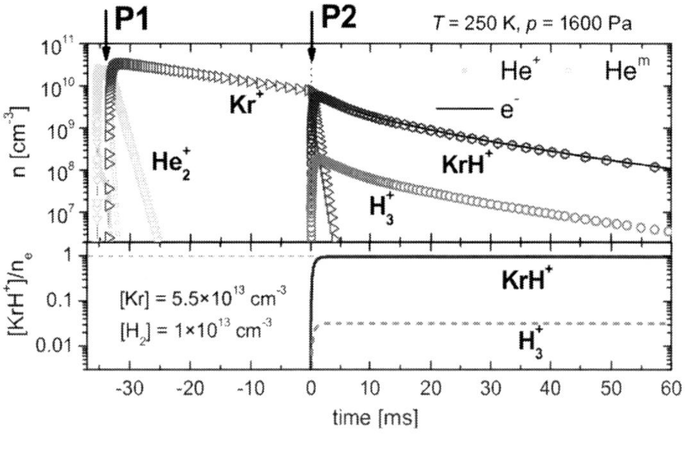

Figure 2. The calculated plasma formation and decay along the flow tube after addition of Kr via port P1 and H_2 via port P2. In the lower panel fractions $[KrH^+]/n_e$ and $[H_3^+]/n_e$ are plotted.

3. Results and discussion

The difference between proton affinities of Kr and H_2 is very small (~0.02 eV) and at used densities of both reactants the forward and backward proton transfer reactions

$$KrH^+ + H_2 \leftrightarrow H_3^+ + Kr \tag{1}$$

are very fast in comparison with recombination. In such conditions the ions KrH^+ and H_3^+ are in thermodynamic equilibrium. In this equilibrium the ratio of ion densities is given by the ratio of densities of H_2 and Kr: $[KrH^+]/[H_3^+] = K_{EQ}\,[Kr]/[H_2]$; where K_{EQ} is the equilibrium constant [15]. It is obvious that in order to study recombination of KrH^+ we had to adjust the densities of [Kr] and $[H_2]$ to achieve the required concentration of ions. To obtain data plotted in Fig. 2 with dominance of KrH^+

ions proper reactant densities were calculated using the above relation. The situation is similar for KrD^+ formation.

As Kr and H_2 are added to helium, KrH^+ and H_3^+ ions will be present in the afterglow and the decay will be controlled by recombination of both ions:

$$\frac{dn_e}{dt} = -\alpha_1[\mathrm{KrH}^+]n_e - \alpha_2[\mathrm{H}_3^+]n_e = -(\alpha_1 f_1 + \alpha_2 f_2)n_e^2 = -\alpha_{\mathrm{eff}} n_e^2 . \tag{2}$$

For the simplicity the diffusion terms are not written, but they are considered in data analysis. The effective rate coefficient, $\alpha_{eff} = (\alpha_1 f_1 + \alpha_2 f_2)$, will be dependent on partial densities of ions (fractions f_1 and f_2) and their rate coefficients (α_1 and α_2). The fractions f_1 and f_2 can be calculated if we can rely on the value of the available equilibrium constant [15]. It is also possible to measure α_{eff} as a function of Kr or H_2 density and to obtain α_1 as a limit for $[H_2] \rightarrow 0$ (equivalent to $f_2 \rightarrow 0$).

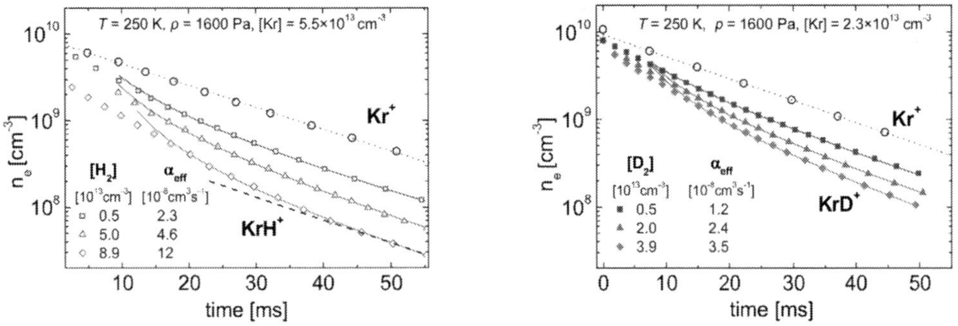

Figure 3. Left panel: The measured decay of KrH^+/H_3^+ dominated plasma along the flow tube (recalculated to the time scale). The corresponding Kr and H_2 densities are denoted. Included is the decay in Kr^+ plasma. **Right panel:** The measured decay of the KrD^+/D_3^+ dominated plasma.

In Fig. 3 are plotted the decay curves measured in KrH^+ and KrD^+ dominated afterglow, obtained for several densities of H_2 and D_2, respectively. The measured decays are very slow and common methods of rate coefficient determination from the slope of a graph of the values $1/n_e(t)$ versus decay time will not be accurate enough. For this reason, an advanced data analysis procedure has been developed [12]. The obtained effective recombination rate coefficients (α_{eff}) are plotted as a function of hydrogen (deuterium) density in Fig. 4. Dependence on the Kr density was measured in separate measurements [16]. The effective rate coefficients plotted in Fig. 4 were obtained at very low Kr densities where α_{eff} is nearly independent on [Kr]. In the final determination of the binary rate coefficients α_{KrH+} and α_{KrD+} the dependence on [Kr] has been also considered.

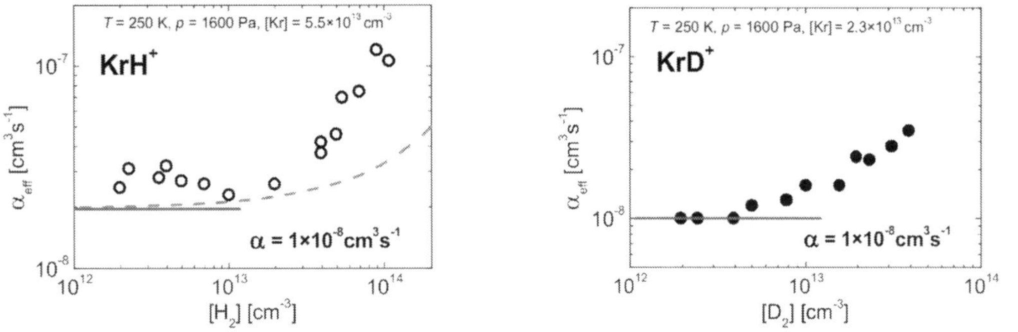

Figure 4. Left panel: The effective recombination rate coefficients measured in KrH^+/H_3^+ dominated plasma. The corresponding Kr density is denoted. The full line indicates the limit of measured data after inclusion of dependence on [Kr]. The dashed line indicates calculated α_{eff}. **Right panel:** The effective recombination rate coefficients measured in KrD^+/D_3^+ dominated plasma.

The obtained binary rate coefficients for recombination of KrH^+ and KrD^+ are $\alpha_{KrH+} = 2\times10^{-8}$ cm^{3}s^{-1} and $\alpha_{KrD+} = 1\times10^{-8}$ cm^{3}s^{-1}. Using the obtained value for α_{KrH+}, the value $\alpha_{H3+} = 1.8\times10^{-7}$ cm^{3}s^{-1} [17] and the equilibrium constant [15] we calculated α_{eff} as a function of [H_2] (for details see e.g. ref. [18]). Obtained dependence is plotted in Fig. 4. The disagreement between measured and calculated data is not influencing the determination of α_{KrH+} but can indicate an inaccurate value of used K_{EQ}.

4. Conclusion

We measured binary recombination rate coefficients $\alpha_{KrH+} = 2\times10^{-8}$ cm^{3}s^{-1} and $\alpha_{KrD+} = 1\times10^{-8}$ cm^{3}s^{-1} in decaying plasma at 250 K.

Acknowledgements: This work is a part of the research plan MSM 0021620834 financed by the Ministry of Education of the Czech Republic and partly was supported by GACR (202/05/P095, 205/05/0390, 202/07/0495) by GAUK 53607 and GAUK 124707

References

[1] Jungen C, Roche A L and Arif M 1997 *Phil. Trans. R. Soc. Lond. A* **355** 1481-1506
[2] Petsalakis I D, Theodorakopoulos G and Buenker R J 2003 *Chem. Phys.* **119(4)** 2004 - 2013
[3] Florescu-Mitchell A I and Mitchell J B A 2006 *Physics Reports* **430** 277
[4] Stromholm C, Semaniak J, Rosen S, Danared H, Datz S, Van der Zande V and Larsson M 1996 *Phys. Rev. A* **54** 3086
[5] Mitchell J B A, Novotny O, Angelova G, LeGarrec J L, Rebrion-Rowe C, Svendsen A, Andersen L H, Florescu-Mitchell A I and Orel A 2005 *J. Phys. B* **38** 693-703
[6] Mitchell J B A, Novotny O, LeGarrec J L, Florescu-Mitchell A I, Rebrion-Rowe C, Stolyarov A V, Child M S, Svendsen A, El Ghazaly M A and Andersen L H 2005 *J. Phys. B* **38** L175-L181703
[7] Gallup G A and Macek J 1977 *J. Phys. B: Atom. Molec. Phys.* **10(9)** 1601-1605
[8] Le Padellec A, Laube S, Sidko O, Rebrion-Rowe C, Rowe B R, Sarpal B and Mitchell J B A 1997 *J. Phys. B: At. Mol. Opt. Phys.* **30** 963-967
[9] Geoghegant M, Adams N G and Smith D 1991 *J. Phys. B: At. Mol. Opt. Phys.* **24** 2589-2599
[10] Novotny O, Plasil R, Pysanenko A, Korolov I and Glosik J 2006 *J. Phys. B: Atomic Mol. Opt. Phys.* **39 11** 2561
[11] Swift J D and Schwar M J R: *Electrical Probes for Plasma Diagnostics* (Iliffe, London, 1970)
[12] Korolov I, Novotný O, Kotrík T, Plašil R, Varju J, Hejduk M and Glosík J 2008 *Contrib. Plasma Phys.* **48** 521-526
[13] Plasil R, Korolov I, Kotrik T, Varju J, Dohnal P, Donko Z, Bano G and Glosik J *present volume*
[14] Ikezoe Y, Matsuoka S, Takabe M and Viggiano A A: *Gas Phase Ion-Molecule Reaction Rates.* Maruzen Co., Japan 1987
[15] Bohme D K, Mackay G I and Schiff H I 1980 *J. Chem. Phys.* **73** 4976-4986
[16] Korolov I, Novotný O, Plašil R, Hlavenka P, Kotrík T, Tichý M, Kudrna P, Glosík J and Luca A 2006 *Czech. J. Phys.* **56** Suppl. B B854-B864
[17] Glosik J, Plasil R, Korolov I, Kotrik T, Novotny O, Hlavenka P, Varju J, Dohnal P, Kokoouline V and Greene C H 2009 *Phys. Rev. A* **79** 052707
[18] Glosik J, Novotny O, Pysanenko A, Zakouril P, Plasil R, Kudrna P and Poterya V 2003 *Plasma Sources Science & Technology* **12(4)** S117-S122

Formation of the ion-pair in electron recombination with H_3^+

Åsa Larson[1], Johanna B Roos[1] and Ann E Orel[2]

[1] Theoretical Chemistry, School of Biotechnology, Royal Institute of Technology, S-106 91 Stockholm, Sweden

[2] Department of Applied Science, University of California, Davis, CA 956161, USA

E-mail: aasa@theochem.kth.se

Abstract. We present a two-dimensional theoretical study of the formation of the ion-pair, $H_2^+ + H^-$, in electron recombination with H_3^+ in its ground electronic and vibrational state. The relevant potential energy surfaces, electronic couplings and autoionization widths are calculated by combining electron scattering calculations with quantum chemistry calculations. The nuclear dynamics on six coupled electronic states is studied by propagating wave packets using the MCTDH (Multi Configuration Time-Dependent Hartree) method. Also the cross section for ion-pair formation in electron recombination with D_3^+ is calculated.

1. Introduction

When an electron recombines with the H_3^+ ion in its ground electronic and vibrational state, several processes are possible. In the first step an excited state of H_3 system is formed. This state can then be stabilized through dissociation into neutral fragments - dissociative recombination - where both the two- (H_2 + H) and three-body (H + H + H) channels are possible. Dissociative recombination is energetically available even at zero collision energy. The H_3 system can also be stabilized through autoionization, *i.e.*, re-emitting the electron. If the system autoionizes into a bound vibrational state of the H_3^+ ion, vibrational excitation occur. If the auoionization goes into the vibrational continuum, the process is dissociative excitation. These reactions require a threshold energy ΔE that will depend on the state of the remaining ion.

In the process of ion-pair formation, the system is stabilized through dissociation into opposite charged fragments. In electron recombination with H_3^+, there are two different ion-pair channels

$$H_3^+ + e^- \longrightarrow \begin{cases} H^- + H_2^+ \\ H^- + H^+ + H \end{cases} \tag{1}$$

The first channel has a threshold energy of $\Delta E = 5.4$ eV and for the second channel the threshold energy is $\Delta E = 8.1$ eV. In the present study, only the first channel (formation of $H_2^+ + H^-$) is included and therefore we will not be able to predict the total cross section for ion-pair formation above 8.1 eV where the second channel becomes energetically available.

Already in 1979, Kulander and Guest [1] calculated the resonant states of H_3 by combining the Configuration Interaction method with Feshbach projection operators. One-dimensional slices of the diabatic potential energy surfaces were calculated where the H-H distance was frozen to

1.65 a_0 and the Jacobi angle θ was set to 90°. They found that the lowest resonant state of H_3 does not cross the ion potential close to its minimum. The resonant state was not able to explain the dissociative recombination cross section at low collision energies. However, when the collision energy is high enough to capture into this resonant state, a peak in the cross section will be seen. Furthermore, they predicted that the lowest resonant state diabatically correlates with the ion-pair limit, $H_2^+ + H^-$, at infinity. However, it will cross and couple to the Rydberg states situated below the ion potential. Therefore, the recombination into the lowest resonant state will produce both neutral fragments as well as the ion-pair. Similar one-dimensional potentials have also been calculated by Michels and Hobbs [2].

More detailed electron scattering calculations were carried out by Orel *et al.* [3] using the Complex Kohn variational method [4]. Using this method, both the resonance energy positions as well as the autoionization widths are obtained. These scattering calculations were carried out in two dimensions where both radial Jacobi coordinates were varied. The Jacobi angle θ was still frozen to 90°. They found that there are four resonant states, two states of $^2A_1$ symmetry and two of $^2B_1$ symmetry. Furthermore, three of these states (the two A_1 states and the lowest B_1 state) have a triple intersection at the equilibrium geometry of the H_3^+ ion. A few scattering calculations were carried out where also the Jacobi angle was varied. It was found that while all resonant states are strongly repulsive with respect to both radial coordinates, they are flat with respect to variations in the angle.

The high energy peak in the cross section for dissociative recombination was first measured in 1993 using the CRYRING ion-storage ring [5]. In this measurement, the neutral fragments were detected and a peak in the cross section centered around 9 eV was observed. The same year, Orel and Kulander [6] used their calculated resonant states to carry out wave packet propagation to predict the high-energy peak in the dissociative recombination cross section. The two radial dimensions were included and it was assumed that everything dissociated into neutral fragments. No couplings between the neutral states were included and the resonant states became flat as soon as they crossed the ion potential. The theoretical study gave a peak in the cross section with a similar shape and position as the one measured in the experiment.

The cross section for ion-pair formation has been measured using an inclined beam experiment [7], a merged beam experiment [8] as well as using the CRYRING [9] and TSR [10] ion-storage ring experiments. The resulting cross sections from these experiments will be displayed in figure 4 below. In the first two experiments, vibrational relaxation is obtained by collisional quenching. The advantage with the ion-storage ring experiments is that the ions are stored in the ring prior to the experiments and the ions will have time to vibrationally relax to its ground vibrational state. In the first three experiments, the H^- fragments were detected and therefore these measurements can not separate between the two channels of ion-pair formation. In the most recent experiments [10], the H_2^+ fragments were measured in coincidence with the H^- fragments and therefore the branching for the two- and three-body ion-pair channels can be determined. However, in this experiment, the absolute cross section was not obtained and therefore, the peak of the total cross section for ion-pair formation was normalized relative to the cross section measured in CRYRING. As can be seen in figure 4, all measurements agree on a magnitude of $2 \cdot 10^{-18}$ cm^2 of the cross section for ion-pair formation. However, the shape of the cross sections differ. The merge-beam experiment shows a sharper onset of the cross section above the threshold energy for ion-pair formation. The recent TSR experiment shows that the two-body ion-pair channel dominates over the three-body channel and some structures are seen in the two-body cross section that are not present in the cross sections for three-body formation.

We have previously carried out an one-dimensional study on the formation of the ion-pair $H_2^+ + H^-$ in electron recombination with H_3^+ [9]. In this study, the Jacobi angle was frozen to 90° (C_{2v} symmetry) and the H-H distance was set to 1.65 a_0. A diabatic representation of the two $^2A_1$ resonant states as well as four Rydberg states of the same symmetry were included.

The molecular dynamics were studied by propagating wave packets on coupled one-dimensional potentials. The calculated cross section for ion-pair formation had a shape and magnitude very different from what has been measured [9]. It had a very sharp threshold were the ion-pair channel opens up and it was followed by a double peak structure that can understood as an interference effect induced by the electronic couplings between the two resonant states. Furthermore, the magnitude of the calculated cross section was about a factor of five larger than all measured cross sections. From this study, we concluded that we have to perform a two-dimensional study of the reaction including all relevant coupled potentials. This is what we intend to do in the present study. The article is outlined as follows: In section 2 it is described how the diabatic representation of the relevant potential energy surfaces, electronic couplings and autoionization widths are obtained. This is followed in section 3 with an outline of how we describe the nuclear dynamics using the wave packet technique. In the following study, we are using the Multi-Configuration Time-Dependent Hartree (MCTDH) method developed in Heidelberg [13] to propagate the wave packets. The numerical details are given in section 3.1. The results are then summarized, where we show the influence on the cross section from the electronic couplings between the ion-pair state and the other neutral states included in the model. We also study ion-pair formation in electron recombination with D_3^+. Atomic units are used throughout the article.

2. Calculation of potentials and couplings

The four lowest resonant states of H_3 are calculated by Orel *et al.* [3] using the Complex Kohn Variational method [4]. In the present study, we include the two lowest resonant states of $^1A_1$ symmetry. In a "quasidiabatic representation", these two resonant states will cross each other in the Franck-Condon region and one of the resonant states will correlate with the ion-pair limit at infinity. The neutral adiabatic states situated below the ion-potential are calculated a with full configuration interaction calculation containing 63 000 configurations. Each H atom is represented by a (8s, 5p, 1d) basis set. The neutral states are diabatized using the R-dependence of the configuration interaction coefficients. The electronic couplings between the two resonant states as well as the couplings between the ion-pair state and the lower Rydberg states are determined using a two-by-two transformation of the adiabatic potentials [11]. In order to determine the couplings between the ion-pair state and the higher Rydberg states, the autoionization width and the effective quantum number of the Rydberg state can be used in a scaling relation [12]. The potentials have to be extrapolated to correct asymptotic forms. In the present study, the two resonant states are included as well as the four lowest Rydberg states. The highest Rydberg state is an "effective state" [11] with a coupling to the ion-pair state that is determined to compensate for the infinite number of Rydberg states not included in the present study. The potential surfaces included in the theoretical model are displayed in figure 1, where the H-H distance is fixed at $r = 1.65$ a_0.

3. Nuclear dynamics

The molecular dynamics is studied by propagating wave packets on the coupled potentials. As will be described below, the MCTDH [13] method is used to integrate the time-dependent Schrödinger equation

$$i\frac{\partial}{\partial t}\overline{\Psi}(t,r,z) = \mathbf{H}\overline{\Psi}(t,r,z). \tag{2}$$

Here $\mathbf{H}$ is the 6×6 Hamiltonian matrix, where the diagonal elements contains the kinetic energy operator plus potentials and the off-diagonal matrix elements contains the electronic couplings to the ion-pair state. Autoionization is included by letting the potentials of the two resonant

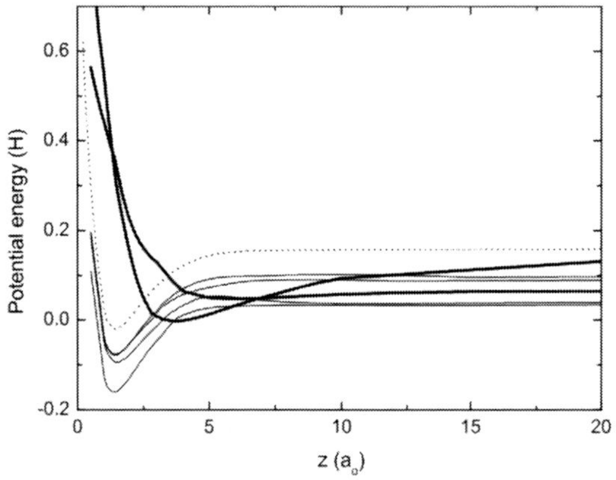

Figure 1. Diabatic potential energy surfaces of H_3^+ (dotted) and H_3 for fixed H-H distance, $r = 1.65$ a$_0$. The resonant states are displayed with thick curves and the Rydberg states are the thin curves.

states be complex above the ion-potential

$$\widetilde{V}_{1,2}(r,z) = V_{1,2}(r,z) - i\frac{\Gamma_{1,2}(r,z)}{2}. \tag{3}$$

This is the so-called local "Boomerang approximation" [14] for treating autoionization. The electron recombination will initiate wave packets on the two resonant states [14]:

$$\Psi_{1,2}(t=0,r,z) = \sqrt{\frac{\Gamma_{1,2}(r,z)}{2\pi}}\chi_{00}(r,z). \tag{4}$$

Here χ_{00} is the ground vibrational state of the H_3^+ ion. The electron capturing into the bound Rydberg states, induced by the non-adiabatic couplings are here neglected.

3.1. MCTDH

The MCTDH method [13] that here is used is a method that is well suited for propagating wave packets in many dimensions. The method is based on the assumption that the time-dependent wave packets are written as a linear combinations of Hartree-products of single-particle functions. For our two-dimensional wave packets we obtain:

$$\Psi(t,r,z) = \sum_{i=1}^{n_r}\sum_{j=1}^{n_z} A_{ij}(t)\phi_i^{(r)}(t,r)\phi_j^{(z)}(t,z). \tag{5}$$

In the present study we include 10 single-particle functions for both the r and z coordinates ($n_r = n_z = 10$) and these single-particle functions are represented in basis functions of discrete variable representation. We are using 200 sine-functions for the r dimension and 500 sine-functions for the z dimension with grids ranging from $0.5 - 10.0$ a$_0$ for the r-dimension and $0.5-20.0$ a$_0$ for z. When the ansatz (5) is inserted into the time-dependent Schrödinger equation, coupled time-dependent differential equations for the coefficients and the single-particle functions are obtained. The wave packets are propagated using the constant mean-field scheme, where the single particle functions are integrated using the Bulirisch-Stoer extrapolation (BS) scheme of order 9 and an error tolerance of 10^{-7}. For the A_{ij} coefficients, a Short Iterative Lanzos algorithm (SIL) of order 15 is applied, with an error tolerance of 10^{-7}. The wave packets are propagated for 60 fs.

The cross section for ion-pair formation is calculated using the technique [15] of integrating the flux absorbed by a Complex Absorbing Potential (CAP), placed in the asymptotic ion-pair channel. Here we placed the CAP at $z_c = 15.0$ a$_0$. The cross section for the individual vibrational states of H_2^+ is calculated by employing projection operators onto these states [15]. Then the total cross section for ion-pair formation is obtained by summarizing contributions from the vibrational states of H_2^+. The flux is integrated with time-steps of $dt = 0.2$ fs. In order to avoid reflections to the edge of the grids, complex absorbing potentials are not only used for the ion-pair state, but also for the other potentials.

In order to use the MCTDH method, the Hamiltonian has to be given in a product form. Here we are using the Jacobi coordinates and the kinetic energy operator is then automatically given in the desired format. However, the calculated potentials, couplings and widths all have to be fitted in the product form

$$V(r,z) = \sum_{i,j} c_{ij} v_i^{(r)}(r) v_j^{(z)}(z), \tag{6}$$

before they are implemented in the MCTDH program. In the present study we are using fits of the surfaces containing up to 49 parameters.

4. Results

In order to obtain a fundamental understanding of the molecular dynamics, we start by studying the reaction by propagating on the ion-pair state alone. Then, we add the couplings to the other resonant state one by one.

4.1. Propagation on the ion-pair state alone

We start by propagating the wave packet on the resonant ion-pair state alone. Autoionization from the resonant state is included. In figure 2, we compare our calculated cross section using the MCTDH method with the cross section we obtain using the one-dimensional model [9] and using a two-dimensional model, where the wave packet is propagated using the Chebyshev propagator [16].

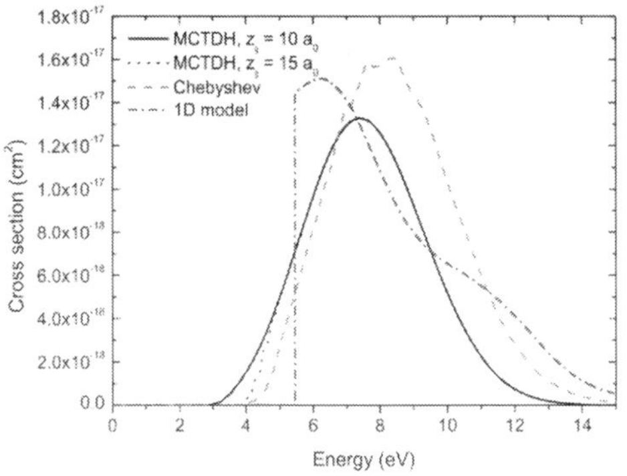

Figure 2. Cross section for propagation on the ion-pair state alone. Comparison between the one- and two-dimensional models and different propagators.

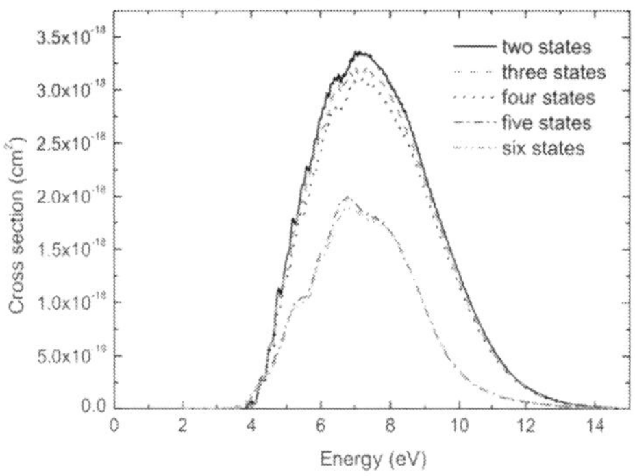

Figure 3. Calculated cross section for ion-pair formation, obtained when different number of states are included in the theoretical model.

The magnitudes of the calculated cross sections are about the same, but the shapes and the positions of the peak differ. The cross section calculated using the one-dimensional model shows a sharp threshold where the ion-pair limit becomes energetically available at 5.4 eV. With the Chebyshev propagator [17] the threshold is not that sharp and the peak of the cross section is centered around 8 eV. When the wave packets are propagating using the MCTDH method, the peak is shifted slightly toward lower energies. This can partly be explained by the fact that we, when using the Chebyshev propagator, used a harmonic oscillator wavefunction for the initial ground vibrational state of the ion. When using the MCTDH method, we instead numerically calculate the vibrational ground state of H_3^+ by carrying out a two-dimensional relaxation on the H_3^+ surface of Meyer *et al.* [18]. As described above, the cross section for ion-pair formation is calculated by integrating the flux absorbed by a complex absorbing potential (CAP). In figure 2, we show the cross sections obtained when the CAP is placed at $z_c = 10$ a_0 and $z_c = 15$ a_0. Since the ion-pair state is a long-range potential, it has not reached its asymptotic energy at these finite distances. Therefore the threshold is shifted toward higher energies when the CAP is moved to larger distances. Except for the threshold behavior, the cross sections calculated with the two CAP distances do not differ. Hypothetically, a CAP placed at infinity should cut the cross section below the threshold energy of 5.4 eV.

4.2. The role of the electronic couplings

In the next step we start to add the couplings between the ion-pair state and the other neutral states of H_3. When we add the second $^2A_1$ resonant state and the couplings between the two states in the Franck-Condon region, the cross section decreases in magnitude with a factor of 0.27. The shape and position of the peak in cross sections does not change. This if different from what we found in the one-dimensional study [9], where the coupling between the two resonant states caused a double peak structure that could be explained by a quantum interference between the two resonant states. When the second dimension is included, this interference effect is smeared out.

We then begin to add the potential energy surfaces of the Rydberg states and the couplings between the Rydberg states and the ion-pair state. When we add the lowest Rydberg state, there is a small decrease of the cross section for ion-pair formation. Adding the next Rydberg state decreases the cross section slightly more, as can be seen in figure 3. The following Rydberg state (state five) has couplings to the ion-pair state both at small and large internuclear distances and we can see that there is a significant drop in the cross section when this state is added to the model. Also some structures in the cross section are induced. The last Rydberg state included in the model (state 6) is the effective Rydberg with the coupling set to compensate for the infinite number of Rydberg states not included in the model. There is only a minor decrease of the ion-pair cross section when the effective state is added. The total effect of including the Rydberg state is to reduce the magnitude of the cross section for ion-pair formation with a factor 0.56. In figure 4 we compare the calculated cross section for the formation of $H_2^+ + H^-$ with the measured cross section of ion-pair formation. Remember that in the first three experiments [7, 8, 9], the H^- were detected and therefore the experiments do not separate between the two- and tree-body reactions of (1). We believe that in order to reproduce the cross section above 8.1 eV, where the three-body ion-pair channel becomes energetically available, higher excited electronic resonant states have to be included in the model.

In the figure, we also show the very recent cross section measured using the TSR ion-storage ring [10]. As mentioned above, in this experiment, the branching ratio for the two- and tree-body channel of ion-pair formation was measured. The total cross section was not obtained and the therefore it has been normalized relative to the peak of the cross section measured in the CRYRING experiment. The two-body channel dominated at lower energies and also some structures are seen in the two-body cross section that are not present in the cross section for

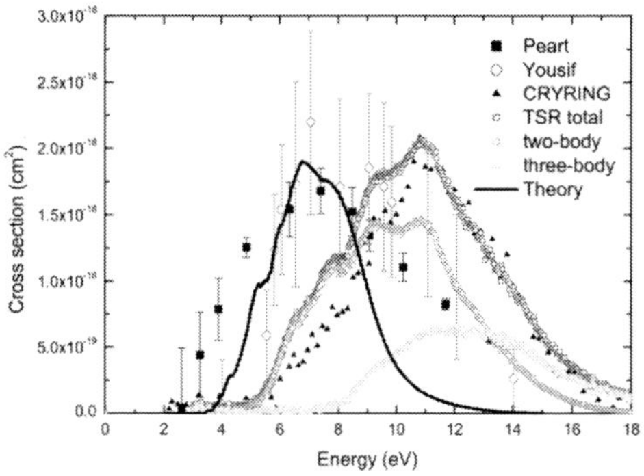

Figure 4. Calculated cross section for formation of H_2^+ + H^- in electron recombination with H_3^+ is compared with measured cross sections.

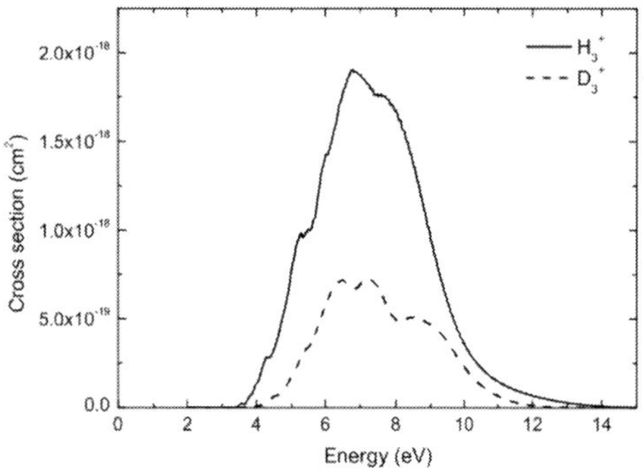

Figure 5. Calculated cross section for ion-pair formation in electron recombination with H_3^+ and D_3^+.

three-body formation. Our calculated cross section has the same magnitude of $2 \cdot 10^{-18}$ cm^2 as all measured cross sections. It is shifted in energy relative to the more recent storage ring experiments. The shape of the calculated cross section agrees better with the measured cross section by Yousif *et al.* [8]. We see some structures in the calculated cross section that resemble the structures observed in the two-body cross section measured in TSR [10], although shifted in energy.

4.3. Isotope effects

The ion-pair formation has only been studied experimentally for the H_3^+ isotope. However, the high-energy peak in the cross section for dissociative recombination has been studied experimentally [19] and theoretically [20] both for H_3^+ and D_3^+. According to both the measurement and the theoretical calculations, the positions and shapes of the high-energy peaks in the DR cross section are very similar, although with different magnitudes. The magnitude of the peak found for H_3^+ is about a factor of 3.1 (1.9 theoretically) larger than the peak found in the cross section for D_3^+.

We have carried out calculations on ion-pair formation with electron recombination with D_3^+. We use the same relevant potentials, couplings and widths as above, but change the reduced masses when we carry out the relaxation to calculate the vibrational wave function of the target ion and when we run the wave packet propagation. The calculated cross section (see figure 5) for D_3^+ has similar shape as the cross section for H_3^+. The structure is slightly more pronounced. The magnitude of the cross section for H_3^+ is a factor of 2.7 larger than for D_3^+. This is a similar ratio as found for the high-energy peak in the DR cross section. This can be understood since the same resonant states are important in both reactions.

5. Conclusions

We have carried out a two-dimensional theoretical study of the formation of $H_2^+ + H^-$ in electron recombination with H_3^+ by including two coordinates and six coupled electronic states. In order to describe the reaction, both radial Jacobi coordinates are crucial. The second dimension will smear out the sharp interference effect observed in a previous one-dimensional study. The role of the Rydberg states is to reduce the cross section. We obtain a cross section that is shifted toward lower energies than found in the more recent storage storage ring experiments [9, 10]. The shape of the cross sections has a good agreement with cross section obtained with the merged beam measurement [8]. The magnitude of the calculated cross section agrees perfectly with all measured cross sections. We believe that in order to obtain the cross section above 8.1 eV, higher excited resonant states have to be included in the model.

The calculated cross section for ion-pair formation in electron recombination with D_3^+ has a similar shape as the corresponding cross for H_3^+. The cross section for H_3^+ is a factor of 2.7 larger and a similar ratio has been observed for the high-energy peaks of the cross sections for DR of H_3^+ and D_3^+.

Acknowledgments

We are grateful to A Wolf and J Hoffmann for providing us with unpublished data and valuable discussions. We would also like to thank H D Meyer for letting us use the MCTDH program. Å Larson acknowledges support from The Swedish Research Council and A E Orel acknowledges support from the National Science Foundation, Grant No. PHY-05-55401.

References

[1] Kulander K C and Guest M F, *J. Phys. B: At. Mol. Opt. Phys.*, **12**, L501 (1979).
[2] Michels H H and Hobbs R H, *Atrophys. J.*, **286**, L27 (1984).
[3] Orel A E, Kulander K C and Lengsfield III B H, *J. Chem. Phys.*, **100**, 1756 (1994).
[4] Rescigno T N, McCurdy C W, Orel A E and Lengsfield III B H, *"The Complex Kohn Variational Method in Computational Methods for Electron-Molecule Scattering"* eds. W. H. Huo and F. A. Gianturco (Plenum, New York (1995)).
[5] Larsson M *et al.*, *Phys. Rev. Lett.*, **70**, 430 (1993).
[6] Orel A E and Kulander K C, *Phys. Rev. Lett.*, **71**, 4315 (1993).
[7] Peart B, Forrest R A and Dolder K, *J. Phys. B: At. Mol. Phys.*, **12**, 3441 (1979).
[8] Yousif F B, Van der Donk P and Mitchell J B A, *J. Phys. B: At. Mol. Phys.*, **26**, 4249 (1993).
[9] Kalhori S *et al.*, *Phys. Rev. A*, **69**, 022713 (2004).
[10] Hoffmann J and Wolf A, Private communications.
[11] Larson Å and Orel A E, *Phys. Rev. A*, **64**, 062701 (2001).
[12] Giusti-Suzor A, Bardsley J N and Derkits C, *Phys. Rev. A*, **28**, 682 (1983).
[13] Worth G A, Beck M H, Jäckle A and Meyer H -D, The MCTDH Package, Version 8.2 (2000), University of Heidelberg, Heidelberg, Germany. Meyer H -D, Version 8.3 (2002). See http://www.pci.uni-heidelberg.de/tc/usr/mctdh/
For a review see: Beck M H, Jäckle A, Worth G H and Meyer H -D, *Phys. Rep.*, **324:1**, 1 (2000).
[14] McCurdy C W and Turner J, *J. Chem. Phys.*, **78**, 6773 (1983).
[15] Jäckle A and Meyer H -D, *J. Chem. Phys.*, **105**, 6778 (1996).
[16] Larson Å, Roos J and Orel A E, *Phil Trans. R. Soc. A*, **364** 2999 (2006).
[17] Tal-Ezer H and Kosloff R, *J. Chem. Phys.*, **81** 3967 (1984).
[18] Meyer W, Botschwina P and Burton P G, *J. Chem. Phys.*, **84**, 891 (1986).
[19] Larsson M *et al.*, *Phys. Rev. Lett.*, **79**, 395 (1997).
[20] Schneider I F and Suzor-Weiner A, *Contrib. Plasma Phys.*, **42**, 578 (2002).

Dissociative Excitation of NeD^+

J. Lecointre[1], V. Ngassam[4], J.J. Jureta[1,2], J.B.A. Mitchell[3], A.E. Orel[4] and P. Defrance[1]

[1] Université Catholique de Louvain, Département de Physique, unité PAMO, Chemin du Cyclotron 2, B-1348 Louvain-la-Neuve, Belgium.
[2] Institute of Physics, P.O.Box 68, 11081, Belgrade, Serbia.
[3] PALMS, 'Equipe d'astrochimie expérimentale', UMR N°6627 du CNRS, Université de Rennes I, F-35042 Rennes, France.
[4] Department of Applied Science, University of California, Davis, Davis CA 95616 USA.

Abstract. Measurements of NeD^+ dissociative excitation cross section were obtained in Louvain-la-Neuve (UCL, Belgium) by using the animated crossed electron-ion beam method. The electronic structure of the system was calculated by using quantum chemistry calculations for the NeD^+ ion and *ab initio* electron scattering calculations using the complex Kohn variational method for the resonant NeD neutral complex formed by the temporary capture of the electron by the ion. The cross section for dissociative excitation are then calculated below the second excitation thresholds of the ion where the process is dominated by a series of doubly-excited resonance states. Theory and experiment are then compared.

1. Introduction

Dissociative excitation (DE) has been much less studied than dissociative recombination (DR). Its higher energy threshold makes it not amenable to plasma techniques and the advantages of merged beam techniques for achieving very low collision energies and the particular importance of DR to low temperature plasma chemistry have meant that the scarce beamtime available at heavy ion storage rings has been mostly devoted to this latter process, as far as electron-molecular ion studies are concerned. It is however particularly attractive for study using crossed beam experiments which cannot go low enough in energy to effectively study DR and yet, not requiring high energy accelerators, can be set up with much less capital and operating expenditure. Indeed, Louvain-la-Neuve in Belgium is one of the facilities where much of the experimental effort in recent years devoted to this subject has been performed. Neon may be used as a coolant gas in the divertor region of ITER to absorb the plasma energy by multiple ionization and excitation [1]. NeH^+ is expected to be formed by collision of Neon atom with vibrationally excited molecular hydrogen in the divertor. We have recently studied the reaction:

$$e^- + NeD^+ \rightarrow NeD^{**} \rightarrow Ne + D^+ + e^- \quad (1)$$

both theoretically using a wave packet propagation method and, experimentally using a crossed electron-ion beam apparatus. The results are reported here after a brief discussion of both the theoretical methods and the experimental setup used.

2. Experimental procedure

In the experiment developed in Louvain-la-Neuve (UCL, Belgium) the animated crossed electron-ion beam method is applied [2]. Ions are extracted from an ECR ion source and accelerated to 12 keV. The fixed-energy molecular ion beam is selected by means of a first double focusing 90° magnetic analyser, additionally focused and purified by a 45° spherical electrostatic deflector. The ion beam is directed into the collision region where it crosses the ribbon-shaped electron beam at right angles. The electron energy (E_e) range extends continuously from the threshold of the initial reaction (excitation or ionization) up to 2.5 keV. Product ions are separated from the primary ion beam by means of a second double focusing 90° magnetic analyser. Due to the transfer of internal potential energy, dissociation fragments exhibit both a broad velocity and a broad angular distribution in the laboratory frame. Selected by the variable analyser slit, the angular acceptance of the magnet (0.1 radian) is large enough to transmit to the detector all ions with the maximum expected kinetic energy release produced by dissociation processes. Product ions are further deflected by a 90° electrostatic spherical deflector and directed onto a channeltron detector. Usually, only a fraction of the signal is detected, so that absolute cross sections are obtained after determination of the product ion transmission. For singly charged fragments, individual contributions of dissociative excitation and dissociative ionization processes are estimated separately [3].

3. Theory

The potential energy and the autoionization width of the resonant states were obtained from electron-scattering calculations using the complex Kohn variational method well described in Ref. [4]. In order to generate an acurate target wave function, we carried out an SCF calculation on the ion to generate a set of molecular orbitals. These orbitals were then used in an all singles and doubles configuration interaction (CI) calculation to obtain the natural orbitals. In order to produce a balanced treatment of the ground and excited states, we took natural orbitals produced from averaging the density matrices for the lowest three states of the ion with the ground Σ state weighted 0.98 and each component of the next Π state weighted 0.01. This was necessary to maintain the relative energy position of the excited state. Three natural orbitals (4σ, $2\pi_x$, $2\pi_y$) were included as well as the occupied orbitals. The basis set was expanded to include 10 additional diffuse orbitals: five s (exponents .299, .1, .033, .011, .0037), four p (exponents .113, .0519, .023, .005) and one d (exponent .170). The two lowest orbitals were frozen and the target states were determined from a full CI in the natural orbital space of six orbitals with the largest occupation numbers.

At the equilibrium geometry four resonance states of Σ symmetry and three of Π resonances are below the first excitation threshold of the ion (18.5 eV). No Δ resonances were found in this energy range. The resonance parameters (autoionization widths and resonance energies) were extracted by fitting the eigenphase sums to a Breit-Wigner form. Fig.1 shows the four Σ (left panel) and three Π (right panel) resonances calculated.

The nuclear dynamics are studied by solving the time dependent Schrödinger equation for each resonant state since they are not coupled to each other, that is:

$$i\frac{\partial}{\partial t}\Psi(R,t) = \left[-\frac{1}{2\mu}\frac{\partial^2}{\partial R^2} + V_{res}(R)\right]\Psi(R,t), \qquad (2)$$

where Ψ represents the amplitude on each of the resonance states and μ is the reduced mass of the system. The resonance potentials appear as local complex potential curves, given by

$$V_{res}(R) = E_{res}(R) - i\frac{\Gamma(R)}{2}, \qquad (3)$$

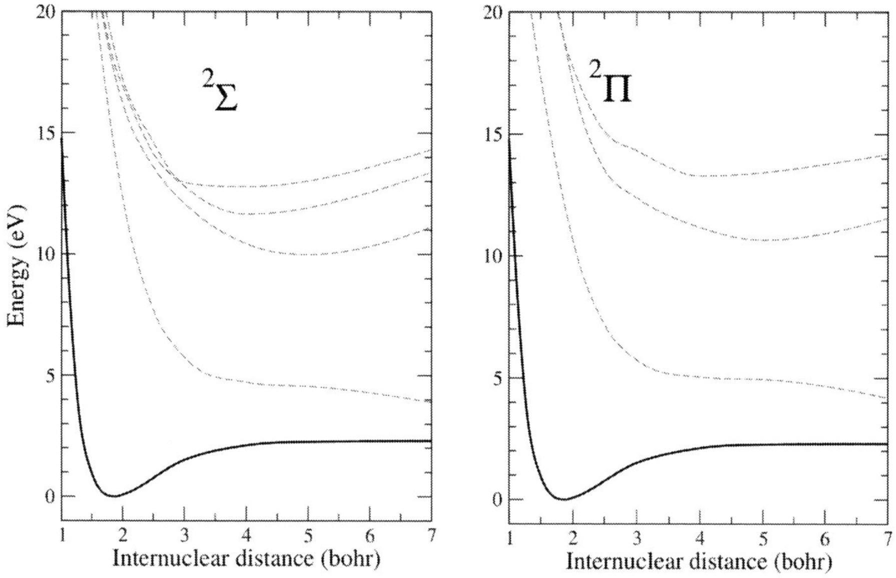

Figure 1. Potential energy curves of Σ (left panel) and Π (right panel) resonance states of NeH. In each panel, the solid curve represents the ground $^1\Sigma$ state NeH^+. The dashed curves represent the resonance states in each symmetry. In these figures, energy is in eV where 1 eV=1.602×10^{-19} J and distance is in bohr where 1 Bohr=5.291772×10^{-11} m.

where $E_{res}(R)$ is the energy position of the resonance potential curve and $\Gamma(R)$ is the autoionization width. The initial wave packet is expressed as

$$\Psi_{\nu_i}(R, t = 0) = \sqrt{\frac{\Gamma_i(R)}{2\pi}}\chi_\nu(R), \tag{4}$$

where $\chi_\nu(R)$ is the initial vibrational wave function of the molecule.

For the initial vibrational level ν_i, the cross section for dissociative recombination via a single resonance state is given by

$$\sigma_{\nu_i}(E) = \frac{4\pi^3}{k_i^2} g \mid \langle\phi_E(R) \mid \Psi_{\nu_i}(R, t)\rangle, |^2 . \tag{5}$$

where $\phi_E(R)$ is the energy normalized wave function given by

$$\phi_E(R) = \sqrt{\frac{\mu}{2\pi k_i(E)}} e^{ik(E)R} \tag{6}$$

4. Results

In the present paper, absolute cross sections are reported for the particular dissociative excitation channel leading to the D^+ formation and results are shown in Fig. 2. The maximum of the DE cross section is found to be $(6.3\pm1.4)\times10^{-17}$ cm^2 (at E_e= 35.1 eV) and two threshold energies are identified at (10.0±0.5) eV and (14.1±0.5) eV. The lowest dissociative states ($B^1\Sigma$ and $b^3\Sigma$) accessible via direct excitation leading to the Ne*(2p^{5}3s, ^{1,3}P)+H^+ pair formation present a high dissociation limit, so that the threshold for dissociative excitation is much higher (around 30 eV). Vibrational excitation of the parent ion cannot be responsible for these observed low

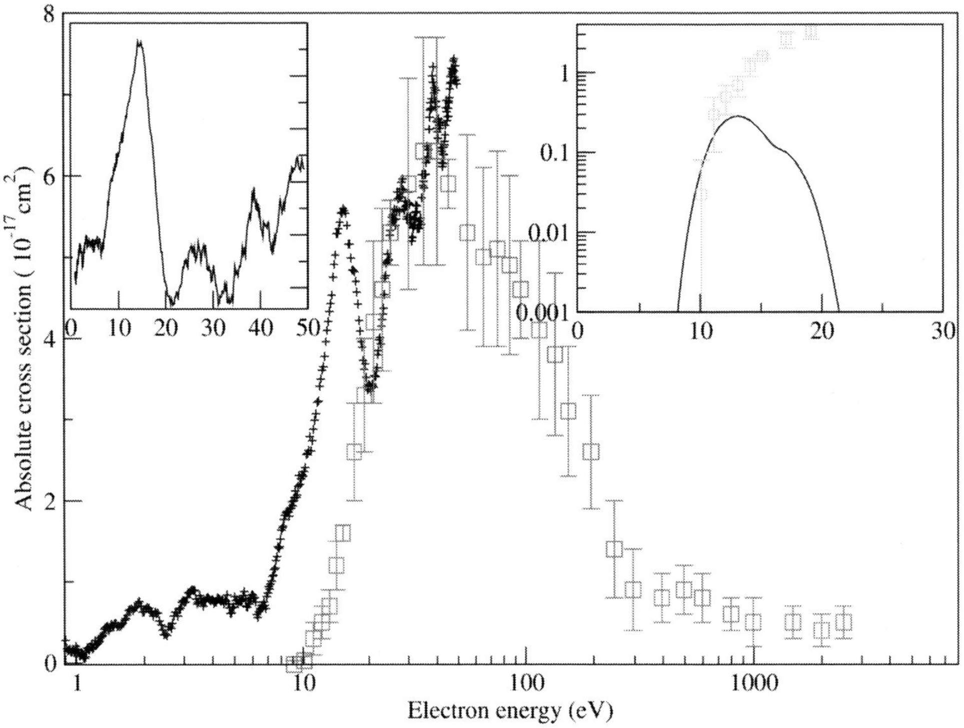

Figure 2. Absolute cross section for dissociative exitation versus electron energy: present measurements for D^+ fragments are represented with opened squares and normalized data of ref [5] are represented with plus sign. The insert on the left shows estimated resonant contributions (see text for details), the insert on the right is a comparison of the data with theory (solid line)

energy thresholds, given that the potential well depth of NeD^+ is only slightly higher than 2.1 eV. Such low experimental thresholds can only be explained by resonant dissociative excitation. The lowest one (10.0 eV) can be attributed to capture into doubly excited states (about 11 eV) followed by autoionization to the $X^1\Sigma$ electronic ground state of NeD^+ (equation 1). The width of the Franck-Condon region may also explain the existence of the two experimental thresholds.

In Fig. 2, present absolute cross sections for dissociative excitation leading to Ne+D^+ products are compared with measurements performed on NeH^+ ions and obtained from an ASTRID storage ring experiment [5]. In that experiment, neutral Ne and Ne+H products are detected but not separately distinguished, so that the corresponding data contain both dissociative recombination and excitation contributions. DR, which is dominant in the low energy range (below 1 eV), cannot be observed in the present experiment as only ions are detected. The storage ring data shown in Fig. 2 have been normalised to the present absolute cross sections, at 20 eV. Several energy thresholds are observed around 7 eV, 20 eV, 33 eV and 42 eV. Below 20 eV, a distinct qualitative difference is clearly observable between the storage ring data and the present ones: the thresholds (10.0 eV and 14.1 eV) do not coincide with the above-listed ones and the broad peak centred on 14 eV is not seen. This is clearly a resonant contribution due to dissociative recombination which is highlighted by subtracting the present results from the storage ring data, assuming that there will be little difference between the NeH^+ data and what one would obtain using NeD^+ ions in the storage ring (left side insert, Fig. 2). The corresponding resonant states evidently lead to formation of two neutral products

(Ne+H), because no resonant signal is observed in the present experiment. Above 20 eV, both experimental results give similar overall shapes for DE, except for the thresholds (33 eV and 42 eV) which cannot be observed in the present experiment due to larger statistical uncertainties.

On the theoretical side, the potential energy curves shown in Fig. 1 were used, as input into the wave packet method previously described, to determine the resonant DE cross section.

The absolute cross section for dissociative excitation of NeD^+ from the ground vibrational state (ν=0) of the ion are shown in the right side insert of Fig. 2. Also shown, opened square, is DE cross section for NeD^+ measured at Louvain-La-Neuve. The theoretical DE cross section for NeD^+ has a peak of 2.83×10^{-18}cm^2 at 13 eV where the overlap between the wave function of the ion continuum and that of the lowest resonant Σ state is maximum. There is also a structure in the DE cross section, located at about 16 eV and due to resonant states that contribute mostly at higher energy. The range of energy in which calculations has been done is smaller compare to experiment. Six excited states of NeD^+ are opened in the energy range of the experiment, and must be explicitly included in the close-coupling expansion. This is beyond the scope of this study. The experimental data from Louvain-La-Neuve shows a threshold in the cross section for DE at 10 eV that is clearly reproduce by the theory. This threshold is due to the fact that even though the DE channel is open at low energy, the cross section is smaller and cannot be detected in the present experiment. As the collision energy increases and the interaction between the lowest Σ resonant state and the ion continuum become stronger; the cross section become significant and the process is detected by the experimental apparatus. These experimental cross sections peak at about 100 eV. This is attributed to the direct DE channels not included in the calculations. In the energy region around the threshold, the present calculation shows a good agreement with the experimental data.

5. References

[1] Mitchell JBA, in *Atomic and Plasma-Material Interaction for fusion*, (Vienna: Internationnal Atomic Energy Agency) **6** 97 (2001)

[2] Defrance P, Brouillard F, Claeys W and Van Wassenhove G, *J. Phys. B: At. Mol. Opt. Phys.* **14** 103 (1981)

[3] Lecointre J, Belic DS, Cherkani-Hassani H, Jureta JJ and Defrance P, *J. Phys. B: At. Mol. Opt. Phys.* **393** 275 (2006)

[4] Rescigno TN, Orel AE, McCurdy CW and Lengsfield BH, in *Computational Methods for Electron-Molecule Collision*, ed Huo WM and Gianturco FA (New York Plenum, 1995)

[5] Novotny O, PhD Thesis, Université de Rennes I (2006)

Dissociative recombination of CF^+: experiment and theory

O Novotný[1], O Motapon[2], M H Berg[1], D Bing[1], H Buhr[1,3], H Fadil[1], M. Grieser[1], J Hoffmann[1], A S Jaroshevich[4], B Jordon-Thaden[1], C Krantz[1], M Lange[1], M Lestinsky[1], M Mendes[1], S Novotny[1], D A Orlov[1], A Petrignani[1], I F Schneider[5], A E Orel[6] and A Wolf[1]

[1] Max-Planck-Institut für Kernphysik, Saupfercheckweg 1, D-69117 Heidelberg, Germany
[2] Department of Physics, Faculty of Science, University of Douala, P.O. Box 24157 Douala, Cameroon
[3] Department of Particle Physics, Weizmann Institute of Science, Rehovot 76100, Israel
[4] Institute of Semiconductor Physics, 630090 Novosibirsk, Russia
[5] Laboratoire de Mécanique, Physique et Géosciences, Université du Havre, 25, rue Philippe Lebon, P.O. Box 540, 76058 Le Havre Cedex, France
[6] Department of Applied Science, University of California at Davis, Davis CA 95616 USA

E-mail: `oldrich.novotny@mpi-hd.mpg.de`

Abstract.
We present results from our recent studies of the dissociative recombination of the CF^+ cation. On one hand, dissociative recombination was measured with 3 MeV CF^+ ions in the heavy-ion Test Storage Ring in Heidelberg, using the twin electron beam configuration with an electron cooler and a separated electron target for collision measurements. In this experiment, the low temperatures of the electron beam provided by a photocathode (temperature in co-moving frame below 1 meV) account for a fast kinetic cooling of the heavy-ion beam and a high resolution in the measured rate coefficients. Fragment imaging measurements show a complete switching of the dissociation route by only a small change of the collision energy and the disappearance of neutral Rydberg product states on crossing the DE threshold. On the other hand, extensive calculations of energy positions and autoionization widths for the doubly excited states of CF between the first and second ionization thresholds have been obtained from electron scattering calculations using the complex Kohn variational method, followed by calculations of the dissociative recombination process with the multichannel quantum defect theory. In preliminary computations, only the first dissociative state in each molecular symmetry, which lies closest in energy to the ion potential at its equilibrium internuclear separation, and thus is dominant for the low-energy dissociative recombination, was included. Although only the direct mechanism of dissociative recombination reaction has been considered in this step, the size and the shape of the DR rate coefficient are already well reproduced.

1. Introduction

Dissociative recombination (DR) is usually the main neutralization channel controlling ionization states and chemical composition in a variety of plasma environments. This also holds for the DR of CF^+ in CF_4 plasmas, commonly used in microelectronics processing [1], where fluorocarbon plasmas are often used to etch silicon surfaces. Although the feed gas is non-reactive, under electron bombardment, it fragments to produce reactive species such as CF,

CF_2 and CF_3 radicals and the corresponding ions. Electron collision cross sections for these transient, reactive species are needed in large-scale simulations of these processing plasmas [2]. As well, it has been recently shown [3] that the DR rate coefficient of CF^+ is a key parameter for the fluorine chemistry in the cold interstellar medium (see also contribution of E. Roueff in this issue). Absolute rate coefficients and their temperature dependencies are therefore needed for meaningful modelling of such plasma environments. The knowledge of further details such as angular dependence of the DR rate coefficient, the branching ratios between dissociative channels, and their energy dependency, contributes to the comprehensive theoretical insight into molecular structure and reaction dynamics.

The rate of recombination of CF^+ with electrons was already studied in a joint experiment at the ASTRID and CRYRING heavy-ion storage rings [4]. At ASTRID the relative DR and DE (dissociative excitation) rate coefficients were obtained for collision energies of up to 12 eV, while at CRYRING the low energy range (up to 0.6 eV), including absolute scaling, was studied.

In this paper we present results of recent experimental and theoretical investigation of the DR of CF^+ with electrons. Experimental studies were carried out at the Test Storage Ring in Heidelberg (TSR). The absolute recombination rate coefficient of CF^+ ions with ultra-cold electrons produced by a GaAs photocathode was obtained for energies of up to 20 eV. The fragment imaging technique was used to describe the reaction dynamics, branching ratios of final neutral states and angular distribution of the products with respect to the ion axis. CF^+ was also used as a "benchmark" ion for testing the cooling performance of the photocathode-equipped electron target for slow beams of heavy ions in the mass range of ~ 30 amu.

We have also performed extensive structure calculations coupled with electron scattering calculations, using the complex Kohn variational method [5], to characterize both the Rydberg states converging to the ground and low-lying excited states of the ion, as well as the resonant states and their autoionization widths as a function of internuclear separation. These data have been used in a Multi-Channel Quantum Defect Theory (MQDT) study of the dissociative recombination. The experimental results are reported here along with the preliminary theoretical calculations.

2. Experiment

The experiment was performed using the TSR in Heidelberg, a heavy-ion storage ring equipped with separated electron-cooler and electron-target devices. The detailed description of the apparatus together with the main principles of storage ring techniques and measurement procedures is given e.g. in [6, 7]; in this section we focus on the special aspects of storing CF^+.

2.1. Fast electron cooling of heavier molecules in the TSR

The energy of the stored CF^+ beam (31 amu) was 3.0 MeV, close to the value defined by maximum magnetic rigidity of the dipole bending magnets of the TSR (1.5 Tm). This beam was produced by stripping CF_3^- to CF^+ in the MPI-K tandem accelerator. The corresponding electron beam energy for matching electron and ion beam velocities is only 53 eV. At such conditions several effects, which are usually negligible or only a small correction for light ions (< 10 amu), start to play an important role. The cross-section of ion collisions with rest gas molecules rises at low energies, which significantly shortens the life time of stored ion beam. The products of charge-transfer reactions then appear as the background in rate and fragment imaging measurements. The longitudinal temperature and density of the electron beam also degrade strongly at low energies [8]. Low electron density, therefore, decreases the measurement rate and signal/background ratio. Higher electron temperatures result in lower energy resolution, and together with a high ion mass and low electron density, prolongate the time needed for full kinematic cooling of the ion beam.

We did not achieve reasonably fast kinematic cooling of the ion beam when using the thermocathode for producing the electron beam. Within the beam lifetime ($\tau \sim 3.6$ s) it was not possible to reach a stable width of the ion beam below 10 mm. Therefore we used the cryogenic GaAs photocathode to produce an ultra-cold electron beam [9] with estimated transverse and longitudinal temperatures of about 1.0 meV and 0.05 meV respectively. The 2D imaging technique [10] was used to study the ion beam profile evolution vs. storage time: The distribution of the centres-of-mass for DR events, projected on to the MCP and the phosphor screen located 12.24 m from the electron-target centre, represents the ion beam profile in the electron-target [11]. The absolute beam width can be extracted when scaling down the projected width and correcting for typical ion spread and dispersion [12]. The typical electron current was 0.3 mA at an electron beam diameter of 13 mm (expansion $\alpha = 20$) which corresponds to an electron density $\sim 3 \cdot 10^6$ cm^{3}s^{-1}. The CF^+ ion beam current right after injection into the TSR was estimated to be a few 100 pA. A transverse cooling time below $\tau_c < 2$ s was achieved demonstrating a high cooling efficiency of the photocathode beam. After only 6 s of pre-cooling very small FWHMs of the centre-of-mass distribution of 1 mm (horizontally) and 0.7 mm (vertically) were measured by an imaging detector. This corresponds to an ion beam widths in the iteraction section of the ring of about 0.5 mm and 0.09 mm, respectively.

2.2. DR and DE rate coefficients

For rate coefficient measurements, an energy sensitive surface barrier detector of 50x50 mm along with the standard cooling-measurement-reference scheme [13] was used. Events appearing at full ion beam energy were assigned to DR while events at energies corresponding to only C or F arriving at the detector were assigned to DE. In both cases, a background corresponding to reactions with rest-gas was subtracted. The absolute scaling factor was obtained in an independent measurement similar to that described in [14]. In short, this method uses the difference between ion beam life times when the electron beam is on or off. It is assumed that the shortening of the life time is exclusively due to DR or DE either by electrons or residual gas and that the rate coefficient does not evolve significantly during storage e.g. due to ro-vibrational relaxation.

Both DR and DE (C^+ + F and C + F^+ channels) rate coefficients measured at the TSR are shown in Figure 1. Below 1 eV the DR rate curve increases faster (roughly $\propto 1/E_{CM}$) than given by the standard scaling $\propto 1\sqrt{E_{CM}}$. Smooth peaks are found at 15 and 30 meV. At higher energies a rich structure appears which is not typical for previous measurements of diatomic molecular ions. Peaks below 4.9 eV correspond to DR pathways via Rydberg states lying below the $a^3\Pi$ excited ionic state (see drop at $\sim$ 4.9 eV) and similar peaks at higher energies correspond to Rydberg states lying below higher ionic excited states [15]. Note the sharp drop of the rate coefficient at 7.75 eV, exactly at the position of the lowest CF^+ dissociation energy (value calculated from thermodynamic data [16]). There is no evolution of the DR rate coefficient between 6 and 22 s of storage time visible within the statistical errorbars.

Both DE channels (Figure 1) do not display significant rate coefficients below their energetic thresholds (7.75 eV for C^+ + F and 13.91 eV for C + F^+); therefore no electronically excited CF^+ is expected to be present in the parent ion beam. Note the very sharp threshold of the C^+ + F channel. There is substantial structure in the DE rate coefficient curve (C^+ + F) which is likely due to vertical transitions into potential curves at higher excited states of CF^+.

Both DR and DE rate coefficents were scaled using the absolute value of the DR rate coefficent measured at near-zero collision energy (limited by electron energy spread), obtaining $\alpha = 8 \cdot 10^{-7}$ cm^{3}s^{-1} at $E_{CM} = 0.1$ meV. The errorbar of the absolute scaling is estimated to be $\sim$30%. The DR rate coefficient curve displays an absolute rate below $E_{CM} = 1$ eV approximately 30% higher than the rate obtained at CRYRING [4] using a DC transformer ion current measurement for absolute scaling. At higher energies, only two broad peaks at

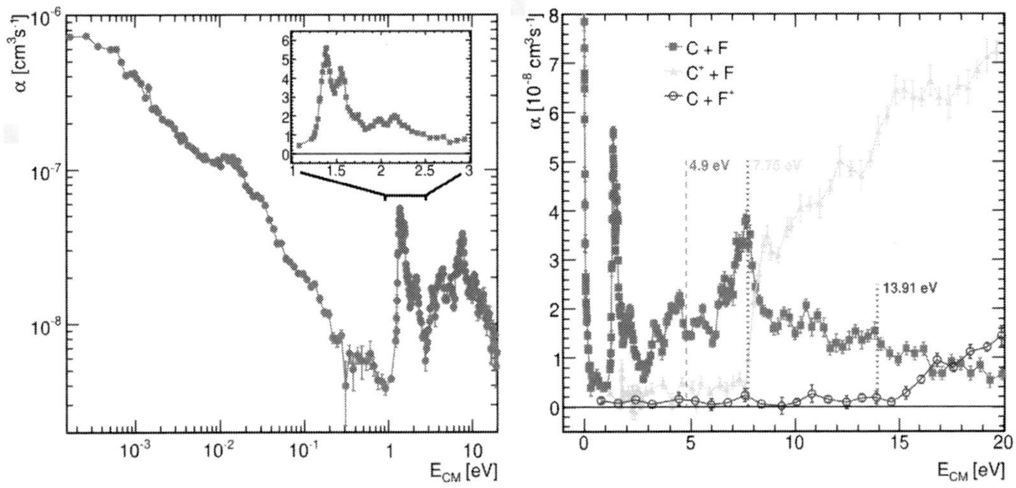

Figure 1. DR and DE rate coefficients vs. E_{CM} in log-log (left) and linear (right) scales. The right panel shows DR (squares) together with DE for C^+ + F (triangles) and C + F^+ (circles). Dotted lines in right panel indicate energetic threshold of particular DE channels with respect to the ground state of CF^+. Dashed line indicates the excitation energy of $a^3\Pi$ of CF^+.

$E_{CM} = 1.5$ eV and $E_{CM} = 7$ eV obtained at ASTRID can be compared to the rich structure obtained at the TSR. The DE rate coefficient (C^+ + F channel) obtained at the TSR has a sharper threshold and a $\sim$ 40% lower absolute value than at ASTRID. The absolute value ratios for DR and DE can be affected by the misinterpretation of DR events as DE events: at high collision energies the kinetic energy release of fragments can cause one fragment, originating from a DR event, to miss the detector. The remaining fragment is then interpreted as a single neutral fragment in a DE event. At the TSR, fragment imaging measurements were performed to show that no more than 5% of DR events are misinterpreted as DE events, even at $E_{CM} > 10$ eV. Simulations also show that the narrow (cold) ion beam at the TSR is essential for minimizing the misinterpretation. There is no information on CF^+ ion-beam width at ASTRID.

2.3. DR fragment imaging

An extensive set of measurements was performed with the fragment imaging technique [10], studying the distribution of projected DR fragment distances on the detector. This allows the assignment of dissociation routes, final states of neutral fragments and their branching ratios for particular interaction energies. Precise fitting of the projected fragment distance distribution [17] was also used to obtain information about the rotational excitation of the parent CF^+ ion, assuming constant DR rate coefficients for particular rotational states and Boltzmann distribution of rotational population at the temperatures given. Very small changes in DR rate coefficient vs. storage time (6-22 s) well support this assumption. Positions of neutral final states marked in the particular plots of fragment distance distributions include the shift due to the rotational excitation of the parent ion determined in the fit and given in the captions.

2.3.1. Near-zero collision energy The results of the measurements at near-zero collision energy (limited by electron beam temperature) is shown in Figure 2. Data were acquired from 3 to 35 s after the injection, where the distributions showed very strong rotational cooling from ~6000 K down to <700 K. At all storage times, only 2 of the 3 energetically allowed final states are observed: $C(^3P) + F(^2P)$ with $E_{KER} = 3.50$ eV and $C(^1D) + F(^2P)$ with $E_{KER} = 2.25$ eV,

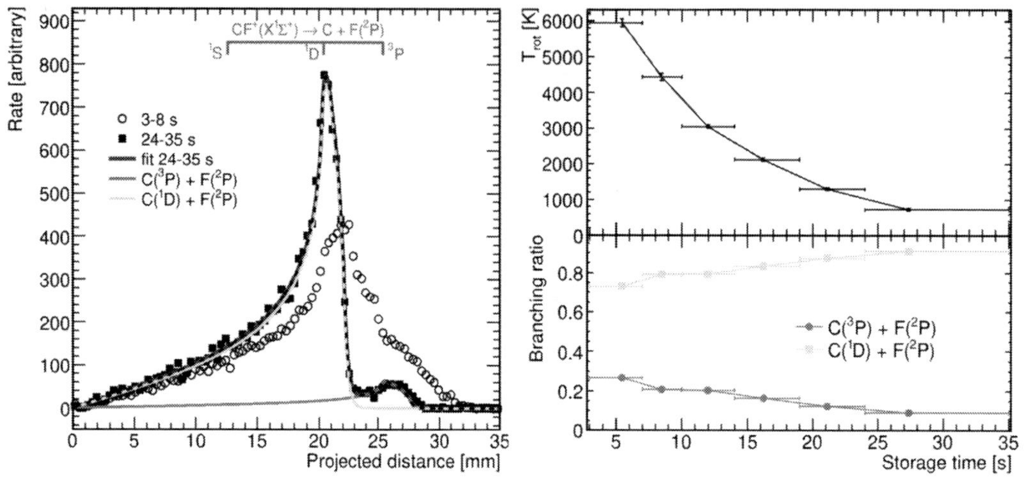

Figure 2. Fragment imaging at near-zero collision energies. **Left:** Distribution of distances of fragments projected on the detector. Experimental data are plotted for two storage time cuts to show the change of the shape due to rotational cooling and the change in branching ratio during ion storage. The fit is shown only for 24-35 s cut with separated curves for two observed final states. The positions of the maxima of the distributions for three final states energetically accessible from the CF^+ ground state at $E_{CM} = 0$ eV and $T_{rot} = 700$ K are marked in the upper part of the graph. **Right top:** Rotational temperature vs. storage time. **Right bottom:** Branching ratio of two observed dissociation routes vs. storage time.

where E_{KER} is the kinetic energy released with respect to the ground state of CF^+. The branching ratio of the lowest final state decreased during the storage time from 28% at 5 s to $< 10\%$ at 30 s. The population of rotational states at the shortest measured storage time (5 s) corresponds to a mean J of $\approx$50 ($T_{rot} \approx 6000$ K). High excitation, therefore, is created due to the CF^+ production method.

2.3.2. Elevated collision energies: 1.37 and 1.53 eV The results of fragment imaging for 1.37 and 1.53 eV collision energies, i. e. on the two highest peaks visible in the inset of Figure 1 (left), are shown in the left part of Figure 3. At 1.37 eV we have observed 100% of the dissociated fragments in the lowest available final state $C(^3P) + F(^2P)$ with an anisotropic angular distribution. At only 0.16 eV higher collision energy (see DR rate curve, Figure 1), the majority of neutral fragments land in the $C(^1D) + F(^2P)$ state with an isotropic angular distribution. In both cases, the obtained rotational temperature exceeds the temperature found for similar storage time at near-zero collision energy by $\sim$40%, which could indicate additional excitation by electrons at elevated collision energies.

2.3.3. Elevated collision energies: 7.48 and 8.56 eV The next two energies for fragment imaging measurement were chosen just below and above the energetic threshold of the $C^+ + F$ DE channel. The results are shown in the right part of Figure 3. It was not possible to fit the projected distance distributions because too many (> 20) final states are accessible at such high collision energies. The signal clearly shows, however, the presence of a Rydberg level ($n \approx 6$) population for DR at E_{CM} below the DE threshold, which disappears when the collision energy exceeds the threshold. On the other hand, the low-n part of the signal remains almost unchanged.

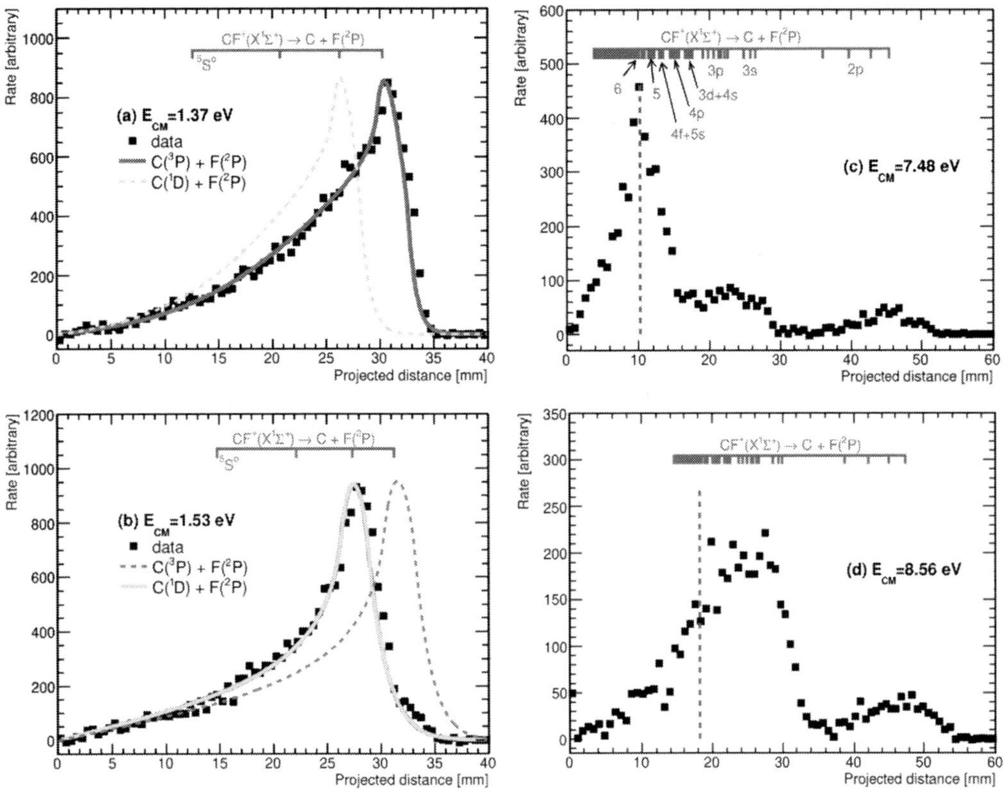

Figure 3. Fragment imaging at elevated collision energies. **Top left:** E_{CM} =1.37 eV, storage time 16-25 s, $T_{rot} \approx 2000$ K. **Bottom left:** E_{CM} =1.53 eV, storage time 10-18 s, $T_{rot} \approx 4000$ K. Full line shows the fit of experimental data, dashed line shows expected shape of "missing" final states in both cases. **Right:** E_{CM} =7.48 eV and E_{CM} =8.56 eV, storage time 10-18 s, showing a Rydberg state contribution with $n \approx 6$ ($\sim$7.2 eV above ground state of CF^+, dashed line) at E_{CM} =7.48 eV (below the DE threshold) and its disappearance for E_{CM} =8.56 eV. In the upper part of all graphs the positions of maximum of distributions of all energetically available final states is provided (for 7.48 eV and 8.56 eV, $T_{rot} \approx 4000$ K is used to compensate the shift due to rotational excitation). See the discussion in the text.

3. Theory

3.1. Calculations of molecular data

The ground state configuration for the ion is [core]$5\sigma^2$ that is, overall $^1\Sigma^+$ symmetry. The Rydberg states of CF converging to this state have configurations [core]$5\sigma^2 nl\lambda$, with l=0, 1, 2, etc. The first excited state of the ion, $a^3\Pi$ has the configuration [core]$5\sigma 2\pi_{x,y}$ and the Rydberg states of CF converging to this state have configurations [core]$5\sigma 2\pi_{x,y} nl\lambda$.

For small internuclear separations, the resonance parameters, which include the resonance energy and the autoionization width, can be obtained from electron scattering calculations. We carried out these electron scattering calculations for low-energy electrons from the ion using the Complex Kohn Variational method [5]. The method has been widely described in the literature [18] and we will only highlight the specifics for this calculation. In this method, the electronic trial wave function of the $(N+1)$-electron scattering system is expanded as

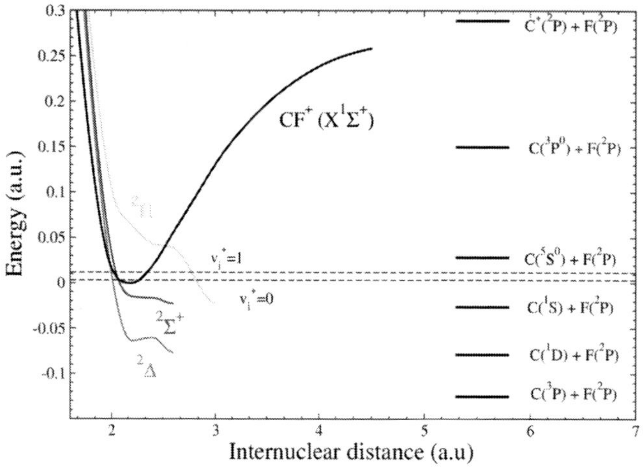

Figure 4. Potential energy curves relevant to low energy DR of CF^+. Ion curve: thick solid line; resonances: color lines.

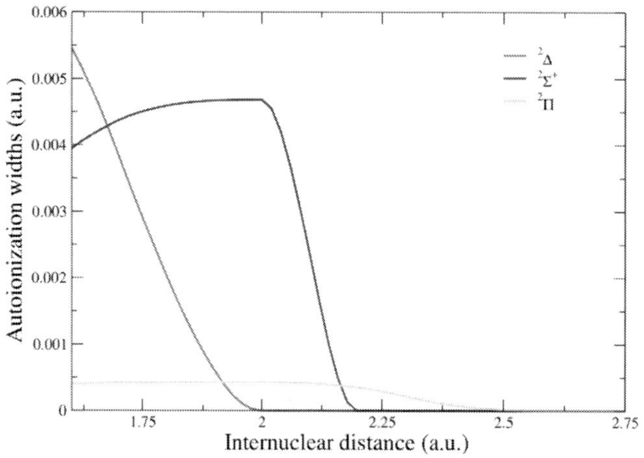

Figure 5. Autoionization widths of the first three resonances.

$$\Psi = \sum_{\Gamma} A[\Phi_{\Gamma}(\mathbf{x}_1...\mathbf{x}_N)F_{\Gamma}(\mathbf{x}_{N+1})] + \sum_{\mu} d_{\mu}\Theta_{\mu}(\mathbf{x}_1...\mathbf{x}_{N+1}) \quad (1)$$

The first sum is denoted as the P-space portion of the wave function and runs over the energetically open target states, where Φ_{Γ} are the N-electron target eigenstates, $\mathbf{x}_i$ denotes space-spin coordinates, A antisymmetrizes the coordinates of the target and scattered electrons. In the second sum, the Θ_{μ} are square-integrable $N+1$ configuration state functions (CSFs) which are used to describe short-range correlations and the effects of closed channels. These functions also contain information about the resonance portion of the scattering wave function. This sum is denoted as the Q-space portion of the wave function. Since the range of scattering energies of interest lies above the first excited state of the ion, three channels were treated in the P-space, the ground Σ and first excited Π states, however, at low scattering energies only one channel may be open.

Journal of Physics: Conference Series **192** (2009) 012021 doi:10.1088/1742-6596/192/1/012021

There are two classes of terms included in the Q-space space. The first class of functions is the set of all $(N + 1)$-electron CSFs that can be formed from the active space of target orbitals. These are terms which will relax any constraints implied by the strong orthogonality between the scattering functions, F_Γ, and the target orbitals. They are generally referred to as "penetration terms" [5]. We also include a second class of terms, "CI relaxation terms", which are the complement $(1 - P)$ of the P-space part of the wave function. This complement, combined with the penetration terms, constitute the correlation part of the trial wave function. To obtain the target state, a set of molecular orbitals were generated from a self-consistent-field calculation (SCF) on the ion, using the TZP basis set of Dunning *et al.* To obtain a more compact representation, a configuration interaction (CI) calculation on the ion, the N-electron system, was carried out consisting of all singles from an active space of 8 orbitals, (the first two orbitals 1σ and 2σ were frozen) 3σ, 4σ, 5σ, 6σ and the two degenerate pairs, $1\pi_{x,y}$ and $2\pi_{x,y}$. To better represent the resonance states of both Σ and Π symmetries, the natural orbitals obtained were then averaged over the three lowest roots, Σ and the doubly degenerate Π, with a weighting of 0.5, 0.25, 0.25 respectively. The final target basis then consisted of 10 natural orbitals with the highest occupation numbers, six σ, and the two degenerate pairs $1\pi_{x,y}$ and $2\pi_{x,y}$ orbitals. The target basis was augmented with five Gaussian functions on the C and seven on the F.

The electron scattering calculations were done in the $(N+1)$-electron system, meaning that 15 electrons are included. The target wave functions used in the calculation were obtained from a full CI in the smaller natural orbital space of six orbitals (the first four were frozen). This forms the first class of terms in the Q-space mentioned above. The variational calculations included spherical harmonics up to l=6, $\mid m \mid$=4. The scattering calculations were performed in all symmetries, and in each case, a cross section and an eigenphase sum was obtained from the scattering calculation. The resonances were well separated in energy, which allowed each eigenphase sum to be fitted to the Breit-Wigner form to obtain the resonance energy and the autoionization width.

The resonance curves compared to the ion ground state, and the corresponding autoionization widths are shown in Figures 4 and 5 for the first states with $^2\Sigma$, $^2\Pi$ and $^2\Delta$ symmetries respectively.

3.2. Dynamics

The dissociation dynamics was performed by using our MQDT-based[19] method[20, 21]. The molecular states obtained as described in the preceding section correspond to a quasi-diabatic representation[22] in which only part of the electronic hamiltonian is diagonalized, within subspaces of electronic states of similar configuration, in order to cope with problems due to the avoided crossings of the potential energy curves. The short-range electronic interactions between states of different subspaces are then found out in terms of an electronic coupling operator $\boldsymbol{\mathcal{V}}$ which couples the ionization channels to the dissociative channels. Starting from $\boldsymbol{\mathcal{V}}$, one builds the short-range reaction matrix $\boldsymbol{\mathcal{K}}$, solution of a Lippmann-Schwinger integral matrix equation:

$$\boldsymbol{\mathcal{K}} = \boldsymbol{\mathcal{V}} + \boldsymbol{\mathcal{V}}\frac{1}{\boldsymbol{E} - \boldsymbol{H}_0}\boldsymbol{\mathcal{K}} \tag{2}$$

$\boldsymbol{H}_0$ being the Hamiltonian operator excluding the interaction potential $\boldsymbol{\mathcal{V}}$. The effects of short-range are valid in the region of small electron-ion and nuclei-nuclei distances, that is the 'A-region' [23] where the Born-Oppenheimer representation is appropriate for the description of the colliding system. There, the energy-dependence of the electronic couplings can be neglected. In the case of weak coupling, a perturbative solution of equation (2) can be obtained. This solution has been recently proven to be exact to second order, in the case of energy-independent electronic coupling [21]. In the external zone, the 'B-region' [23], corresponding to large electron-core distances, the Born-Oppenheimer model is no longer valid for the ionization channels, and

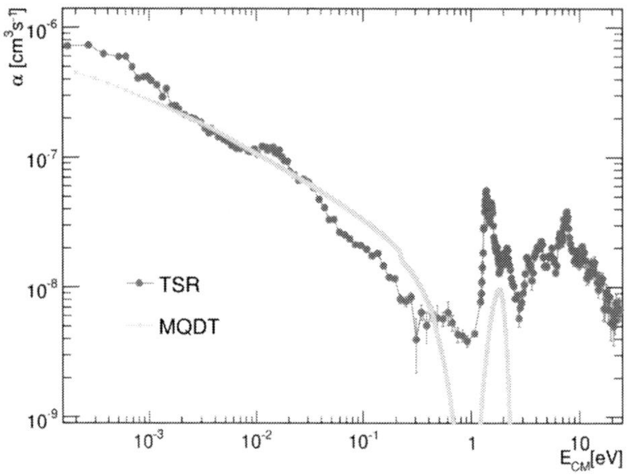

Figure 6. Maxwell anisotropic rate coefficients compared to TSR experiment for $v_i^+ = 0$.

a close-coupling representation, in terms of 'molecular ion + electron', is more appropriate. This corresponds to a frame transformation which involves projection coefficients, which are the building blocks of the 'generalized' scattering matrix $\boldsymbol{X}$, accounting for open and closed channels. Finally, the open, closed and open-closed blocks of $\boldsymbol{X}$are used eventually to build-up the 'proper' scattering matrix $\boldsymbol{S}$.

For a molecular ion initially in the level v_i^+ and recombining with an electron of energy ε, the cross section of capture into *all* the dissociative states d_j of the same electronic symmetry Γ is given by:

$$\sigma^{\Gamma}_{\mathrm{diss}\leftarrow v_i^+} = \frac{\pi}{2}\cdot\frac{\rho^{\Gamma}}{2}\cdot\frac{1}{\varepsilon}\cdot\sum_j\sum_l |S^{\Gamma}_{d_j\leftarrow lv_i^+}|^2 \quad (3)$$

Here ρ^{Γ} is the ratio between the multiplicities of the neutral system and the ion, and l label the partial waves in the wavefunction of the incident electron. In this work, we did not include the Rydberg bound states responsible for the *indirect* DR process. Only the lowest resonances with $^2\Sigma$, $^2\Pi$ and $^2\Delta$ symmetries, which should dominate the low-energy recombination cross section with a strong direct contribution, were included in the calculation.

3.3. Results and discussion

The anisotropic Maxwell rate coefficients for both experiments and theory are reported on Figure 6. The agreement is very good up to an electron energy close to 1 eV. At higher energies the discrepancies observed can be accounted for by higher dissociative curves that have not been considered in the present treatment.

Acknowledgments

We thank the TSR accelerator group, for efficient support during the beamtimes. OM acknowledges financial support from the French Ministère des Affaires Etrangères through the Service de Coopération et d'Action Culturelle of the French Embassy in Yaoundé (CAMEROON). AEO acknowledges support from the National Science Foundation Grant No. PHY-05-55401. IFS acknowledges financial support from CPER, CRIHAN, CNRS (PCMI, GdR "Cataplasme" and "ARCHES"), JSPS and Max Planck Society. HB acknowledges partial support from the German Israeli Foundation for Scientific Research and Development (G.I.F.) under Grant I-900-231.7/2005 and by the European Project ITS LEIF (HRPI-CT-2005-026015).

References

[1] V Georgieva A B and Gijbels R 2003 *J. Appl. Phys.* **94** 3748
[2] Denpoh K and Nambu K 2000 *Japan. J. Appl. Phys.* **39** 2804
[3] Neufeld D A *et al.* 2006 *Astron. Astrophys.* **454** L37–L40
[4] Novotny O *et al.* 2005 *J. Phys. B: At. Mol. Opt. Phys.* **38** 1471–1482
[5] Rescigno T N, Lengsfield B H and McCurdy C W 1995 *Modern Electronic Structure Theory* vol 1 (Singapore: World Scientific) p 501
[6] Kreckel H *et al.* 2005 *Phys. Rev. A* **95** 263201
[7] Sprenger F *et al.* 2004 *Nuc. Instr. and Meth. A* **532** 298–302
[8] Orlov D A *et al.* 2006 *Non-neutral plasma physics VI* (*AIP conference proceedings* vol 862) (AIP) p 274
[9] Orlov D A *et al.* 2005 *J. Phys.: Conf. Ser.* **4** 290
[10] Amitay Z *et al.* 1996 *Phys. Rev. A* **54** 4032
[11] Wolf A *et al.* 2004 *Nuc. Instr. and Meth. A* **532** 69
[12] Orlov D A *et al.* 2007 *Proc. Workshop on beam cooling and related topics (COOL'07)* ed Hasse R W and Schaa V R W pp 230–233 URL `online at JACoW (http://www.jacow.org/cl07)`
[13] Al-Khalili A *et al.* 2003 *Phys. Rev. A* **68** 042702
[14] Pedersen H B *et al.* 2005 *Phys. Rev. A* **72** 012712
[15] Petsalakis I D and Theodorakopoulos G 2000 *Chem. Phys.* **254** 181–186
[16] Linstrom P and Mallard W (eds) 2005 *NIST Chemistry WebBook* (*NIST Standard Reference database* no 69) (Gaithersburg MD, 20899: National Institute of Standards and Technology) URL `http://webbook.nist.gov`
[17] Amitay Z *et al.* 1999 *Phys. Rev. A* **60** 3769
[18] Orel A E 2005 *J. Phys.: Conf. Ser.* **4** 142
[19] Jungen C *Molecular Applications of Quantum Defect Theory* (Bristol: Institute of Physics Publishing)
[20] Giusti A 1980 *J. Phys. B: At. Mol. Opt. Phys.* **13** 3867
[21] Ngassam V *et al.* 2003 *Eur. Phys. J. D* **26** 165
[22] Sidis V and Lefebvre-Brion H 1971 *J. Phys. B* **4** 1040
[23] Jungen C and Atabek O 1977 *J. Chem. Phys.* **66** 5584

Seventh International Conference on Dissociative Recombination (DR2007) IOP Publishing
Journal of Physics: Conference Series **192** (2009) 012022 doi:10.1088/1742-6596/192/1/012022

Spectroscopy and dissociative recombination of the lowest rotational states of H_3^+

A Petrignani[1], H Kreckel[2], M H Berg[1], S Altevogt[1], D Bing[1], H Buhr[3], M Froese[1], M Grieser[1], J Hoffmann[1], B Jordon-Thaden[1], C Krantz[1], M B Mendes[1], O Novotný[1], S Novotny[1], D A Orlov[1], S Reinhardt[1,4], and A Wolf[1]

[1] Max-Planck-Institut für Kernphysik, Saupfercheckweg 1, D-69117 Heidelberg, Germany
[2] Columbia University, 550 West 120th Street, New York, NY 10027, USA
[3] Department of Particle Physics, Weizmann Institute of Science, 76100 Rehovot, Israel

E-mail: Annemieke.Petrignani@mpi-hd.mpg.de

Abstract. The dissociative recombination of the lowest rotational states of H_3^+ has been investigated at the storage ring TSR using a cryogenic 22-pole radiofrequency ion trap as injector. The H_3^+ was cooled with buffer gas at ~15 K to the lowest rotational levels, (J, G)=(1,0) and (1,1), which belong to the ortho and para proton-spin symmetry, respectively. The rate coefficients and dissociation dynamics of $H_3^+(J, G)$ populations produced with normal- and para-H_2 were measured and compared to the rate and dynamics of a hot H_3^+ beam from a Penning source. The production of cold H_3^+ rotational populations was separately studied by rovibrational laser spectroscopy using chemical probing with argon around 55 K. First results indicate a ~20% relative increase of the para contribution when using para-H_2 as parent gas. The H_3^+ rate coefficient observed for the para-H_2 source gas, however, is quite similar to the H_3^+ rate for the normal-H_2 source gas. The recombination dynamics confirm that for both source gases, only small populations of rotationally excited levels are present. The distribution of 3-body fragmentation geometries displays a broad part of various triangular shapes with an enhancement of ~12% for events with symmetric near-linear configurations. No large dependences on internal state or collision energy are found.

1. Introduction

The H_3^+ ion is of utmost theoretical and astrophysical importance. It is the simplest polyatomic ion and thereby constitutes an important benchmark ion for the theory of both molecular structure and molecular collision dynamics. It is also a key ion in the universe as it initiates a network of reactions leading to the formation of complex molecules. The dissociative recombination (DR) of H_3^+ plays a crucial role as destruction mechanism for the H_3^+ ions [1, 2] and as production mechanism for the neutral hydrogen fragments. The DR reaction has been extensively researched over the past decades [3, 4] and with the knowledge on the reaction increasing, the focus is moving to more and more detailed investigations. Internal-state dependences of the DR reaction, at zero as well as at elevated electron energies, are now being studied [5, 6]. Disagreements on the magnitude of the DR rate coefficient of H_3^+ have been dominating the scene for quite some time [7]. Presently, the differences that have been observed

[4] Present address: Max-Planck-Institut für Quantenoptik, Garching, Germany

are attributed to different internal-state distributions of the ions and consent between storage-ring experiments has been reached [8, 9]. Furthermore, previous discrepancies with afterglow studies seem to have been resolved recently and the latest theoretical calculations show good overall agreement with experiment [10, 11, 12]. However, the predicted rate coefficient still shows local discrepancies up to an order of magnitude around several electron energy regions. Again, differing internal-state distributions are given as possible explanation. In the cold interstellar space, the difference of the DR rate between the two lowest rotational levels, (J, G)=(1,0) and (1,1) at an energy difference corresponding to kT=33 K only, may be of particular importance. Since for H_3^+ these two levels are equivalent with the ortho and the para symmetry of the total nuclear spin, respectively, we shortly denote the two levels as ortho- and para-H_3^+. Previous experiments indicate that the DR rate coefficient of para-H_3^+ could be larger than that of ortho-H_3^+ for specifically the low electron energies prevailing in the cold interstellar medium [9]. In these experiments, para-H_2 was used as precursor gas and although the rotational populations of the H_3^+ ion beam could not be measured, an enhancement of the para-H_3^+ population is expected in this case. The higher para-H_3^+ rate was until recently in contradiction to theory, which predicted the opposite [11, 13]. However, new theoretical results now state a roughly ten times higher rate for para-H_3^+ at low temperature [12].

The DR of cold H_3^+ has two fragmentation channels for low-energy electrons, a 3-body breakup containing only ground-state atoms and a 2-body breakup that includes the H_2 product molecule,

$$\begin{aligned} H_3^+(J,G) + e^-(E_d) \quad &\rightarrow H(1s) + H(1s) + H(1s)\,, \\ &\rightarrow H_2(v,J) + H(1s) \end{aligned}$$

Here, the cold initial H_3^+ ion is given to be in a low rotational state of its ground vibrational state with J the total angular momentum associated with the motion of the nuclei and $G = |K - l|$, where K is the projection of J onto the molecular axis and l is the vibrational angular momentum. The product atoms are in their ground state, whereas the H_2 molecule may be rovibrationally, (v, J), excited. Further dissociation channels become possible when additional energy, E_d, is introduced via the electrons. At typical molecular-cloud temperatures of 10-60 K, H_3^+ populates the two to three lowest rotational levels, (J, G)=(1,0), (1,1), and (2,2), of the vibrational ground state, where the (2,2) level belongs to the para spin symmetry.

We present a two-fold investigation of cold H_3^+. The first focus is on the production and characterisation of rotationally cold H_3^+ populations using normal-H_2 and para-H_2 as parent gas. In order to probe the different rotational distributions, rovibrational spectroscopy on the H_3^+ ions in a 22-pole radiofrequency ion trap is implemented. The second focus is on the DR of $H_3^+(J, G)$ populations with altered ortho-to-para ratio. High-resolution rate coefficients up to 0.4 eV are given and additionally compared to a hot H_3^+ beam to investigate rotational heating. The issue of internal excitation is addressed by fragment imaging. For the first time, the dissociation of H_3^+ were studied using the $3D$ imaging technique [14], and 3-body breakup from both $2D$ and $3D$ imaging are presented.

2. Experimental Setup

The cold H_3^+ samples are prepared in a 22-pole radiofrequency (rf) ion trap setup, as shown in the inset of Figure 1. The ions are produced in a rf storage ion source [15] through electron-impact ionisation of H_2 and subsequent collisions of the H_2 with the hence produced H_2^+. The H_3^+ ions are then guided through a rf quadrupole to the 22-pole rf ion trap [16]. Here, the ions are trapped radially by the 22 poles and axially by an entrance and exit electrode. The number of stored ions may be varied from a few to several millions of ions, whereas the storage time may be varied from ms to several seconds. During storage, the ions are buffer-gas cooled with helium

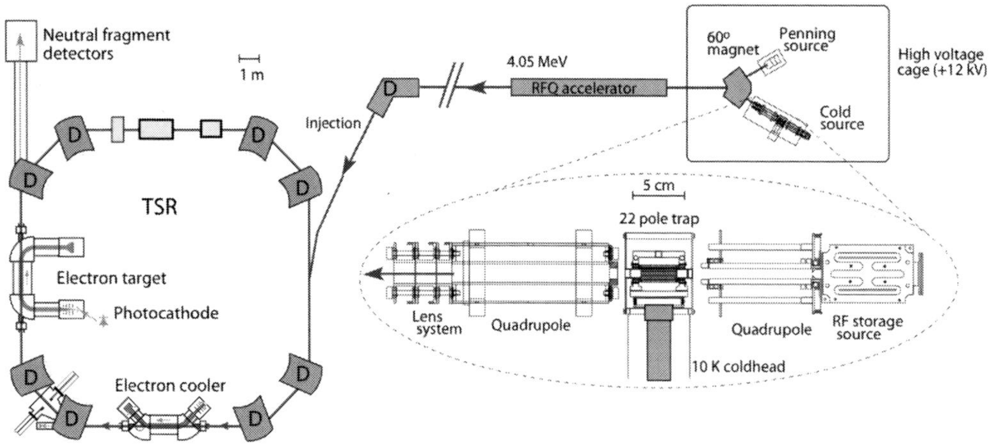

Figure 1. The 22-pole trap setup used as injector for the ion storage ring TSR. The H_3^+ ions are prepared in the trap, accelerated to 4.05 MeV, and injected into the ring. A cold electron beam, produced with a liquid-nitrogen cooled photocathode [19], is merged with the ion beam in the electron target. The neutral fragments are detected either with a surface barrier counting detector or a position-sensitive imaging system [20, 21].

at a chosen ambient temperature of 10-60 K. For more details on the 22-pole trap setup, the reader is referred to Refs. [17, 18]. In the present work, the setup is used as a stand-alone device for probing the $H_3^+(J,G)$ populations implementing rovibrational spectroscopy and it is used as an ion injector for the storage ring TSR to the study of the DR of the cold H_3^+ (see Figure 1). Unfortunately, the conditions required for the spectroscopy measurements are different from those of the DR measurements. As such, care has to be taken when deploying information from the laser-probed populations for the TSR measurements.

2.1. Spectroscopy Experiments

To verify the buffer-gas cooling inside the 22-pole trap and to probe the rotational-state populations, an action spectroscopy scheme is used. To this purpose, rovibrational spectroscopy in the infrared is implemented as H_3^+ lacks any stable electronically excited states. As a sensitive probe of the vibrational transitions, a laser-induced chemical reaction with argon as reactant is used. The charge exchange reaction,

$$H_3^+(v=0) + Ar \rightarrow ArH^+ + H_2 - 0.57 \text{ eV} \tag{1}$$

is endothermic by 0.57 eV for ground-state H_3^+ ions. However, if at least two vibrational quanta are excited in H_3^+, the reaction becomes energetically possible. Once the ions in the trap are cooled down to the chosen temperature, a tunable infrared diode laser of ~1400 nm is used to trigger transitions from the lowest rotational states of H_3^+ to the third harmonic of the bending-mode vibration. The laser-excited ions will react with the abundant argon in the trap to form ArH^+, which is trapped as well and used as spectroscopic signal. The ions are extracted from the trap, analysed by a mass spectrometer, and directed to a scintillation detection system capable of counting single ions with near-unity efficiency. These spectroscopy measurements are performed at a minimum temperature of 55 K to keep the argon in the gas phase. For more details, see Ref. [17].

The action spectroscopy scheme has been used before to probe cold H_3^+ produced with normal-H_2 [17]. Here, also the H_3^+ rotational distribution produced with para-H_2 is presented.

Moreover, important improvements have currently been implemented to enhance the cooling of the H_3^+ in the trap as well as to enhance the ArH^+ spectroscopy signal. Gas purities, laser power (increased by a factor of 10), and frequency resolution have now been refined. As a result, the typical ArH^+ count at resonance is 1.5-2 per trap filling of 300 H_3^+ ions, an order of magnitude higher than before. The number densities of the helium and argon gas inside the trap have been optimised to $\sim 10^{14}$ cm^{-3} and $\sim 10^{12}$ cm^{-3}, respectively. The hydrogen number density, $\sim 10^{10}$ cm^{-3}, and the trap storage time, 250 ms, were unchanged. Recently, new results on the action spectroscopy to find near-visible transitions of H_3^+ have been published demonstrating the high sensitivity of the method [22].

2.2. Dissociative Recombination Experiments

The DR measurements were performed with the 22-pole trap as injector of the cold ion beam for the storage ring TSR. The trap temperature was set to 15 K in order to simplify the DR measurements, as only the lowest para and ortho rotational states, (1,0) and (1,1), are populated at that temperature. The number of H_3^+ ions had to be hugely increased to the maximum trap filling of around $2 \cdot 10^6$ in order to supply a sufficient DR count rate [23]. The H_3^+ ortho-to-para ratio is dominantly determined by collisions with H_2, therefore the H_2 number density (10^{10} cm^{-3}) was kept the same as in the spectroscopy experiments. The helium number density was increased to $\sim 10^{15}$ cm^{-3} to compensate for the slightly reduced trap storage time of 100 ms. After this cooling time, the ions were released to be accelerated and stored in the TSR at an ion-beam energy of 4.05 MeV. The transmission of the ion beam was about 10%, resulting in very low counting rates (in a later storage-ring experiment the transmission could be improved to 50%). The ring storage time was set to 10 s during which the ion beam was merged in the electron target with an ultracold electron beam produced by a liquid-nitrogen cooled photocathode [19]. The transversal and longitudinal temperatures of this electron beam were $T_\perp = 1$ meV and $T_\parallel = 30$ μeV, respectively. The electron density was roughly $9 \cdot 10^5$ cm^{-3}. Rate-coefficient curves were measured with a new large surface barrier detector of 10x10 cm^2 at collision energies between 0 and 0.4 eV by varying the electron detuning energy, E_d. The dynamics of the DR reactions were investigated at $E_d = 0$ and 6 meV through the $2D$ [20] and, as a first for H_3^+, the $3D$ [14, 21] imaging technique. The electron cooler was set to 0 eV collision energy to continuously cool the ion beam and, for the first 0.5 s of storage, the electron target was additionally set to 0 eV in order to speed up the initial translational phase-space cooling of the beam. This initial cooling time was deliberately kept low in order to minimise possible internal heating-effects from the toroidal regions, where the electron beam is no longer parallel to the ion beam and elevated detuning energies may cause rotational excitation of the ions. Only for the 0-eV imaging measurements, the electron cooler was switched off after this time, whilst the target remained at 0 eV.

3. Results

3.1. Rovibrational Spectroscopy

Figure 2 shows the observed ArH^+ signal for the transitions from the three lowest rotational states present at 55 K to the third vibrational level. The upper figure results from H_3^+ ions produced with normal-H_2, the lower from H_3^+ ions produced with para-H_2. We are currently still studying possible effects of the change of the precursor gas on the absolute normalisation of the signal. The relative normalised intensities for a given normal-H_2 or para-H_2 measurement, however, are reproducible and well under control. The intensities are normalised to the laser power of $\sim$12 mW, to a trap filling of 300 ions, and to an ArH^+ lifetime of 25 ms, such that they may be coarsely compared to the previously published spectroscopy measurement [17].

The experimental changes mentioned before have now lead to improved signal strengths giving rise to nicely observable Gaussian profiles. The cooling of the ions also proved to be

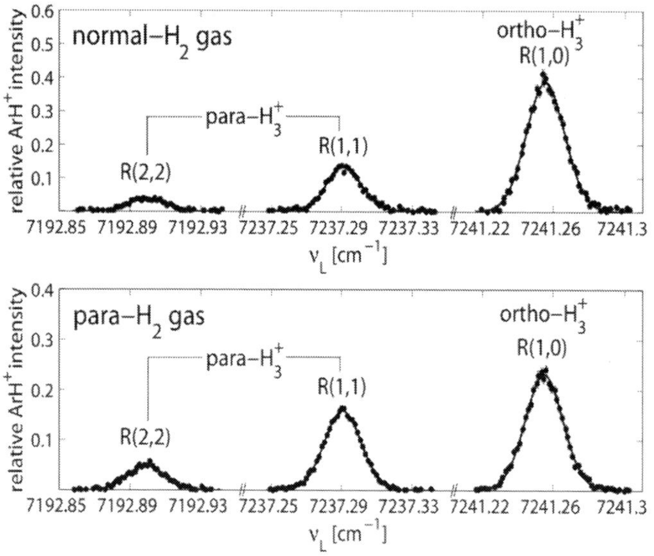

Figure 2. The measured absorption profiles for the three observed transitions in the $(0, 3^1) \leftarrow (0, 0^0)$ vibrational overtone band of H_3^+ produced with normal-H_2 (upper panel) and para-H_2 (lower panel). The signal is derived by averaging over many trap fillings.

more efficient at the optimised helium pressure and much lower temperatures are observed than in 2004 [17]. The profile widths of the respective transitions denoted by $R(J, G)$ – where R stands for the $(J+1)$ branch and J, G stand for the rotational quantum numbers of the lower state – are determined by Doppler broadening, which conforms a temperature of ~67 K for all transitions. The line-intensity ratios are determined by the given laser intensity, the Einstein coefficients, and the initial-level populations. Assuming the theoretical Einstein coefficients of 8.9, 4.9, and 8.2 s^{-1} for $R(1, 0)$, $R(1, 1)$, and $R(2, 2)$, respectively [24], the populations and hence ortho-to-para and para-to-para ratios can be extracted. The "para rotational temperature" is derived from the ratio between the two para transitions, $R(1, 1)$ and $R(2, 2)$, and matches the translational temperature of ~67 K for both ion-source gases. Thermal equilibrium seems to have been reached in both cases. The "ortho-to-para temperature" is derived from the ratio between the ortho transition, $R(1, 0)$, and the para transitions, $R(1, 1)+R(2, 2)$. A significant decrease of the ortho-H_3^+ line relative to the para lines is visible when using para-H_2 as parent gas. The ortho-to-para temperature correspondingly decreases, namely from ~140 K to ~36 K. This clearly shows that under certain circumstances the ortho-to-para ratio can become non-thermal and can deviate significantly from the translational and rotational temperature of an ion sample. The population for the normal-H_2 parent gas conforms an ortho-to-para ratio of 60:40 using the above mentioned Einstein coefficients, whereas the H_3^+ sample produced with para-H_2 shows the opposite, an ortho-to-para ratio of 40:60. It seems that the use of para-H_2 as parent gas leads to enrichment of the rotational states belonging to the para-H_3^+ symmetry.

3.2. Dissociative Recombination Rate Coefficients

As mentioned before, the 22-pole trap was used as an injector for the storage ring TSR, where the DR rate of the cold H_3^+ populations was investigated between 0 and 0.4 eV collision energy. Also, to investigate rotational heating, the DR rate of a hot H_3^+ beam produced with a Penning source was measured. Due to low count rates for the cold ion beams (only 10% transmission), insufficient statistics could be gathered around the 10-eV resonance. Therefore, the rate curves have been normalised to the 12-meV resonance. This resonance is at an energy low enough for sufficient statistics, however, presumably at an energy high enough for the normal- and para-H_3^+ rates to be the same, as observed in Ref. [9]. The hot H_3^+ rate curve, which does contain enough statistics, is normalised to the 10-eV resonance. Both normalisations are to the photocathode

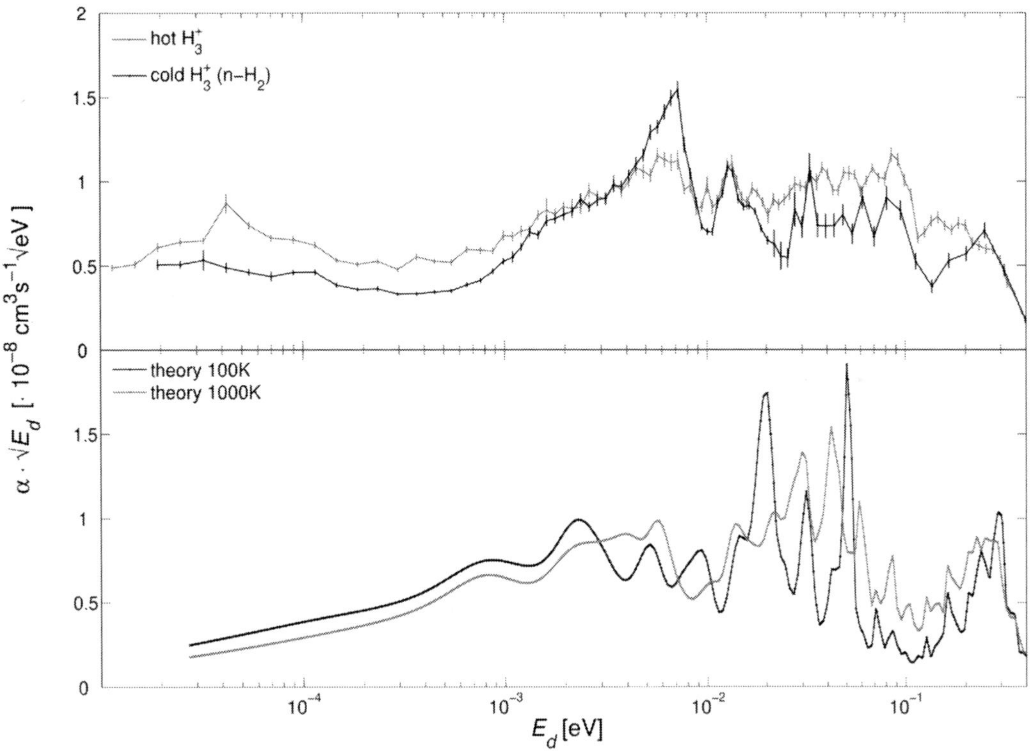

Figure 3. The upper frame shows the measured reduced rate coefficients of the cold (black) and hot (grey) H_3^+ produced with normal-H_2 in the 22-pole trap and the Penning source, respectively. The lower frame shows the reduced theoretical rates for 100 K (black) and 1000 K (grey), folded with the experimental electron temperature from Ref. [12].

measurement of Ref. [9] as it bears a similar energy resolution. Figures 3 and 4 show the reduced rate coefficients, $\alpha \cdot \sqrt{E_d}$, with α the energy-dependent rate coefficient that is proportional to $\sqrt{E_d}$. Any structures observable on the reduced rate are related to resonant behaviour. Figure 3 (upper frame) displays the measured rates of the hot and cold H_3^+ produced with normal-H_2. Overall, the hot H_3^+ rate is higher than the cold H_3^+ rate, except for the resonance around 7 and 270 meV. Although the temperature of the H_3^+ beam from the Penning source is on the order of 10^3 K (see Section 3.3), resonant behaviour is still clearly visible. Figure 3 (lower frame) shows the theoretical predictions for 100 K and 1000 K as calculated by Ref. [12]. Similar to the experiment, the hot H_3^+ rate is overall higher, however, at very low energies, the opposite behaviour is predicted. Several structures appear both in theory and in experiment. The intensities of the theoretical rates seem to be systematically lower than those measured, leading to a better agreement between our cold H_3^+ beam and the 1000 K theoretical rate. This raises the question whether heating effects in the storage ring might have considerably warmed up the initially cold ion beam. The fragmentation dynamics presented in the next section, however, demonstrate that this is not the case.

Figure 4 displays the reduced rates of the cold H_3^+ ion beams produced with normal- and para-H_2 (upper frame). The H_3^+ rate using para-H_2 exhibits slightly more structure below 5 meV collision energies. Otherwise, the rate coefficients are quite similar and show no sign of the predicted order of magnitude difference in DR rate between para-H_3^+ and ortho-H_3^+ [12].

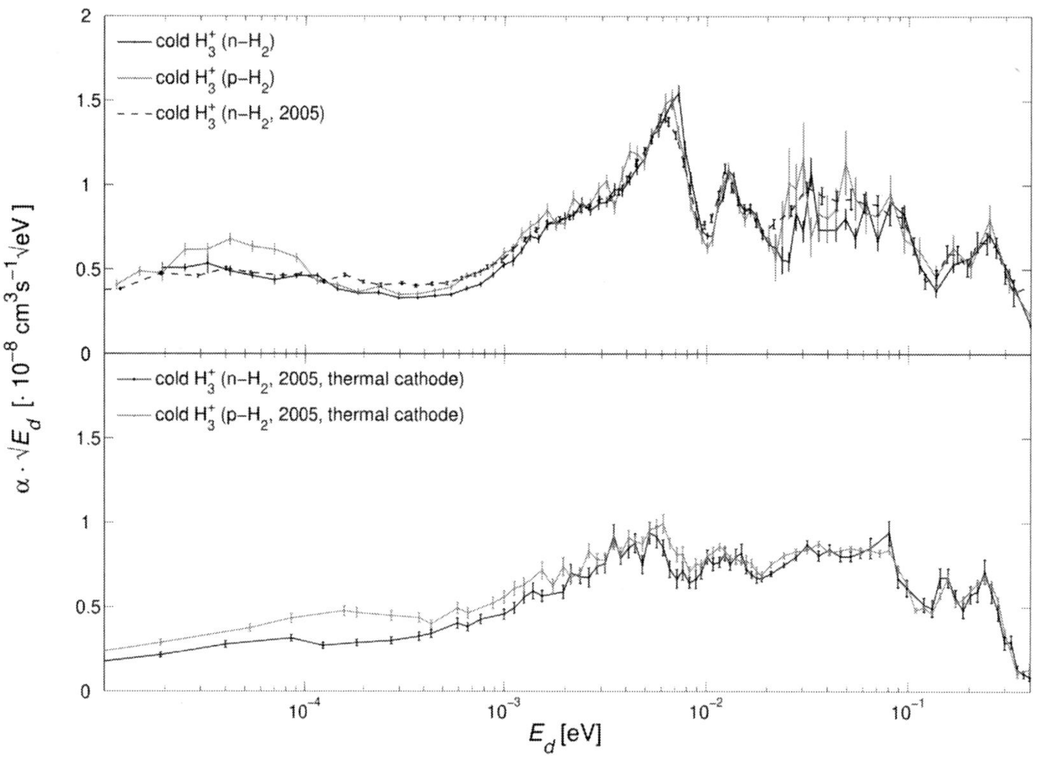

Figure 4. The upper frame shows the high-resolution reduced rates of H_3^+ produced with para-H_2, present data (grey), and normal-H_2, present data (solid black) and previous data (dashed black) [9]. The lower frame shows the previously measured lower-resolution reduced rates of H_3^+ produced with para-H_2 (grey) and normal-H_2 (black) [9].

The two H_3^+ populations in the storage ring produced with normal- and para-H_2 are unknown and may very well differ less than was the case in the spectroscopy measurements, however, a DR rate that is an order of magnitude faster for para-H_3^+ than for ortho-H_3^+ should manifest itself even at small increases in the para-H_3^+ contribution. No large dissimilarities between the two source gases are observed that would point towards the predicted ten times more rapid dissociation of para-H_3^+. Still, nearly identical $H_3^+(J,G)$ populations in the storage ring for both source gases cannot be excluded.

Also shown in figure 4 is the previous photocathode measurement for normal-H_3^+ (upper frame) and the previously measured H_3^+ rates for the normal- and para-H_2 source gases as measured with the thermal cathode (lower frame) [9]. The photocathode data clearly exhibit more pronounced resonances as the resolution of the liquid-nitrogen cooled photocathode ($T_{e,\perp} = 1$ meV) [19] is much better than that of the thermal cathode ($T_{e,\perp} = 4$ meV). A small resonance is revealed around 17 meV and below 5 meV structure is unfolded. The increased structure visible for the para-H_2 source gas raises the H_3^+ rate slightly, in accordance with the previous thermal-cathode experiment, i.e., assuming the normalisation on the 12-meV resonance is valid. Further analysis is still required. It is noted that the internal distributions in the ring may also differ between the present and previous DR experiments, both in ortho-to-para ratio for the H_3^+ produced with para-H_2 as in the achieved internal temperature that may give rise to a differing contribution from excited rotational levels irrespective of source gas.

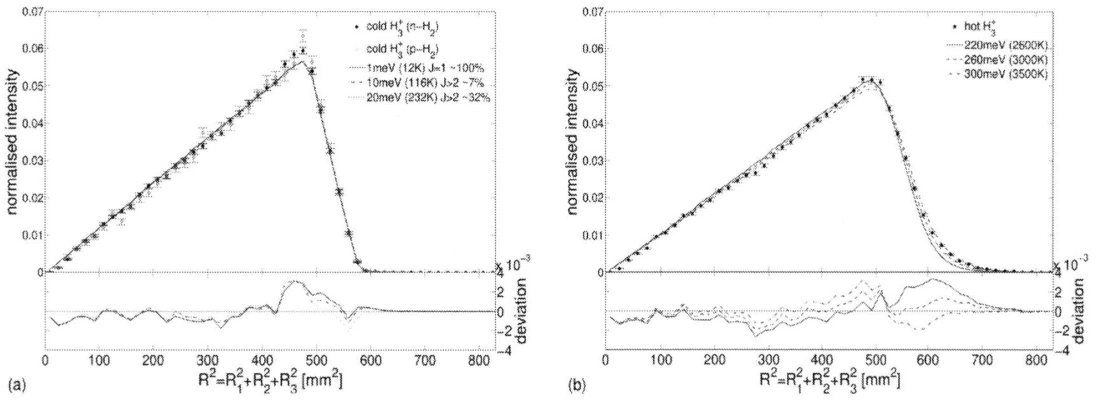

Figure 5. The R^2 distributions of the H_3^+ beams for (a) the cold H_3^+ produced with normal- and para-H_2 in the 22-pole trap and (b) the hot H_3^+ produced with normal-H_2 in the Penning source. All black curves are Monte-Carlo simulations of isotropic dissociations at the given temperatures, considering the allowed rotational levels and their probability in a Boltzmann distribution. All distributions are normalised to unit area. The deviation of the measured distributions of H_3^+ produced with normal-H_2 from the simulations are plotted in the respective lower frames.

3.3. Dissociative Recombination Dynamics

The breakup dynamics of the DR of H_3^+ and their possible (J, G) state-dependences have been studied as well. Imaging data have been taken for both the 2-body and 3-body fragmentation channel. Here, only dissociation into three H atoms is treated. Figure 5a) shows the projected distance distributions of our cold H_3^+ beams from the 22-pole trap as acquired by the $2D$ imaging [20], normalised to unit area. The distance plotted is $R^2 = R_1^2 + R_2^2 + R_3^2$ with R_i the distance from particle i to the centre-of-mass in the $2D$ detection plane. Forward Monte-Carlo simulations for isotropic dissociations at three different rotational temperatures, normalised to unit area, are plotted to reveal a rough estimate of the temperature of the ion beam and anisotropy effects. To clarify differences, the deviation of the distribution for the normal-H_2 source gas from the simulations is displayed in the lower frame. The discrepancy at the peak of the distribution could indicate anisotropy and further simulations are needed to investigate the observed feature. Figure 5b) shows the R^2 distribution of the hot H_3^+ from the Penning source together with forward simulations at three high temperatures. Below, the difference of the measured data with the simulations is shown. Again, no large anisotropies can be observed.

The best temperature fit is determined from the least-squares minimum between the falling slope of the data and the simulation, with the simulation normalised to the maximum of the measured data (not displayed) in order to disregard the anisotropy effects. The hot H_3^+ distribution appears to have a temperature of around 3000 K. For the cold H_3^+, the least-squares deviation is smallest at the lowest temperature and increases with rising temperature. Although a significant contribution of the J=2 level cannot be excluded, the temperature is most certainly below room temperature, which demonstrates that the ion beams are indeed cold. The apparent difference between experiment and theory at low temperatures can therefore not be explained by considerable heating of the ion beam.

Figure 6a) gives the Dalitz plot of the hot H_3^+ beam from the Penning source for 0-eV collision energy, for the first time acquired by the $3D$ imaging technique [14]. The geometries of the corresponding dissociations are displayed in the lower sextant. As proof-of-principle for the $3D$ imaging, the events dissociating near-perpendicular to the detector are displayed in figure 6b). Near-perpendicular events with $cos(\theta)$>0.8 are selected, where θ is the angle between the

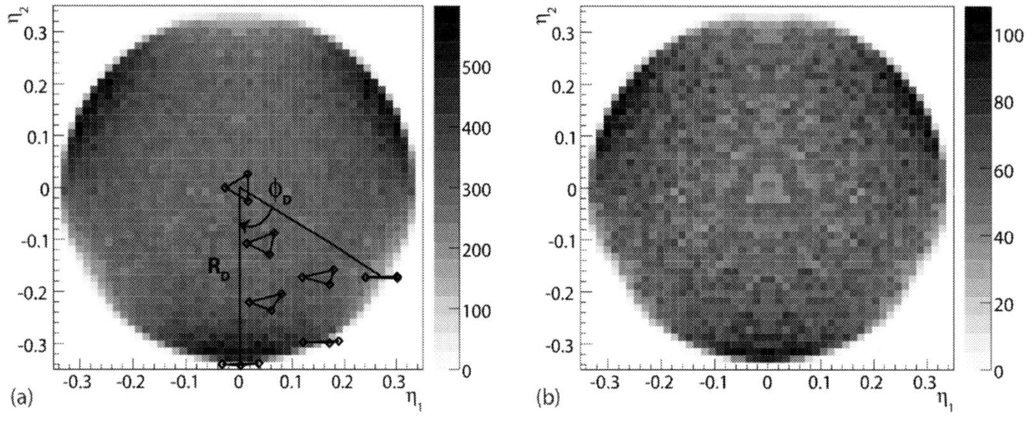

Figure 6. The $3D$ imaging data of hot H_3^+ from the Penning source for a) all dissociations and b) the near-transversal dissociations, i.e., dissociations with $cos(\theta)$>0.8, where θ is the angle between the normal of the molecular plane and the beam axis.

normal of the molecular plane and the beam axis. The tranversal $3D$ Dalitz plot is indeed similar to the total $3D$ Dalitz plot. The $3D$ imaging data are also consistent with the Dalitz plots obtained previously by Monte-Carlo reconstructions of $2D$ projected distance distributions [20].

To investigate the distributions for the different H_3^+ populations in more detail, the Dalitz coordinates have been transformed to polar coordinates (as illustrated in figure 6a) to be able to integrate over the angle and the radius. The integrated radius describes the amount of (non-)linearity and the integrated angle the deformation from the isosceles/equilateral triangle. Figures 7a) and b) display the integrated-coordinate distributions at 0 eV. A uniform background, i.e., isotropic distributed geometries, can be observed superimposed with ~12% anisotropic dissociations, which exhibit a preference towards linear breakup, independent of the H_3^+ population. As discernable in the Dalitz plot, the near-linear dissociations peak around the symmetric linear breakup and drop to (even slightly below) the level of the isotropic background for the asymmetric linear geometries. The integrated-coordinate distributions show no significant internal-state dependence. Figures 7c) and d) display the distributions for the hot and the cold H_3^+ at 6 meV together with the 0-eV distributions of the hot H_3^+ for comparison. The breakup of the cold H_3^+ produced with para-H_2 at 6 meV was not measured. The fragmentation dynamics of the hot H_3^+ at 6 meV resemble those at 0 eV. The cold H_3^+ breakup exhibits a small difference, peaking slightly more around the symmetric linear dissociations. No large energy dependence is observed.

4. Conclusions

The rovibrational spectroscopy shows that the use of para-H_2 as parent gas can indeed enrich the para contribution of the H_3^+ population. In the storage-ring experiment, where the para enrichment cannot be confirmed, no increase in the DR rate has been found when using para-H_2 as source gas that may support the prediction of a 10 times higher para-H_3^+ rate than ortho-H_3^+ rate. The trend of increasing overall rate with increasing temperature as predicted by theory is, however, confirmed by the present data. Furthermore, the H_3^+ ions prepared

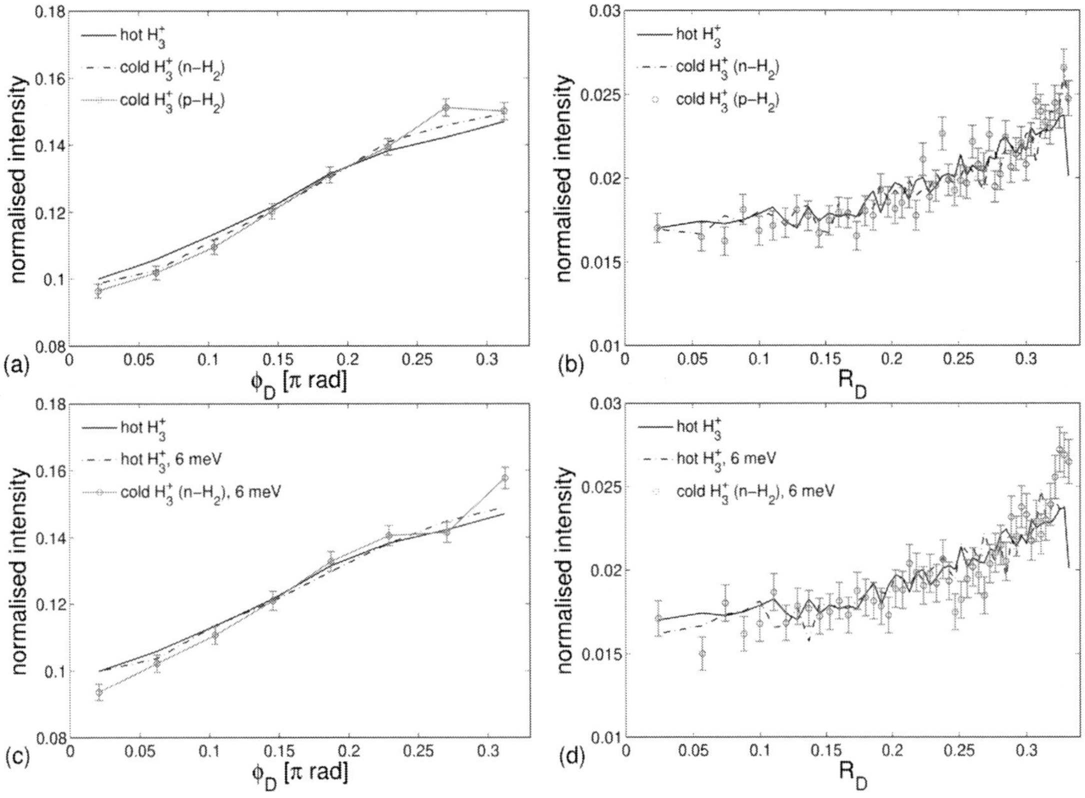

Figure 7. The projected Dalitz polar coordinates at a,b) 0 eV and c,d) 6 meV for the given H_3^+ populations. For comparison, the hot H_3^+ data at 0 eV are displayed again with the 6 meV distributions. As illustrated in figure 6, R_D ranges from the equilateral triangle to linear dissociation geometries and ϕ_D ranges from asymmetric to symmetric geometries. To improve clarity, only the errorbars of the distribution with the lowest statistics are shown.

in the 22-pole trap and transferred to the TSR are really rovibrationally cold, featuring a temperature that is distinctly lower than 300 K. As such, heating effects cannot account for the discrepancy in intensity between the theoretical and experimental rate-coefficient curves. The 3-body fragmentation dynamics of the DR of H_3^+ show no large dependences on the internal-state distribution of the initial H_3^+ nor on electron energy. A fraction of 12% of the H_3^+ ions are found to dissociate in symmetric near-linear geometries, while 88% have a uniform distribution of geometries.

Acknowledgments

HB acknowledges partial support from the German Israeli Foundation for Scientific Research and Development (G.I.F.) under Grant I-900-231.7/2005 and by the European Project ITS LEIF (HRPI-CT-2005-026015). We are grateful to the TSR accelerator group for their support during the experiments.

References

[1] McCall B J 2005 *Astrochemistry: Recent Successes and Current Challenges Proceedings IAU Symp. No. 231* ed Lis D C, Blake G A and Herbst E pp 165–174
[2] Oka T 2006 *PNAS* **103** 12235–12242

[3] Larsson M 2000 *Phil. Trans. R. Soc. Lond. A* **358** 2433–2444
[4] McCalll B J 2006 *Phil. Trans. R. Soc. Lond. A* **364** 2953–2963
[5] Wolf A, Kreckel H, Lammich L, Strasser D, Mikosch J, Glosík J, Plašil R, Altevogt S, Andrianarijaona V, Buhr H, Hoffmann J, Lestinsky M, Nevo I, Novotny S, Orlov D A, Pedersen H B, Terekhov A S, Toker J, Wester R, Gerlich D, Schwalm D and Zajfman D 2006 *Phil. Trans. R. Soc. Lond. A* **364** 2981–2997
[6] Greene C H and Kokoouline V 2006 *Phil. Trans. R. Soc. Lond. A* **364** 2965–2980
[7] Oka T 2000 *Dissociative Recombination: Theory, Experiment and Applications IV* ed Larsson M, Mitchell J B A and Schneider I F (Singapore: World Scientific) pp 13–24
[8] McCall B J, Huneycutt A J, Saykally R J, Djuric N, Dunn G H, Semaniak J, Novotny O, Al-Khalili A, Ehlerding A, Hellberg F, Kalhori S, Neau A, Thomas R D, Paal A, Österdahl F and Larsson M 2004 *Phys. Rev. A* **70** 052716
[9] Kreckel H, Motsch M, Mikosch J, Glosík J, Plašil R, Altevogt S, Andriananrijaona V, Buhr H, Hoffmann J, Lammich L, Lestinsky M, Nevo I, Novotny S, Orlov D A, Pedersen H B, Sprenger F, Terekhov A S, Toker J, Wester R, Gerlich D, Schwalm D, Wolf A and Zajfman D 2005 *Phys. Rev. Lett.* **95** 263201
[10] Glosik J, Korolov I, Plasil R, Novotny O, Kotrik T, Hlavenka P, Varju J, Greene C H, Kokoouline V and Mikhailov I A 2007 (*Preprint* `arXiv/0710.2339`)
[11] Kokoouline V and Greene C H 2003 *Phys. Rev. A* **68** 012703
[12] dos Santos S F, Kokoouline a V and Greene C H 2007 *J. Chem. Phys.* **127** 124309
[13] Kokoouline V and Greene C H 2005 *Phys. Rev. A* **72** 022712
[14] Strasser D, Urbain X, Pedersen H B, Altstein N, Heber O, Wester R, Bhushan K G and Zajfman D 2000 *Rev. Sci. Instrum.* **71** 3092–3098
[15] Teloy E and Gerlich D 1974 *Chem. Phys.* **4** 417–427
[16] Gerlich D 1995 *Phys. Scr.* **T59** 256–263
[17] Mikosch J, Kreckel H, Wester R, Plašil R, Glosík J, Gerlich D, Schwalm D and Wolf A 2004 *J. Chem. Phys.* **121** 11030
[18] Kreckel H, Mikosch J, Wester R, Glosík J, Plašil R, Motsch M, Gerlich D, Schwalm D, Zajfman D and Wolf A 2005 *Dissociative Recombination: Theory, Experiment and Applications VI* ed Wolf A, Lammich L and Schmelcher P (London: Institute of Physics and IOP Publishing) pp 126–133
[19] Orlov D A, Weigel U, Schwalm D, Terekhov A S and Wolf A 2004 *Nuclear instruments and methods in Physics Research A* **532** 418–421
[20] Strasser D, Lammich L, Kreckel H, Krohn S, Lange M, Naaman A, Schwalm D, Wolf A and Zajfman D 2002 *Phys. Rev. A* **66** 032719
[21] Nevo I, Novotny S, Buhr H, Andrianarijaona V, Altevogt S, Heber O, Hoffmann J, Kreckel H, Lammich L, Lestinsky M, Pedersen H B, Schwalm D, Wolf A and Zajfman D 2007 *Phys. Rev. A* **76** 022713
[22] Kreckel H, Bing D, Reinhardt S, Petrignani A, Berg M H and Wolf A 2008 *J. Chem. Phys.* **129** 164312
[23] Kreckel H 2003 *Internal excitations of stored triatomic hydrogen molecular ions* Ph.D. thesis University Heidelberg
[24] Bianca M Dinelli S M and Tennyson J 1992 *J. Mol. Spectrosc.* **153** 718–725

Measurements of EEDF in recombination dominated afterglow plasma

R Plasil[1], I Korolov[1], T Kotrik[1], J Varju[1], P Dohnal[1], Z Donko[2], G Bano[2], J Glosik[1]

[1] Charles University, Faculty of Mathematics and Physics, Department of Surface and Plasma Science, Prague, Czech Republic
[2] Research Institute for Solid State Physics and Optics, Hungarian Academy of Sciences, Budapest, Hungary

E-mail: radek.plasil@mff.cuni.cz

Abstract. Electron energy distribution functions (EEDF) have been measured in decaying plasma in Flowing Afterglow Langmuir Probe (FALP) experiment. The measurements have been carried out in diffusion and recombination governed plasmas used for studies of recombination of KrD^+ and H_3^+ ions.

1. Introduction

Recombination of ions with electrons and diffusion are two important processes controlling degree of ionisation in low temperature plasmas. Typical low temperature plasma is the afterglow plasma, which is formed in an active discharge, and decays after termination of ionisation or after moving out from the region with ionisation. If gas is flowing through a discharge region, e.g. along a tube, it can take plasma away from the discharge and the plasma will then decay along the flow of gas. Such "Flowing Afterglow" (FA) is used to study relaxation processes, ion-molecule reactions and electron – ion recombination. In principle decaying plasma in FA is not in thermodynamic equilibrium, nevertheless it can be very close to thermodynamic equilibrium. Validity of the assumption of equilibrium has to be discussed for particular experimental conditions and verified by measuring plasma parameters. Further we will discuss FALP experiment (Flowing Afterglow with Langmuir Probe) and experimental conditions used for recombination studies when helium is used as a buffer/carrier gas. In FALP experiments it is usually assumed that afterglow plasma is thermalised, $T = T_e = T_{ion} = T_{He}$. We will show that such assumption is not always fulfilled. The clear misbalancing factor in helium afterglow is a presence of internally excited long-lived particles formed in He discharge, in ion-molecule reactions, and in recombination. In the majority of FALP experiments Ar is added to helium flow immediately downstream from the discharge region to quench helium metastables ($He(2^3S)$ and $He(2^1S)$) formed in the discharge. After addition of Ar these metastables are removed from the plasma by Penning ionisation and plasma relaxes rapidly, this process is clear and well described [1].

We will study processes connected with recombination and with formation of recombination dominated plasma (demonstrated on Ar^+, Kr^+, KrD^+ and H_3^+). In these processes energetic electrons and heavy excited particles can be formed. In Penning ionisation and in superelastic collisions these excited particles can transfer their internal energy to electrons of already cold decaying plasma. The generation of energetic electrons can influence the electron energy distribution function (EEDF) [2] and the decay of the plasma. Further plasma decay depends on the balance between generation, recombination, thermalisation and diffusion of electrons.

2. Experiment

The principle of the FALP method used in the present experiments is shown in Fig. 1. Helium buffer gas flows (at pressure p_{He} = 1600 Pa) through the glass discharge tube and He_2^+ dominated plasma is generated by microwave excitation. The flowing buffer gas then drives plasma along the flow tube, where reactant gases required for the formation of studied ions can be introduced via injection ports (P1 and P2 are 3 and 35 ms downstream from the discharge, respectively).

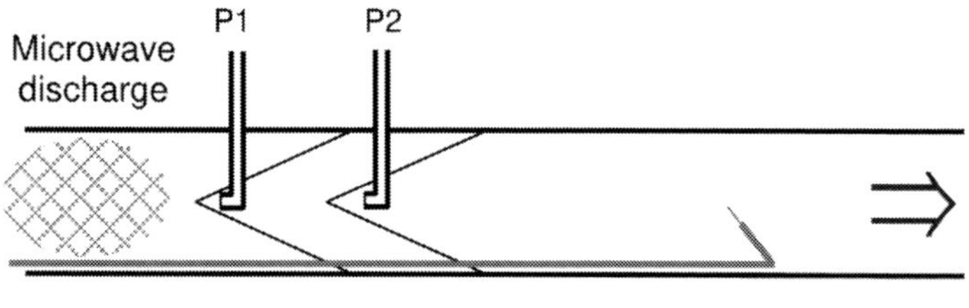

Figure 1. Principle of FALP method. He buffer gas carries the decaying plasma from discharge region along the flow tube. The reactant gases are added via ports P1 and P2. The plasma decay is monitored downstream from P2 (t_{P2} = 0).

The plasma parameters (including EEDF) are determined by measuring probe characteristics (I_{Probe} versus U_{Probe}) with axially movable Langmuir probe. Example of electron energy probability function (EEPF) obtained in Ar^+ dominated late afterglow plasma (at t = 45 ms) in He/Ar mixture is plotted in Fig. 2. The plasma is not recombining and decays only due to diffusion losses. Plotted is "uncorrected" EEPF calculated from the second derivative of probe current (I_{Probe}") using Druyvesteyn formula and "corrected" EEPF [3, 4]. The EEPF (f_{EEPF}) is connected with corresponding EEDF (f_{EEDF}) by the formula $f_{EEDF}(\varepsilon) = f_{EEPF}(\varepsilon)\cdot\sqrt{\varepsilon}$. Advantage of EEPF is that for Maxwellian EEDF the corresponding EEPF is exponential, i.e. in semilogarithmic plot it is a straight line. The measured EEPF is nearly Maxwellian over the whole covered range of electron energies. The obtained electron temperature is T_e = 360 K i.e. it is slightly higher than T_{He}, $\Delta T_e \sim$ +110 K. The increase can be caused by presence of excited Ar^+ ions in $^2P_{1/2}$ spin state, with excitation energy $\Delta\varepsilon$ = 0.18 eV. These ions formed in Penning ionisation can transfer energy to electrons in superelastic collisions.

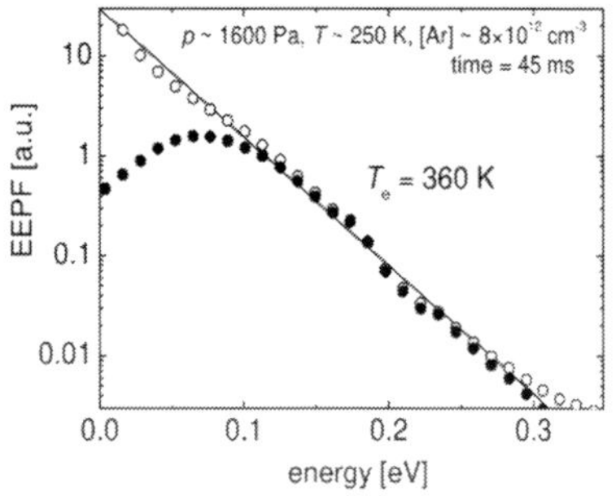

Figure 2. Electron energy probability function (EEPF) obtained from measured characteristic of Langmuir probe. The straight line corresponds to Maxwellian electron energy distribution. T_e is given by the slope of the plot. Note that the corrected (open symbols) and uncorrected (full symbols) EEPFs differ only in low energy region and the character of the EEPF is not changed for $\varepsilon > 0.1$eV. That is why we will not use corrections. We will expect Maxwellian distribution in low energy region with temperature given by Maxwellian body of measured EEDF.

Part of the increase of T_e is probably also due to systematic error of the measurements due to the fluctuation of plasma potential, potential of the probe and inhomogeneity of the probe surface. We suppose that the systematic increase given by "apparatus effect" of the technique used is $\delta T_e \sim$ +70 K. The evolution of EEPF (normalised to electron density) in Ar^+ dominated afterglow is shown in Fig. 3. In the inset evolutions of electron density (n_e) and electron temperature (T_e) are plotted. The exponential decay of n_e corresponds to diffusion losses. Note that T_e is higher at low decay time and it is decreasing towards higher decay time, from 450 K to 350 K, this corresponds T_e =380-280 K after subtraction of $\delta T_e \sim$ +70 K.

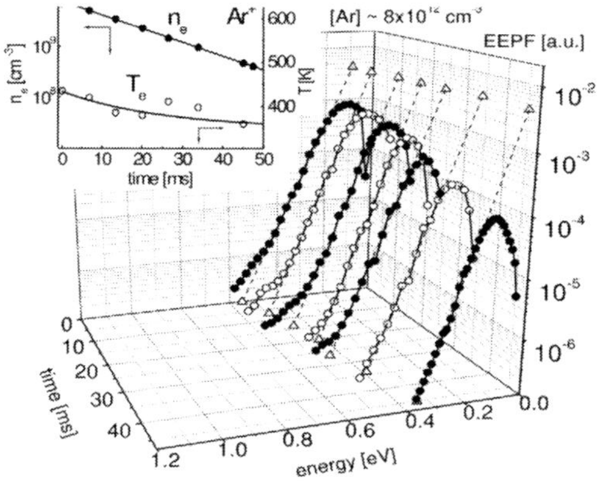

Figure 3. The evolutions of electron energy probability function (EEPF), electron density (n_e) and electron temperature (T_e) along the flow tube in Ar^+ dominated plasma.

3. Measurements of EEDF in recombining plasma

If krypton is added instead of Ar to the helium via P1, then in following ion molecule reactions and Penning ionisation Kr^+ dominated plasma is formed. Kr^+ ions in two spin states, $^2P_{3/2}$ (ground state) and $^2P_{1/2}$ are present in such plasma. At the position of port P2 and further downstream Maxwellian EEDF was observed but the electron temperature is higher than in Ar^+ dominated plasma. The measured T_e decreases during afterglow from 600 K to 450 K. From this $\delta T_e \sim +70$ K can be attributed to the already discussed "apparatus effect", but higher increase is due to transfer of energy from the metastable particles. We suppose, that the increase is connected with presence of $Kr^+(^2P_{1/2})$, which can in superelastic collision transfer excitation energy ($\Delta\varepsilon = 0.67$ eV) to electrons, and in consequent electron – electron collisions the energy is transferred to electron gas. Because the plasma is not recombining we do not see other source of energy, which can heat electrons along the flow tube. The electron density gradients and gradients of electric potential are comparable with Ar^+ dominated plasma, so they cannot be source of the difference in the temperatures. Because of high electron density in non-recombining plasma ($n_e \sim 10^{10}\,cm^{-3}$) and low energy of "produced" electrons (≤ 0.67 eV) Maxwellisation is very fast.

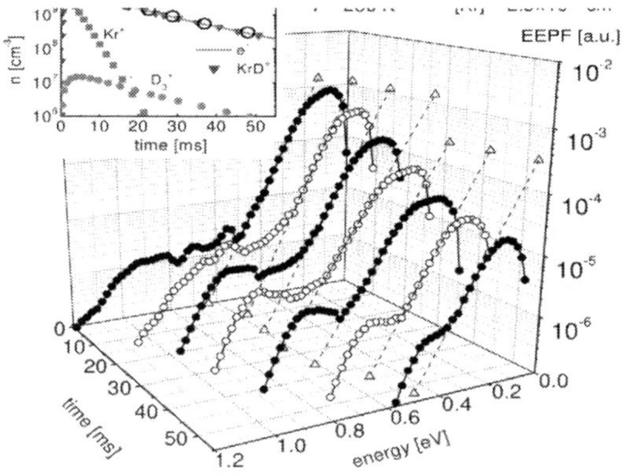

Figure 4. The evolution of EEPF in KrD^+ dominated recombining plasma. The measured electron density and the calculated ion composition are plotted in the inset.

The situation is different if deuterium is introduced via the port P2 to the above-described Kr^+ dominated afterglow plasma (with Maxwellian EEDF). The used density of D_2 is low in comparison with Kr density, $[D_2]/[Kr] \sim 0.1$. In the sequence of ion-molecule reactions ions KrD^+ and D_3^+ are formed. The densities of these two types of ions are in equilibrium and the ratio $[KrD^+]/[D_3^+] = K_{Eq}[Kr]/[D_2]$, where $K_{Eq}(T_{He})$ is corresponding equilibrium constant. Because of the low value of $[D_2]/[Kr]$ the KrD^+ ions are dominant in the decaying plasma. The formation of KrD^+ from $Kr^+(^2P_{3/2})$ is exoergic by ~1.2 eV, from $Kr^+(^2P_{1/2})$ by ~ 1.9 eV. The excess energy gained from exothermicity of reactions can be stored in internal excitation of KrD^+ and in superelastic collisions transferred to electrons. The measured EEPFs (normalised to electron density) are plotted in Figure 4. The production of energetic electrons is evident. The variation of the high-energy part of EEPF along the flow tube is coupled with decrease of electron density and "cut off mechanism" [5]. The high-energy group of electrons does not exceed 2% of all electrons. The body of EEPF is Maxwellian and can be characterised by the temperature. The high-energy group does not directly influence the measurements of recombination rate coefficient ($\alpha \sim 2\times10^{-8}\,cm^3s^{-1}$); nevertheless it influences the electron temperature of the recombining plasma. The measured experimental data have been compared with EEDF obtained by numerical calculations assuming source of energetic electrons (assuming $\varepsilon_{fast} = 0.7$ eV). The numerical simulation describes the temporal evolution of the EEDF in the presence of electron – electron (Coulomb) collisions and electron – helium atom collisions. Monte Carlo approach is used for electron – neutral collision modelling. The preliminary results are in agreement with measured EEDF, further calculations are in progress to improve the model.

If hydrogen is added via port P2 to Ar^+ dominated plasma described in experimental section (see also Fig. 3), H_3^+ dominated plasma is formed within few centimetres from the port. In this plasma recombination is fast loss process ($\alpha \sim 2\times10^{-7}\,cm^3s^{-1}$, see accompanying paper in present volume). The measured evolution of EEPF together with the calculated ionic composition and measured decay of electron density are plotted in Fig. 5. In the first 10 ms the EEPF exhibits a high energy group, during the further decay the deviation disappears and electrons are nearly thermal; nevertheless the temperature is slightly higher than in Ar^+ dominated plasma. This observation is in qualitative agreement with our recent model of multicollisional character of H_3^+ recombination. Further studies are in progress.

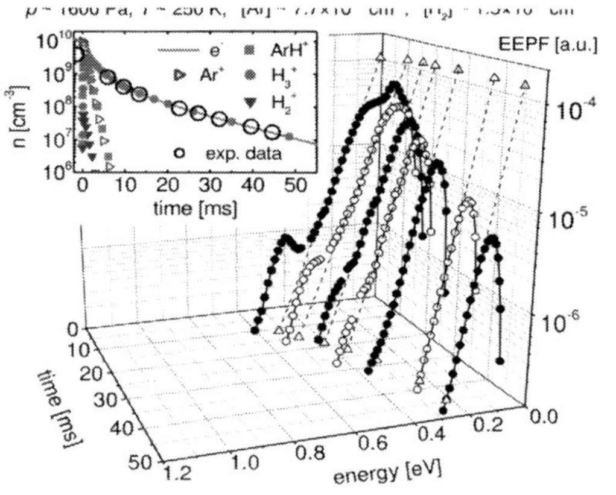

Figure 5. The evolution of EEPF in H_3^+ dominated afterglow plasma. The measured electron density and calculated ion composition are plotted in the inset.

4. Conclusion

We measured EEDF and its evolution during diffusion and recombination governed afterglow in FALP. The experimental conditions were identical with those in the case of recombination studies of H_3^+ and KrD^+. The low-energy part of EEDF is close to Maxwellian and the corresponding temperature can be used to characterise electron gas in the decaying plasma. The heating due to inelastic processes, causing increase of electron temperature, is observed. Only very small fraction of electrons is not Maxwellian (<2%), so the measured rate coefficient can be considered as thermal, corresponding to temperature of the body of the EEDF. The calculation of EEDF under assumption of production of fast electrons supports measured data.

Acknowledgements: This work is a part of the research plan MSM 0021620834 financed by the Ministry of Education of the Czech Republic and partly was supported by GACR (202/05/P095, 205/05/0390, 202/07/0495) by GAUK 53607, GAUK 124707 and GAUK 143307. ZD has been supported by OTKA grants T-48389 and IN-69892.

References

[1] Glosík J, Bánó G, Plašil R, Luca A and Zakouril P 1999 *Int. J. Mass Spectrom.* **189** 103–113
[2] Kolokolov N B, Kudrjavtsev A A and Blagoev A B 1994 *Physica Scripta* **50** 371–402
[3] Gorbunov N A, Grochola A, Kruk P, Pietruczuk A and Stacewicz T 2002 *Plasma Source Sci. Technol.* **11** 492–497
[4] Korolov I, Plašil R, Kotrík T, Dohnal P, Novotný O and Glosík J 2008 *Contrib. Plasma Phys.* **48** 461-466
[5] Arslanbekov R R, Kudryavtsev A A and Tsendin L D 2001 *Phys. Review E* **64** 016401

Journal of Physics: Conference Series **192** (2009) 012024 doi:10.1088/1742-6596/192/1/012024

Anisotropic fragmentation in low-energy dissociative recombination

S. Novotny[1], H. Rubinstein[2], H. Buhr[1,2], O. Novotný[1], J. Hoffmann[1], M.B. Mendes[1], D.A. Orlov[1], M.H. Berg[1], M. Froese[1], A.S. Jaroshevich[1,3], B. Jordon-Thaden[1], M. Grieser[1], C. Krantz[1], M. Lange[1], M. Lestinsky[1], A. Petrignani[1], I.F. Schneider[4], D. Shafir[2], F.O. Waffeu Tamo[4], D. Zajfman[2], D. Schwalm[1,2] and A. Wolf[1]

[1]Max-Planck-Institut für Kernphysik, D-69117 Heidelberg, Germany
[2]Weizmann Institute of Science, Rehovot 76100, Israel
[3]Institute of Semiconductor Physics, 630090 Novosibirsk, Russia
[4]University of Le Havre, 76058 Le Havre, France

E-mail: steffen.novotny@mpi-hd.mpg.de

Abstract. On a dense energy grid reaching up to 75 meV electron collision energy the fragmentation angle and the kinetic energy release of neutral dissociative recombination fragments have been studied in a twin merged beam experiment. The anisotropy and the extracted rotational state contributions were found to sensitively depend on energy. Both show pronounced variations on a likewise narrow energy scale as the rotationally averaged rate coefficient. For the first time angular dependences described by Legendre polynomials higher than 2nd order could be deduced. Moreover, a slight anisotropy at zero collision energy was observed which is caused by the flattened velocity distribution of the electron beam.

1. Introduction

In low-energy dissociative recombination (DR) two pathways of compatible strength, the direct and the indirect mechanism [1], compete with each other and lead to interference effects resulting in rich resonant structures of the DR cross section [2]. These occur as both of them can often access the same neutral doubly excited dissociative state, which is formed by purely electronic energy exchange in the direct process and through intermediate ro-vibrationally excited Rydberg states in the indirect process. Non-adiabatic coupling mechanisms between the electronic and nuclear motion in the indirect process further cause the DR reaction to become particularly sensitive on the ro-vibrational state of the molecular ion [3, 2].

On the experimental side the resonant energetic structures in the DR cross section have so far mainly been studied in event-by-event counting experiments which have revealed for the deuterated hydrogen molecular ion HD^+ pronounced narrow resonances down to meV electron collision energies [4]. In the present experiment we have focused for the first time on the nature of the resonances appearing in HD^+ DR at energies below ~ 75 meV through measurements of the fragment kinetic energy release (KER) and the angular distribution. Thereby the fragment angular distribution can be related to the angular dependence of the electron capture reaction, i.e. its dependence on the angle relative to the molecular axis. However, the fragment direction reflects the initial orientation of the molecule only if its rotation proceeds slowly in comparison to the dissociation time; only in this case the axial recoil approximation [5, 6, 7] is justified.

The measurements presented here have been made possible by the new twin-merged beam setup with nearly monochromatic electrons at the heavy ion storage ring TSR in Heidelberg, Germany. Fragment imaging experiments could thus be performed under stable ion beam conditions and variable low electron collision energies. The use of a photocathode electron source [8] provided an electron beam with low energy spread for high precision measurements down to the meV range. These conditions enabled us to obtain both the KER as well as the product angular distribution on a dense energy grid. The rotational levels of the molecular ions contributing to the resonances could then be extracted from the KER and revealed similar as the measured angular distribution likewise narrow variations as a function of energy as the DR rate coefficient (see also [9]).

The HD^+ molecular ion has been chosen for a number of theoretical and experimental studies and has become, due to its simple structure, a benchmark system in DR research. Moreover, the stored infrared active HD^+ ion cools down to the vibrational ground state within hundreds of ms [10] and reaches an equilibrium rotational state distribution mainly determined by the surrounding room temperature after only a few seconds. At low electron collision energies E_d the HD^+ DR process is then dominated by the doubly excited neutral HD potential curve $(2p\sigma_u)^2$ of ${}^1\Sigma_g^+$ symmetry which leads to only one final state configuration, [H($n=1$)+D($n=2$)] or [H(2)+D(1)], for $E_d < 1.14$ eV [11]. This simplifies the analysis of the KER and the HD^+ ion thus becomes an ideal study case on the DR angular dependence at low collision energies.

2. Experimental

2.1. Setup

In the present experiment, an HD^+ ion beam is produced in a standard Penning ion source and accelerated to 1.44 MeV by the HSI accelerator [12] before injection into the storage ring TSR [13]. Compared to earlier experiments the ion beam is merged with electron beams in two separate sections of the storage ring (Fig. 1). In the electron-target section (ETS) [14], optionally equipped with a thermionic- or photocathode, the electrons are tuned to well defined energies E_d independently of the electron cooler, where an electron beam of constant acceleration voltage precisely defines the ion beam energy. The electron cooler of the TSR yields an electron beam with a density of 1.6×10^7 cm^{-3} and thermal electron energies of $kT_\perp \sim 10.0$ meV and

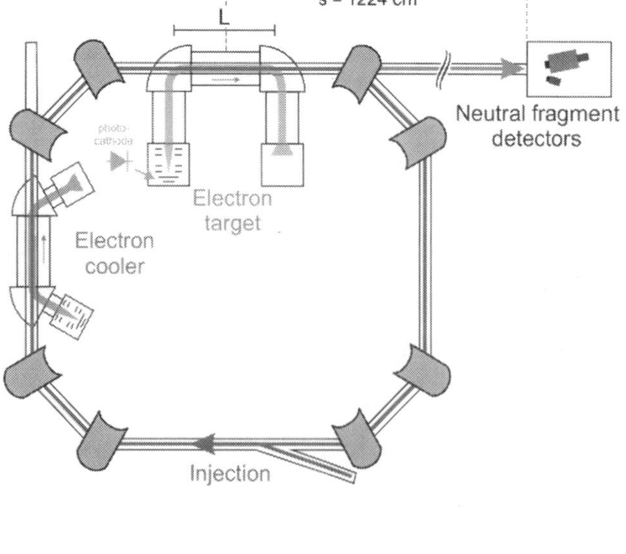

Figure 1. (color online) The twin-merged beam setup at the TSR consisting of the electron-cooler and electron-target section (ETS). Neutral fragments are recorded downstream of the ETS by counting or fragment imaging detectors.

$kT_{\parallel} \sim 0.1\,\mathrm{meV}$ transverse and parallel to the beam direction, respectively. In the ETS the energy resolution of the detuning energy E_d in the co-moving reference frame is mainly limited by the thermal electron velocities. The GaAs photocathode in the ETS provides an electron beam with thermal energies of $kT_{\perp} \sim 0.5\,\mathrm{meV}$ and $kT_{\parallel} \sim 0.03\,\mathrm{meV}$ [15]. In the experiments conducted with the photocathode two different electron densities are used, $n_{e,l} = 0.63 \cdot 10^6\,cm^{-3}$ and $n_{e,h} = 1.27 \cdot 10^6\,cm^{-3}$, which are referred to as the low and high electron density throughout the paper. For some measurements the thermionic cathode is used in the ETS, yielding $kT_{\perp} \sim 2.0\,\mathrm{meV}$ and $kT_{\parallel} \sim 0.045\,\mathrm{meV}$ at $n_e = 2.79 \times 10^6\ \mathrm{cm}^{-3}$.

Neutral fragments stemming from DR events are recorded downstream of the ETS either by a counting or imaging detector (Fig. 1). For high resolution fragment imaging a detector [16] has been placed at $\bar{s} = 12.24$ m from the center of the merging region of the ETS, where fragments impinging on a 80-mm diameter microchannel-plate (MCP) at a rate of < 1 kHz produce light spots on the attached phosphor screen. A CCD camera records the light spots at a frame rate of up to 30 s^{-1}. In order to ensure that only fragments stemming from a single DR event are recorded in each camera frame, the phosphor screen is switched off as soon as the first fragment arrives, but leaving enough time for the detection of the second particle. The transverse distance D between the two fragments is then determined from the fragment positions of a double-hit event with a resolution of $< 100\,\mu\mathrm{m}$. For DR rate measurements an energy sensitive surface barrier detector can be placed temporarily ~0.5 m in front of the MCP detector.

After each injection the ion beam is stored for 20 s. During the first 7 s of the total storage time period the electron cooler and the ETS are both operated at velocity-matching values to phase-space cool the ion beam. Taking advantage of the twin-merged beam setup the photocathode in the ETS is then detuned to non-zero electron collision energies for 7–17 s from injection before being reset to cooling conditions in the final 3 s. The electron cooler beam remains at zero relative velocity, thereby constantly defining and stabilizing the ion beam energy. The initial 7 s precooling period ensures common ion beam conditions for the subsequent storage time, i.e. with respect to phase-space cooling and the initial molecular ion state distribution. Reference imaging measurements at the end of the initial 7 s and during the last 3 s are acquired, probing the rotational state distribution of the ions through DR at zero detuning energy. The first rotational probe is used to confirm that the same equilibrium state distribution for all measurements is reached before the ETS energy is detuned, while the final reference measurement confirmed the absence of any significant rotational excitation by electrons at non-zero collision energy. In case of the use of the thermionic cathode the total storage time was 15 s and the ETS could be operated continuously at detuning energy without affecting the ion beam quality.

2.2. 2D fragment imaging

Fragments originating from a DR event arrive after a flight distance s at the detector. Their transverse distance D is then determined through $D = s\delta_J \sin\theta$ [17] by the fragmentation angle θ relative to the beam direction in the co-moving frame and the maximum laboratory emission angle δ_J. Latter contribution depends on the molecular ion rotational state J through the KER, which is defined by the initial molecular ion and final fragment states as well as the electron collision energy. Averaged over the interaction region in the ETS with $s_1 \leq s \leq s_2$ ($s_2 - s_1 = L = 1.3$ m) two essential properties of the fragmentation process influence the observed transverse fragment distance distribution $F(D)$. Towards the high distance edge, limited by the maximum distance $s_2\delta_J$, the sensitivity of $F(D)$ on the KER increases and is thus dominated by the rotational level contributions. The shape of the total $F(D)$ spectrum at lower distances contains the information on the angular fragmentation pattern. Figure 2 shows the measured transverse distance distributions at several electron collision energies. Both properties, the rotational state contributions as well as the shape related to the angular distribution, vary distinctly as functions of energy, even for an energy step as small as 2 meV (Fig. 2(c) - (d)).

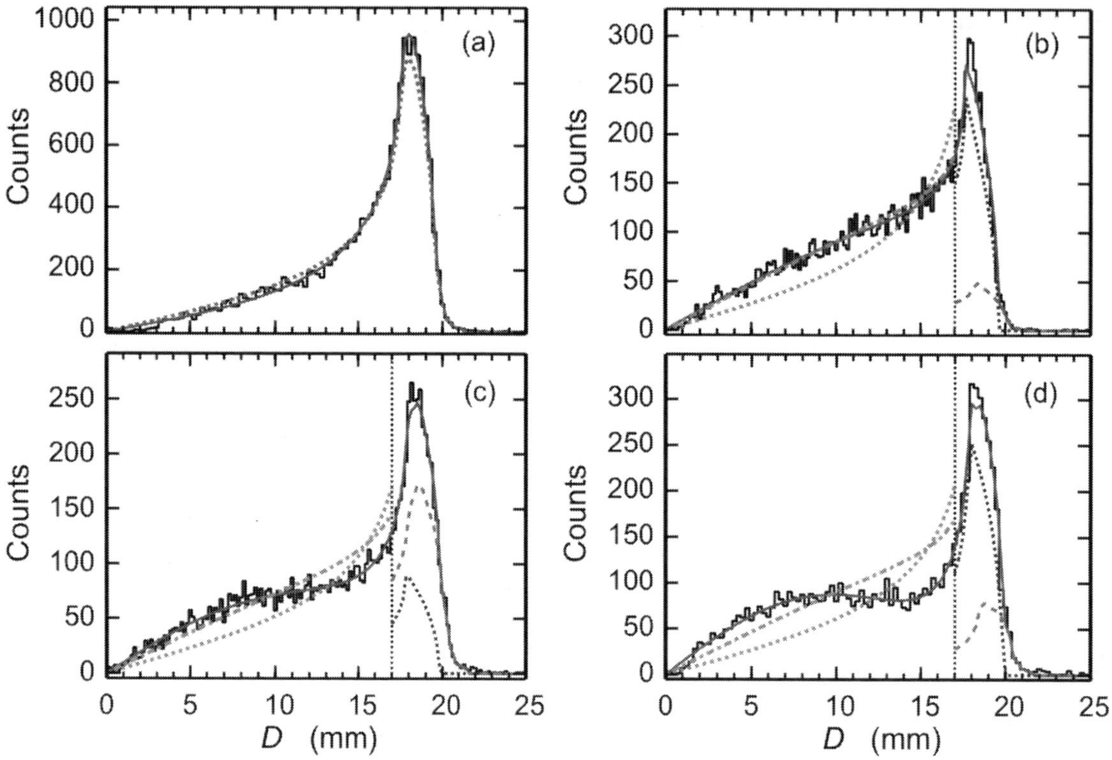

Figure 2. (color online) Transverse distance distributions $F(D)$ for $E_d = 0$ meV, thermal cathode (a), $E_d = 8$ meV (b), $E_d = 25$ meV (c) and $E_d = 27$ meV (d), all using the photocathode, with fits applying Eq.(3) (solid lines, green). The measured distributions and their best fits are compared in (a) to an isotropic distribution (dotted, red), whereas in (b,c,d) the relative contributions of Legendre polynomials of order $n \leq 2$ (dash-dotted, green) to the total fits are shown for $D < 17$ mm, while for $D > 17$ mm the relative rotational state contributions of $J \leq 1$ (dotted, blue) and $J \geq 2$ (long-dashed, red) are displayed.

In order to analyze the $F(D)$ spectra quantitatively, the angular distribution $W(\theta)$ of the fragments in the co-moving frame relative to the beam direction is described by Legendre polynomials

$$W(\theta) = \sum_n a_n P_n(\cos\theta) \tag{1}$$

with Legendre coefficients a_n of order n. Since the transverse distance D does not allow to distinguish forward and backward dissociation, i.e. θ from $\pi - \theta$, only even Legendre polynomials have to be considered and the total distribution is normalized such that $a_0 = 1$. For each polynomial $P_n(\cos\theta)$ and rotational state contribution the transverse distance distribution $F_{J,n}(D)$ is computed by integrating over all event distances s. The resulting equations for $n = 0$ and 2 have been derived earlier [17]. For the present analysis it is necessary to consider also the

$n = 4$ Legendre polynomial which results in

$$F_{J,4}(D) = \begin{cases} \frac{35\,D}{32\,L\,\delta_J^2}\left(\frac{\Gamma_1^3}{s_1^4} - \frac{\Gamma_2^3}{s_2^4}\right) + \frac{15\,D}{64\,L\,\delta_J^2}\left(\frac{\Gamma_2}{s_2^2} - \frac{\Gamma_1}{s_1^2}\right) + \frac{9}{64\,L\,\delta_J}\left(\arccos\frac{D}{\delta_J\,s_2} - \arccos\frac{D}{\delta_J\,s_1}\right) \\ \qquad \text{for } 0 \le D \le \delta_J\,s_1 \\ -\frac{35\,D\,\Gamma_2^3}{32\,L\,\delta_J^2\,s_2^4} + \frac{15\,D\,\Gamma_2}{64\,L\,\delta_J^2\,s_2^2} + \frac{9}{64\,L\,\delta_J}\arccos\frac{D}{\delta_J\,s_2} \\ \qquad \text{for } \delta_J\,s_1 \le D \le \delta_J\,s_2 \\ 0 \qquad \text{otherwise} \end{cases} \tag{2}$$

with Γ_i $(i = 1, 2)$ defined as $\sqrt{s_i^2 - (D/\delta_J)^2}$ and L describing the interaction length. The small rotational energy differences allow only to deduce rotationally averaged angular distributions, that is we describe the total, normalized distance distribution $F(D)$ by

$$F(D) = \sum_J b_J \sum_n a_n F_{J,n}(D) \tag{3}$$

with rotational state contributions b_J, which are proportional to the population as well as to the DR rate coefficient of the rotational state J.

The parameters b_J and a_n are extracted from a least-squares fit to the data. In a first step, assuming constant, J-independent DR rate coefficients, the rotational weighting factors b_J are related to each other through a Boltzman distribution and effective rotational temperatures T are deduced from the fits as a measure of the J states contributing at a given detuning energy E_d. In a second step, the parameters b_J are fitted individually with the exception of those for the two lowest rotational levels, $J = 0$ and 1, which are forced in the fit to $b_0 = b_1 = b_{01}/2$ as the projected distance distributions for the lowest two rotational states cannot be distinguished due to their small energy difference. This enables us in the further analysis to extract the relative contribution b_{01} of the two lowest rotational levels. In contrast to earlier fragment imaging experiments the fits require Legendre polynomials up to order $n = 4$, while contributions from higher orders were found to be negligible and were therefore set to zero in the final analysis. In the examples shown of Fig. 2 the best fits are displayed together with some of their individual components. Note that the low-J contribution b_{01} increases by more than a factor of two between 25 meV and 27 meV, i.e. within a small step of only 2 meV. Moreover, the necessity to include the 4th order Legendre polynomial is stressed by comparing the total fit result to the contribution involving only the lowest two Legendre polynomials.

3. Fragmentation kinematics

3.1. Rotational state contributions at nearby DR resonances

In the range below ~ 0.1 eV the DR cross section of HD^+ is believed to be strongly dependent on the initial molecular ion rotational state [3, 2] resulting in the pronounced structures observed in the measured rotationally averaged DR rate $\tilde{\alpha}(E_d)$ (Fig. 3(a)) [4]. This is convincingly corroborated by the effective rotational temperatures T displayed in Fig. 3(b), which were extracted from the fragment distance distributions as discussed above. We find that between 7 s and 17 s T varies strongly as a function of energy on a similar narrow scale as $\tilde{\alpha}(E_d)$ and mostly deviates from the average state contribution T probed by reference measurements at zero detuning energy (shaded bars for the measurement at high and low electron density, respectively). Note, that the effective rotational temperatures deduced at detuning energies $E_d > 8$ meV do not depend on the different target electron densities employed in the measurements nor on the different initial rotational state distribution reached during the first

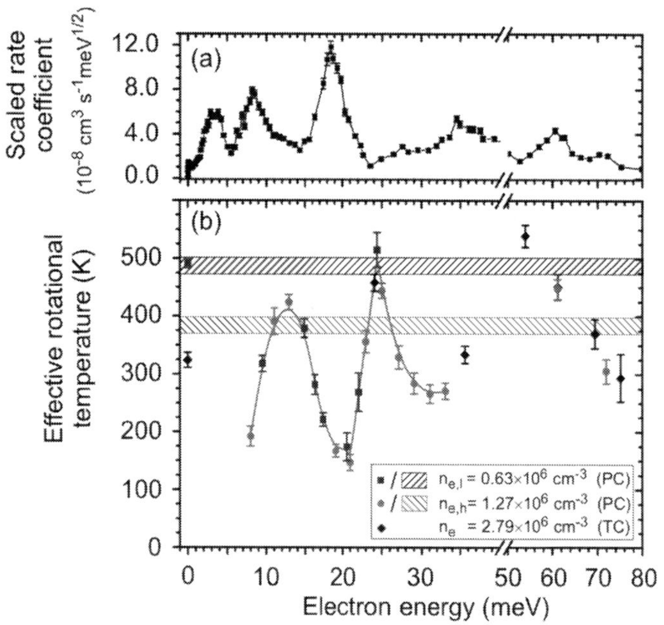

Figure 3. (color online) (a) Scaled, rotationally averaged DR rate coefficient $\tilde{\alpha}(E_d)E_d^{1/2}$ and (b) effective rotational temperature as functions of the detuning energy E_d for the different measurements performed (PC: Photocathode, TC: thermionic cathode). The two bars extending over the full energy range mark the effective rotational temperatures measured at zero detuning energy between 6.5 s and 7 s, i.e. after the initial cooling phase and before jumping to the detuning energy E_d.

7 s of combined electron cooling; this clearly shows that the observed variations of T with E_d are caused by the varying contribution of low and high rotational states to the DR rate. At zero detuning energy, on the other hand, the influence of superelastic collisions (SEC) between rotational states, as reported in [18, 19], is reflected in the electron density dependent equilibrium temperatures obtained after initial cooling.

To obtain a more detailed view on the rotational states contributing to the DR rate and on the angular fragment distribution we fitted the projected distance distributions with Eq.(3) treating b_{01}, the individual b_J for $J > 1$, and the Legendre coefficients a_2 and a_4 as free parameters (Fig. 4). The results show that at certain detuning energies molecular ions in the lowest two rotational states contribute with more than 80% to the recorded DR events (Fig. 4(a)). Furthermore the b_{01} fraction is found to change very rapidly within only a few meV.

3.2. Anisotropy at non-zero detuning energy

As shown in Fig. 4, the Legendre coefficients obtained from the analysis applying Eq.(3) yield similar narrow energy dependent structures as $\tilde{\alpha}(E_d)$, and peaks in the a_2 and a_4 coefficients generally coincide with enhanced b_{01} factors. The 4th order coefficient a_4, observed for the first time in DR angular distributions, turns out to depend particularly sensitive on the collision energy E_d and to decrease with E_d approaching zero eV. The a_2 parameter, on the other hand, oscillates slightly around a value of 0.8 for energies above 8 meV.

Within the axial recoil approximation, which requires the time scale for dissociation to be much smaller than the rotational period, the fragment angular correlation is governed by the partial waves of the incoming electrons leading to dissociation. As the d partial wave is expected to dominate [20, 9], a_4 coefficients of the order of $a_4 \approx 2.5$ are expected within this approach independent of the amount of the s wave contribution which is present as well. While the necessity to include Legendre polynomials of up to the order of 4 in the analysis of the projected distance distributions is affirming the importance of d waves in the entrance channel, the a_4 coefficients are found to be considerably smaller than expected. In particular one notes, that minima in the measured low-J fraction, i.e. an enhancement of contributions from highly rotating molecules, tend to coincide with minima in the Legendre coefficients, suggesting a loss

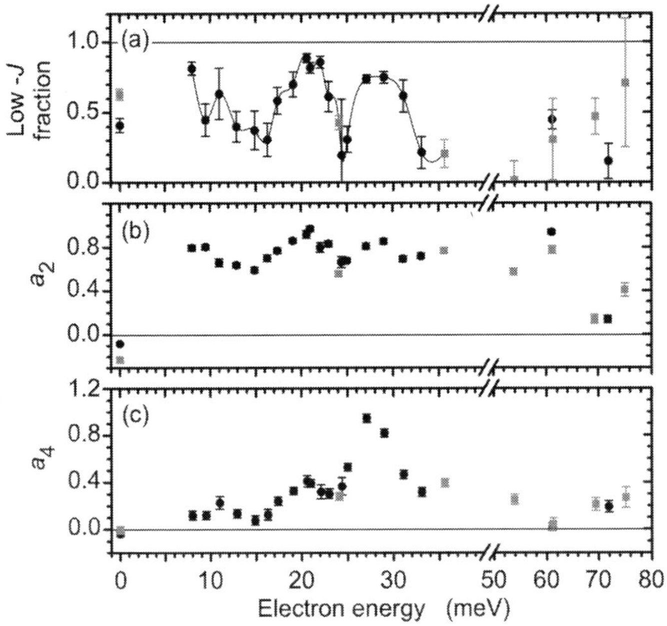

Figure 4. (a) Low-J fraction b_{01} (symbols connected to guide the eye), and (b, c) Legendre coefficients a_2 and a_4 as functions of the detuning energy E_d (see also [9]). The photocathode is used except for the grey symbols indicating use of the thermionic cathode.

of anisotropy due to a sizable rotation of the neutralized molecule before dissociation. While the direct mechanism of DR is expected to be fast, the indirect DR process [1], involving ro-vibrational excited neutral Rydberg states and thought to be responsible for the pronounced resonance structure in the DR rate coefficient, could well lead to delays in the dissociation comparable to the rotational period, thereby violating the non-rotation assumption of the axial recoil approximation.

MQDT-calculations of the angular fragment distributions from the DR of HD^+ including ro-vibrational couplings are presently being performed [21]; so far, first results obtained in the axial recoil approximation show reasonable agreement with the data at selected electron energies but generally predict too large anisotropies, indicating again the need to expand the theoretical description beyond the axial recoil limit to arrive at a more detailed understanding of the fragment angular distributions observed in the DR of HD^+.

3.3. Anisotropy at zero detuning energy

For a spherical symmetric energy distribution around the detuning energy E_d the fragment angular distribution has to be isotropic for symmetry reasons if the detuning energy is adjusted to $E_d = 0$. However, due to the flattened velocity distribution of the electron beam, the spherical symmetry is violated and the fragment distribution may be anisotropic despite of E_d being zero.

This effect could be observed for the first time in our present measurements (Fig. 2(a)). As expected, the anisotropy is found to depend on the transversal electron beam temperature as $T_\perp >> T_\parallel$; while the data collected with an electron beam from the photocathode, characterized by $kT_\perp \approx 0.5$ meV is still compatible with an isotropic distribution, a small but statistically significant negative a_2 coefficient was observed when using an electron beam produced from the thermionic source, which exhibited a perpendicular temperature of $kT_\perp \approx 2.0$ meV. Note also that the negative sign of the extracted a_2 coefficient is consistent with the results obtained at non-zero detuning energies considering the change from a predominantly longitudinal electron approach to a case where the electrons are approaching mainly from the transverse direction.

4. Conclusion

Fragment imaging combined with a twin-merged beam setup opens up a new observational window on the kinematic nature of the DR process. High resolution and energy controlled measurements on a narrow grid of energies enable the study of the DR product kinematics down to a few meV.

For HD^+ this technique made it possible to focus on different aspects of the ro-vibrational capture resonances in the region of up to $\sim$ 75 meV electron collision energy. Firstly, the KER allowed us to determine the contribution of the lowest two rotational states of the HD^+ ion in the recorded DR events. As a function of energy the b_{01} fraction varies strongly, similar to the structure of the rotationally averaged DR rate coefficient $\tilde{\alpha}(E_d)$. Since the probe measurements confirm a common initial rotational state distribution and exclude different ion beam conditions, the collision energy dependent variations in the low-J fraction thus reflect the sensitivity of the HD^+ DR process on the initial ro-vibrational state. Secondly, the measured fragment angular distribution reflects within the axial recoil approximations directly the dependence of the electron capture on the orientation of the molecular axis with respect to the incoming electrons. The angular distributions observed here require for the first time 4th order Legendre contributions for a complete description expected if d waves are contributing in the entrance channel leading to DR. However, the a_4 coefficients are considerably smaller than expected and vary together with the a_2 coefficients on a similar energy scale as the DR rate coefficient $\tilde{\alpha}(E_d)$. The partial correlation of enhanced contributions of highly rotating molecules and reduced Legendre coefficients might indicate the break-down of the axial recoil approximation.

At zero detuning energy the flattened velocity distribution of the electron beam on the fragment distribution was studied. With an electron beam produced by the thermionic cathode of the electron target, which generates electrons at a higher transverse temperature as the photocathode source, a slight anisotropic fragment angular distribution could indeed be observed as expected in the presence of an asymmetric electron velocity distribution.

D. Schwalm acknowledges support by the Weizmann Institute through the Joseph Meyerhoff program. This work has been funded in part by the German Israeli Foundation for Scientific Research (GIF) under Contract No.I-707-55.7/2001.

References

[1] J N Bardsley 1968 *J. Phys. B: At. Mol. Opt. Phys.* **1** 365–380
[2] I F Schneider *et al* 1997 *J. Phys. B: At. Mol. Opt. Phys.* **30** 2687
[3] H Takagi 1993 *J. Phys. B: At. Mol. Opt. Phys.* **26** 4815
[4] A Al-Khalili *et al* 2003 *Phys. Rev. A* **68** 042702
[5] R N Zare 1967 *J. Chem. Phys.* **47** 204
[6] T F O'Malley and H S Taylor 1968 *Phys. Rev.* **176** 207
[7] S L Guberman 2004 *J. Chem. Phys.* **120** 9509
[8] D A Orlov *et al* 2004 *Nucl. Instrum. Methods A* **532** 418
[9] S Novotny *et al* 2008 *Phys. Rev. Lett.* **100** 193201
[10] Z Amitay *et al* 1998 *Science* **281** 75
[11] D Zajfman *et al* 1995 *Phys. Rev. Lett.* **75** 814
[12] M Grieser *et al* 1993 *Nucl. Instrum. Methods A* **328** 160
[13] D Habs *et al* 1989 *Nucl. Instrum. Methods B* **43** 390
[14] F Sprenger *et al* 2004 *Nucl. Inst. Meth. A* **532** 298–302
[15] D A Orlov *et al* 2005 *J. Phys.: Conf. Ser.* **4** 290–295
[16] I Nevo *et al* 2007 *Phys. Rev. A* **76** 022713
[17] Z Amitay *et al* 1996 *Phys. Rev. A* **54** 4032
[18] D Shafir *et al* 2009 *Phys. Rev. Lett.* **102** 223202
[19] D Shafir *et al this conference*
[20] A Giusti-Suzor, J N Bardsley and C Derkits 1983 *Phys. Rev. A* **28** 682
[21] F O Waffeu Tamo 2008 *Dissociative Recombination in the Plasmas of Atmospheric and Energetic Interest* Ph.D. thesis University of Le Havre and University of Douala unpublished

An Ultra Cold Photoelectron Gun for the Heidelberg TSR Target Section

C Krantz[1], D A Orlov[1], J Hoffmann[1], M Lestinsky[1], O Novotny[1], A S Jaroshevich[2], A S Terekhov[2] and A Wolf[1]

[1] Max Planck Institute of Nuclear Physics, Saupfercheckweg 1, 69117 Heidelberg, Germany
[2] Institute of Semiconductor Physics, Lavrenteva 13, 630090 Novosibirsk, Russia

E-mail: `claude.krantz@mpi-hd.mpg.de`

Abstract. We present a cryogenic photoelectron gun, developed for the target section of the Test Storage Ring (TSR) of the Max Planck Institute of Nuclear Physics (MPIK). Cooled to cryogenic temperatures by liquid nitrogen, the photocathode source provides an electron gas with an initial thermal energy spread of around 10 meV. The beam optics of the target section reduce the electron temperatures to much lower values in the comoving frame of the beam. Recently the photocathode source has seen signi cant improvements regarding its reliability. By controlling several cathode degradation mechanisms, including cryosorption, vacuum degrading leak electron currents, and backstream of ionised restgas particles, cathode lifetime and currents have been subject to substantial improvements. Presently the photoelectron gun can deliver currents of up to 1 mA at lifetimes of about 24 h. The ability of the photoelectron beam to cool slow, heavy molecular ion beams was demonstrated by cooling a 3 MeV CF^+ beam in the TSR. At an electron cooling energy of only 53 eV and a perveance-limited current of 0.34 mA, a cooling time below 2 s has been achieved, with a very small transverse relative momentum spread of $2.5 \cdot 10^{-5}$ and nal ion beam cross-section of $0.5 \times 0.1\,\mathrm{mm}^2$.

1. The TSR Target Section

The Test Storage Ring of the MPIK is presently the only heavy-ion storage ring equipped with two independent merged electron-ion beam facilities. The so-called electron cooler is normally operated in velocity-matched mode, such that the stored ion beam is continuously phase-space cooled. The second, probe electron beam, hereafter referred to as the electron target, can be detuned in velocity with respect to the ions, and allows electron-ion recombination experiments with unprecedented energy resolution.

To achieve the highest possible energy resolution, the electron target can be equipped with a cryogenic photoelectron gun instead of a thermocathode. The heart of the cryogenic gun is a p-doped GaAs (Cs/O) photocathode. Cooled by liquid nitrogen to about 100 K, it delivers an electron gas with an initial thermal energy spread of around 10 meV, roughly corresponding to the lattice temperature of the cold GaAs crystal [2].

The beam optics of the electron gun greatly enhance the energy resolution in the comoving frame of the electron beam [3]. Adiabatic magnetic expansion by eld ratios of typically 20 to 40 reduces the kinetic energy spread transverse to the beam direction according to

$$k_B T_\perp = \frac{k_B T_c}{\quad} \,, \tag{1}$$

Journal of Physics: Conference Series **192** (2009) 012025 doi:10.1088/1742-6596/192/1/012025

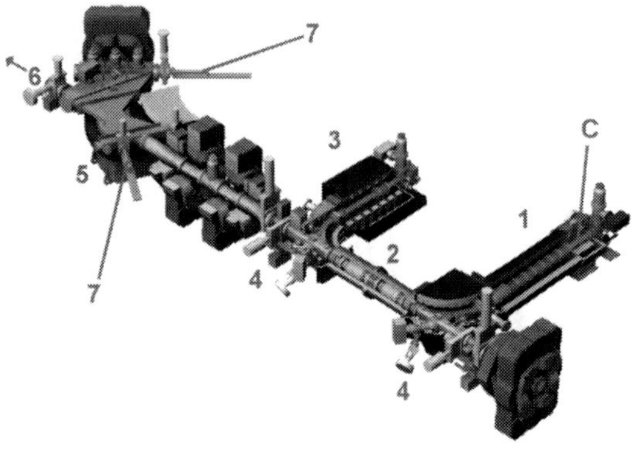

Figure 1. The TSR Electron Target and adjacent ring elements: C – cathode position, 1 – accelerator section, 2 – interaction section, 3 – collector section, 4 – toroids merging and demerging the electron/ion beams, 5 – bending dipole magnet, 6 – neutral counting and imaging detectors, 7 – charged fragment detectors.

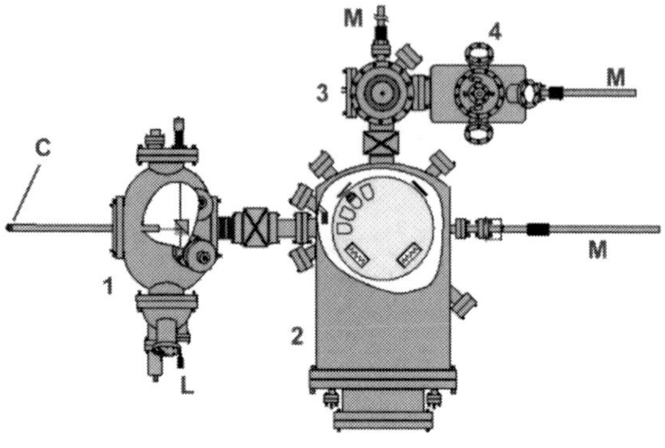

Figure 2. Schematic drawing of the photocathode vacuum setup (top view): 1 – gun chamber, 2 – preparation chamber, 3 – loading chamber, 4 – atomic hydrogen chamber, M – magnetic manipulators, C – cathode position in the gun, L – excitation laser port.

where T_c is the cathode temperature. Values of $k_B T_\perp$ as low as 0.5 meV have been observed [4]. Subsequent adiabatic acceleration of the electron cloud kinematically reduces the logitudinal velocity spread in the comoving frame of the beam according to

$$k_B T_\parallel \approx \frac{(k_B T_c)^2}{eU} + C\frac{e^2 n^{1/3}}{4 \quad _0} . \tag{2}$$

The rst, kinematic, term in eq. (2) describes the decrease in longitudinal temperature of the beam related to an acceleration by a voltage U. Potential energy relaxation by plasmon emission in the accelerated electron beam leads to a rise in longitudinal temperature described in the second term of eq. (2). $C \approx 1$ denotes the acceleration constant and n the electron particle density. For the TSR target equipped with the photoelectron gun, a value $k_B T_\parallel$ of 22 μeV has been measured [5].

In low energy operation ($U \lesssim 100$ V) the maximum extractable current I is limited by the gun perveance P, which is of around 1 μPerv for the photoelectron gun:

$$I = PU^{3/2} = 1\,\text{mA}\ \frac{P}{1\,\mu\text{Perv}} \left(\frac{U}{100\,\text{V}}\right)^{3/2} . \tag{3}$$

Guided by a magnetic eld of usual inductance 0.04 T, the cold electron beam is then merged with the stored ion beam in an interaction section of 1.4 m length (c.f. g. 1). Afterwards, the beams are demerged and the electrons dumped onto a Faraday cup in the so-called collector section. Neutral and charged recombination products are separated from the stored beam in the next bending magnet of the TSR downstream of the electron target. Recombination processes can be investigated using counting and imaging detectors.

2. The Photoelectron Source

The heart of the ultracold gun is a transmission mode GaAs photocathode. A surface of p-doped GaAs, activated by a thin (about 1 monolayer) layer of cesium and oxygen, is characterised by a surface state of e ective Negative Electron A nity (NEA), i.e. the vacuum level lies energetically below the conduction band minimum in the bulk of the crystal. Thanks to the NEA, electrons photoexcited from the valence to the conduction band may escape into the vacuum without further energy supply.

2.1. Photoelectron Emission from GaAs

Excited to the conduction band, electrons quickly thermalise into a Boltzmann distribution, whose spread corresponds to the temperature of the crystal bulk T_c. Unfortunately, the emission process itself is characterised by further energy and momentum relaxation through electron scattering at surface states, such that the transverse and longitudinal electron energy distributions in vacuum expand to approximately the span of the NEA which is of 0.1 to 0.2 eV [6]. However, in normal cathode operation the power of the excitation laser beam is set so high that electrons accumulate in front of the cathode surface. Eventually, a space charge builds up, e ectively preventing electrons with longitudinal energies lower than about the conduction band minimum to be emitted from the cathode. It has been shown [4] that the e ective initial energy spread of the electron gas again corresponds approximately to the temperature of the GaAs bulk, of around $k_B T_c = 10$ meV.

Acting as a bu er, the space charge also de nes the e ective quantum e ciency of the photoelectron gun which is of about 1% when the space charge potential corresponds to the conduction band minimum [4]. The quantum yield (QY) measured in current-limited mode is much higher – 20 % or more for freshly activated cathodes, with a record value of 35 %, measured at 635 nm wavelength in re ection mode. During operation in the electron gun, the overall QY of cathodes slowly decreases due to various mechanisms leading to destruction of the Cs/O layer. Thus after a given time of operation a GaAs sample has to be replaced by a newly activated one.

2.2. The Photocathode Setup

The photocathode vacuum setup is designed to allow a fast, all-in-vacuum cathode replacement procedure which takes only half an hour, thereby ensuring a quasi-continuous operation of the electron target. It consists of four distinct vacuum chambers (c.f. g. 2). Transfer of cathodes among chambers or into the electron gun is done using magnetically coupled manipulators. The so-called gun chamber is anged to the target's acceleration section and thus connected to the TSR vacuum. It contains the actual electron gun and features a laser window for transmission-mode excitation of the GaAs samples as well as high voltage and liquid nitrogen feedthroughs. Its base pressure varies around $2\ \ 10^{\ 11}$ mbar, depending on the overall vacuum conditions in the target and in the TSR. The preparation chamber, separated from the gun by a UHV valve, has a base pressure of $5\ \ 10^{\ 12}$ mbar. It can contain up to four cathode samples and features facilities for cesium/oxygen activation as well as ovens for heat-cleaning of samples. In addition the setup features a chamber dedicated to atomic hydrogen treatment of cathode samples after several cycles of heat-cleaning, activation and use [1]. Exposure to free hydrogen radicals removes any traces of activation layer from the GaAs and restores the initial quality of the crystal surface. Last, the setup features a loading chamber, where samples can be removed from or inserted into the setup without disturbing the high vacuum [2].

3. Photocathode Performance in Target Operation

The main drawback of GaAs photocathodes is their limited lifetime, especially in high current operation. While all QY degradation is eventually due to destruction of the Cs/O monolayer on the GaAs surface, we have identi ed several e ects that can lead to such degradation. Carefully monitoring and controlling these e ects, we signi cantly increased the maximum usable extraction current of the photoelectron gun to about 1 mA. This is su cient for a large number of experimental needs, especially for slow molecular ion beams, where electron currents are limited by gun perveance anyway (c.f. eq. (3)). As a benchmark for the performance of the photoelectron target at such low energies, a slow beam of CF^+ was cooled by an electron beam of energy 53 eV and current 0.34 mA [7].

3.1. Extraction Currents and Lifetimes

At low current drain (several nA), the so-called dark-lifetime of a photocathode is limited by the partial pressures of various chemical agents in the UHV restgas. In the case of GaAs photocathodes H_2O and CO_2 are especially known to be dangerous [8]. For cathodes operated in the TSR electron target, dark lifetimes are of around one week or longer.

Too strong cooling by liquid nitrogen radically reduces cathode lifetimes, as the cold surface adsorbs restgas such as CH_4, CO and CO_2 [4]. Undercooling of a photocathode results in an instantaneous, steep drop in the emission current, which can be partially restored by reduction of the cooling power. Formerly the cathode temperature was tuned in a rather indirect way by the liquid nitrogen supply line pressure. In 2007, a nitrogen ow controller was installed, allowing very reliable tuning of the cooling power, thus undercooling is not an issue anymore. Normally used cathode temperatures are estimated to be about 100 K.

At normal operation currents (0.5 to 1 mA) electron emission from the cathode itself degrades the vacuum conditions, as leak currents lead to desorption from gun electrodes and chamber walls. Since the magnetic guiding eld e ectively prohibits electron propagation transverse to the beam direction, the major part of these leak currents consist of electrons that travel to the collector as normal, but, upon impact onto the faraday cup, produce secondary electrons which can drift back towards the gun and impinge next to the cathode. These leak currents may thus be minimized using the beam optics of the collector section alone, which do not a ect the experimentally relevant position of the main electron beam in the interaction section.

Another degradation channel is caused by electron impact ionisation of the target restgas. During target operation, restgas molecules can be positively ionised inside the electron beam. Trapped in the magnetic guiding eld, they may drift towards the gun section where they are accelerated and focused onto the cathode surface by the electron beam optics. This high energetic ion bombardment leads to a very localised destruction of the Cs/O layer which is most

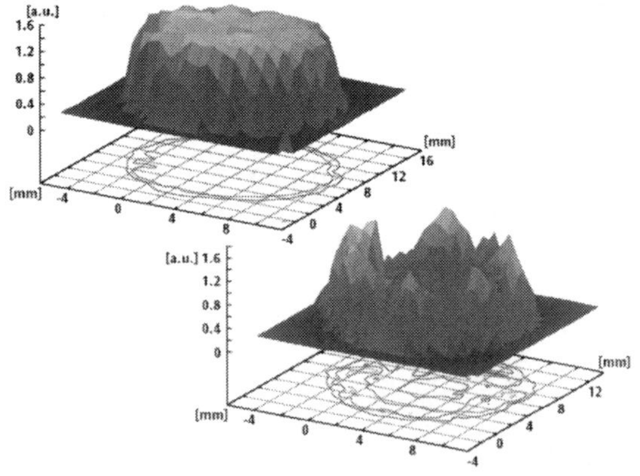

Figure 3. Electron beam pro les, depicting the current density distribution over the beam cross-section. Shown are a freshly activated cathode (top) and a heavily degraded one (bottom).

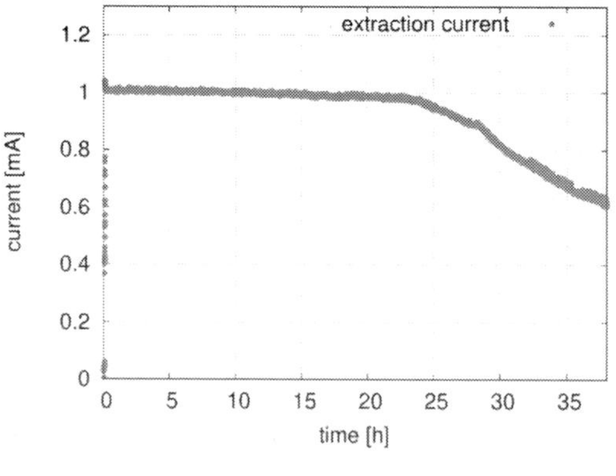

Figure 4. Extraction current of the photoelectron gun versus time. The current is stable at 1 mA for approximately 23 h, then the cathode would be replaced in normal operation.

prominent in the center of the cathode surface, as depicted in g. 3. Ion trapping in the target can be prevented using an electric clearing eld in the collector section. Additionally, a small positive potential threshold can be added to the acceleration structure of the electron gun. As initial kinetic energies of ionised particles are small, even a very weak positive potential step of 5 to 10 V, which does not a ect the high energetic electron beam, is su cient to prevent ion drift towards the electron gun.

By careful control of the above mentioned cathode degradation channels, we were able to increase the photoelectron gun's maximum extraction current from about 0.2 mA to 1 mA, while maintaining cathode lifetimes of typically 24 h (c.f. g. 4). At currents below 1 mA lifetimes of several days have been observed.

3.2. Cooling of Slow Heavy Ion Beams

As can be seen from eq. (2), the quality of an electron beam with respect to its longitudinal temperature degrades at low energies eU as the rst term in (2) rises in magnitude. Additionnally, the maximum current that can be extracted from the cathode decreases according to eq. (3). The perveance P of the photoelectron gun is of around 1 μPerv, i.e. at electron energies below 100 eV, the maximum extraction current, and thus the electron density n, is limited by perveance rather than by cathode degradation.

A slow beam of CF^+, stored in the TSR at a total kinetic energy of 3 MeV (around 90 keV/u), was used as a benchmark for the performance of the photoelectron target at low acceleration voltage [7]. Usage of the TSR's electron cooler for this beam gave unsatisfactory results, with cooling times of around 15 s and a nal width of the stored CF^+ beam of typically a few millimeters. Thus, in contrast to the usual operation scheme, the electron target rather than the electron cooler was used to cool the ion beam. The electron cooling energy, matching the velocities of electrons and ions, was only 53 eV, thereby limiting the extraction current to about 340 μA. The electron beam was magnetically expanded by a eld ratio of $= 20$ to a diameter of about 13 mm, which corresponds to an electron density $n \approx 3$ 10^6 cm ^{3}. A guiding eld

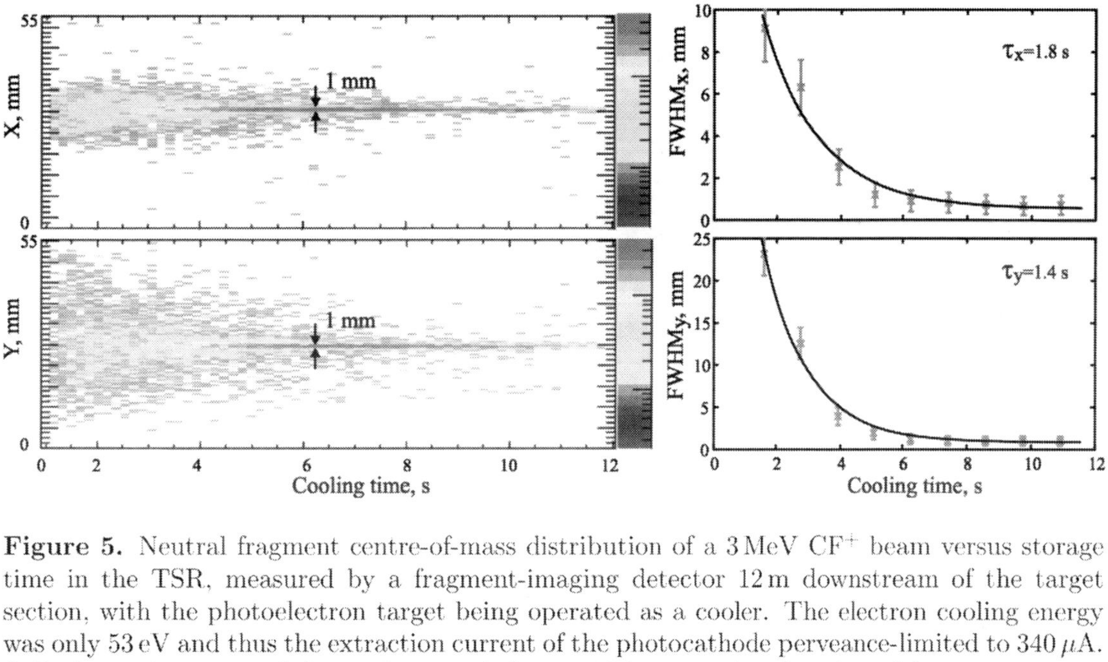

Figure 5. Neutral fragment centre-of-mass distribution of a 3 MeV CF^+ beam versus storage time in the TSR, measured by a fragment-imaging detector 12 m downstream of the target section, with the photoelectron target being operated as a cooler. The electron cooling energy was only 53 eV and thus the extraction current of the photocathode perveance-limited to 340 μA. Still, the cooling time of the ion beam is below 2 s. The gure is taken from [9].

inductance of 0.04 T was used [9].

The CF^+ ion current injected into the storage ring was estimated to be around 100 pA. At such low currents, the TSR's regular beam diagnostics cannot be used anymore. Thus the molecular fragment imaging detector was used to analyse the evolution of the transverse emittance of the ion beam during electron cooling. 12 m downstream of the position of the electron target, the detector observed the centre-of-mass distribution of the neutral fragments emerging from Dissociative Recombination of CF^+. The spread of this centre-of-mass distribution allows to determine the transverse momentum spread and cross-section of the ion beam in the interaction section [9].

The measurement is shown in gure 5. After 6 seconds, the width of the neutral fragment centre-of-mass distribution measured at the detector is of 1 mm in the horizontal (X) and 0.7 mm in the vertical (Y) direction transverse to the beam. This corresponds to a transverse relative momentum spread of the ion beam in the interaction section of around 2.5 10 ^{5} along both directions. The width of the beam was found to be 0.5 mm along the horizontal and 0.09 mm along the vertical direction. As shown in gure 5, the cooling time was below 2 s for both dimensions, demonstrating a very good cooling performance of the photoelectron beam even at low energies and currents [9].

4. Summary

The target section combined with the cryogenic photoelectron gun features unprecedented energy resolution, as has been demonstrated in both atomic and molecular recombination experiments at the TSR. Transverse and longitudinal temperatures of the electron beam can be as low as 0.5 meV and 22 μeV respectively.

The performance and reliability of the photocathodes have recently seen signi cant improvements. Carefully controlling various degradation mechanisms, we increased the usable current limit from around 0.2 mA to 1 mA, while keeping lifetimes of 24 hours or more, which is by now su cient for practically all experimental needs.

Equipped with the photoelectron gun, the target's cooling capabilities were found to be very good, even at electron beam energies of 53 eV, where the longitudinal beam temperature increases and the electron current is perveance-limited to about 0.3 mA, i.e. signi cantly lower than the maximum current allowed from the point of view of cathode reliability.

References

[1] Orlov D A *et al.* 2009 *J. Appl. Phys.* **106**, 054907
[2] Orlov D A *et al.* 2004 *Nucl. Instr. and Meth. A* **532**, 418
[3] Sprenger F *et al.* 2004 *Nucl. Instr. and Meth. A* **532**, 298
[4] Orlov D A *et al.* 2005 *J. Phys. Conf. Series* **4**, 290
[5] Lestinsky M *et al.* 2008 *Phys. Rev. Lett.* **100**, 033001
[6] Orlov D A *et al.* 2001 *Appl. Phys. Lett.* **78**, 2721
[7] Novotny O *et al.* **this issue:** Dissociative Recombination of CF^+: Experiment and Theory
[8] Wada T *et al.* 1990 *Jpn. J. Appl. Phys.* **29**, 2087
[9] Orlov D A *et al.* 2007 *Proceedings of COOL 07* (http://www.jacow.org/cl07), 230

Rotational state effect and fragmentation of small polyatomic molecular ions

V Zhaunerchyk,[1] W D Geppert,[1] M Hamberg,[1] M Kaminska,[1,3] E Vigren,[1] A Al-Khalili,[1] S Rosen,[1] M Danielsson,[1] F Österdahl,[1] V Bednarska,[2] A Petrignani,[2] E Bahati,[4] M E Bannister,[4] M R Fogle,[4] C R Vane,[4] M Larsson,[1] W J van der Zande[2] and R D Thomas[1]

[1] Department of Physics, Albanova University Centre, Stockholm University, SE-106 91 Stockholm, Sweden
[2] Institute for Molecules and Materials, Radboud University Nijmegen, PO Box 9010, NL-6500 GL Nijmegen, The Netherlands
[3] Institute of Physics, Jan Kochanowski University, Swietokrzyska 15, PL25406, Kielce, Poland
[4] Physics Division, Oak Ridge National Laboratory, P. O. Box 2008, Oak Ridge, Tennessee 37831-6377, USA

vz@physto.se

Abstract. In the paper we report the first experimental observation of rotational state effects in dissociative recombination of H_2^+. We also report the branching fractions from the DR of BH_2^+, N_3^+ and O_3^+ and the dynamics occurring in the full fragmentation channel are discussed.

1. Introduction

Dissociative recombination (DR) is a process in which a molecular ion recombines with an electron and subsequently dissociates into neutral fragments. The DR reaction first time attracted attention in 1931 when it was put forward to be a depletion mechanism for ionized oxygen molecules in the terrestrial ionosphere in attempt to explain aurora and night airglow [1]. Beyond such atmospheric applications, DR is also of great interest for cold astrophysical environments such as diffuse and dense interstellar clouds where H_3^+ is by far the dominant molecular ion species [2,3]. Due to the overwhelming importance of the DR process in all low-temperature and low-density plasmas great deal of attention has been addressed to its study.

The earlier theoretical treatments of the DR reaction were devoted to the simplest molecular ion, H_2^+ [4,5]. The total cross section was calculated from the interference between the direct and indirect mechanisms [6,7]. The calculations showed that the contribution from the direct mechanism is more significant for vibrationally excited states due to the poor overlap between the ground state and dissociative $^1\Sigma_g^+(2p\sigma_u)^2$ curve. The results also showed that contribution from the indirect mechanism involving pre-dissociative Rydberg states appeared as narrow dips in the cross section. The first experimental DR investigation of H_2^+ became possible due to the development of a single-pass merge-beam machine enabling high energy resolution measurements to be performed [8]. The experimental results supported the theoretical calculations regarding positions of resonances and the behavior of the indirect mechanism, i.e., narrow dips in the cross section were observed [9]. It is noted that the rotational population of the ions was experimentally uncontrolled and therefore must have reflected the thermal distribution. Moreover, the theoretical calculations did not take into account the molecular rotation since it was believed to be unimportant. The question of the influence of the rotational motion was first put forward by Takagi, who theoretically predicted a significant rotational effect [10]. Existing contradictions between the latest calculations reported by Takagi and earlier theoretical and experimental results has led us to investigate rotational effects in the DR of H_2^+ at the storage ring CRYRING.

It is well established that the DR rate coefficient at room temperature is usually on the order of 10^{-7} cm^3s^{-1}, while the fragmentation behaviour is not completely understood. An early model developed by Bates argued that fragmentation favours the least rearrangement of valence bonds [11]. However, three-body break-up has been observed to be a main route for many polyatomic ions. Moreover, observation of four-body break-up was reported, though it was not a dominant channel [12]. A systematic investigation into the fragmentation of XH_2^+ ions (X=H, C, N, O and P) has revealed a propensity for the three-body break-up [13-17]. To shed more light on the DR fragmentation of XH_2^+ ions we chose to investigate the next ion from this series; BH_2^+. The investigation of the homonuclear triatomic molecular ions N_3^+ and O_3^+ has also been undertaken.

2. Experiment

The experiment was performed at the heavy-ion storage ring CRYRING located at the Manne Siegbahn Laboratory, Stockholm. For investigation of the rotational state effects, para- and normal-H_2 were separately introduced into a supersonic-expansion ion source at a pressure of 2 atm [18]. The rotational temperature of the produced ions is 30 K which means that para-H_2^+ occupies only the rotational ground state (J=0) while the population of normal-H_2^+ corresponds to 25% of J=0 and 75% of J=1. In order to suppress contribution from vibrationally excited states ion-molecular reaction involving Ne + $H_2^+(v>1) \rightarrow NeH^+ + H$ was employed by introducing Ne into the ion source; Ne:H_2 90%:10%. For production of BH_2^+, O_3^+ and N_3^+ a cold plasma ion source was used. Before injection into the ring H_2^+ ions were pre-accelerated to 600 keV by RFQ, while BH_2^+, O_3^+ and N_3^+ ions were injected into the ring at the energy of 40 keV. All ions were further accelerated by a RF cavity to final energies of 8, 3.25, 1.96 and 2.25 MeV, respectively. The interaction between the ion and electron beams occurred over a length of 0.85 m in the electron cooler. The neutral products of the DR reaction were separated from the ion beam by a bending magnet and left the ring tangentially where they were detected with a silicon detector. For measurement of the branching fractions from the DR of the BH_2^+, O_3^+ and N_3^+ ions a well-known grid technique was utilized [19].

3. Experimental results and discussion

If the rotational population of H_2^+ was not changed during extraction, acceleration and storage of the ion beam, the conditions in the ion source would mean that para-H_2^+ and normal-H_2^+ occupied the lowest rotational levels. Moreover, due to the super-elastic collisions in the electron cooler, further de-excitation of vibrationally excited states could be expected [20]. Measurements were made after 3 s and 17 s of storage. It is noted that due to the failure of the ion current measurement, only relative rate coefficient was obtained. To avoid logarithmical scaling and to make DR resonances more pronounced the measured rate coefficient was multiplied with $1/\sqrt{E}$. In order to reduce statistical scattering in the raw data, averaging procedure was utilized (figure 1). The rate coefficients measured after storing the ion beam for 17 s for both para- and normal-H_2^+ contain sharp peaks, while the rate coefficients measured after 3 s are smooth. This fact confirms that ion beam for different storage times involved different vibrational populations and due to the efficient super-elastic collisions the rate coefficient obtained after 17 s is expected to contain predominantly contribution from the vibrational ground state. The peak at ≈6 meV is a para-feature, while the peak at ≈20 meV is an ortho-feature. Higher energy peaks are present in both para- and normal-spectra. The results show that contribution from

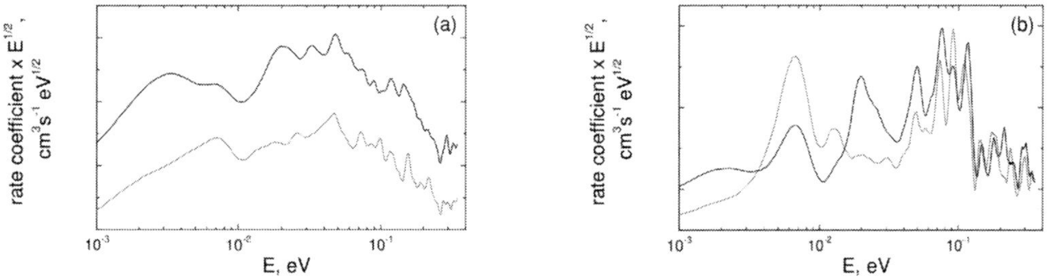

Figure 1. The rate coefficients multiplied by square root from interaction energy for both para- and normal-H_2^+ are shown by gray and black curves for storage time of 3 s (a) and 17 s (b).

indirect mechanism appears as windows where the rate coefficient is enhanced. Even if the absolute rate coefficient contains narrow dips, they are smeared out due to the experimental energy resolution and not observed. Our results disagree with the earlier single-pass merged-beam experiment by Van der Donk *et al.*, in which the energy resolution was worse than in CRYRING, but narrow dips in the cross section were obtained [9].

Investigation into fragmentation of BH_2^+ revealed the following fractions: n(BH+H)=0.56±0.03, n(B+H$_2$)=0.09±0.02 and n(B+H+H)=0.35±0.05. It is noted that this is the first observation for the fragmentation of XH_2^+ ions which is not dominated by the three-body break-up. In order to explain such phenomenon the suggestion of Vikor *et al.* was utilized: the importance of the three-body break-up increases with available energy [21]. The validity of the suggestion is seen from figure 2. It seems

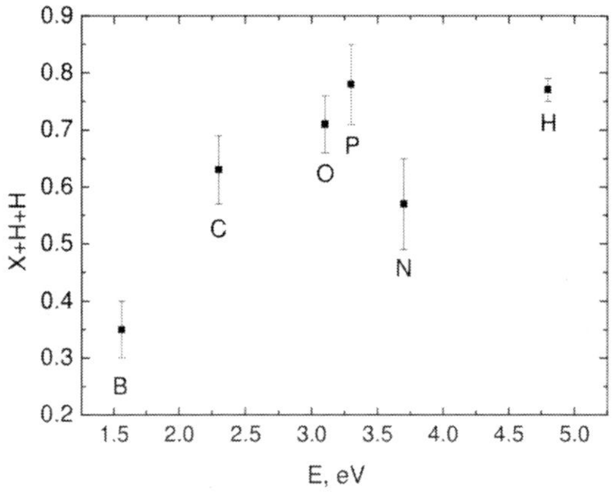

Figure 2. Fraction of three-body break-up in DR of XH_2^+ ions versus exothermicity [13-17].

to be inappropriate to compare ions from series of XH_2^+ with other ions, for instance, fragmentation of H_3O^+ is dominated by OH+H+H break-up, where the kinetic energy available in this channel is only ≈1.3 eV [19].

In the DR of N_3^+ and O_3^+ two-body and three-body channels are exothermic. It is surprising that fragmentation of O_3^+ almost exclusively proceeds through the three-body break-up (0.94±0.03), while for N_3^+ this channel constitutes only 0.08±0.03. So far, such a high fraction of three-body break-up has been observed for $D_5O_2^+$ [22]. The result for $D_5O_2^+$ might be expected since fragmentation involves ruptures of cluster bonds, while in O_3^+ the cleavage is of valence bonds. The suggestion of Vikor *et al.* is valid for X_3^+ ions studied so far as well: N_3^+ – 0.08 (0.7 eV), H_3^+ – 0.77 (4.8 eV) and O_3^+ – 0.94

(6.27 eV). Another explanation for different fragmentation pattern observed in the DR of O_3^+ and N_3^+ is based on consideration of bond strength in the molecules. N_3^+ is quite strongly bound and the lowest dissociation limit, $N^+ + N_2$, lies 3.53 eV above the ionic ground states [23], while for O_3^+ the dissociation energy into the lowest limit, $O + O_2^+$, is 0.6 eV [24]. It is likely that the likelihood of the bond rupture increases with the weakness of the bond.

References

[1] Kaplan J 1931 *Phys. Rev.* **38** 1048
[2] Geballe T R and Oka T 1996 *Nature (London)* **384** 334
[3] McCall B J, Geballe T R, Hinkle K H and Oka T 1998 *Science* **279** 1910
[4] Giusti-Suzor A, Bardsley J N and Derkits C 1983 *Phys. Rev. A* **28** 628
[5] Nakshima K, Takagi H and Nakamura H J 1987 *J. Chem. Phys.* **86** 726
[6] Bates D R 1950 *Phys. Rev.* **78** 492
[7] Bardsley J N 1968 *J. Phys. B* **1** 365
[8] Auerbach D, Cacak R, Caudano R, Gaily T D, Keyser C J, McGowan J M, Mitchell J B A and Wilk S F J 1977 *J. Phys. B* **10** 3797
[9] Van der Donk P, Yousif F B, Mitchell J B A and Hickman A P 1991 *Phys. Rev. Lett.* **67** 42
[10] Takagi H 1993 *J. Phys. B* **26** 4815
[11] Bates D R 1986 *Astrophys. J.* **306** L45
[12] Geppert W, Ehlerding A, Hellberg F, Kalhori S, Thomas R D, Novotny O, Arnold S T, Miller T M, Viggiano A A and Larsson M 2004 *Phys. Rev. Lett.* **93** 153201
[13] Datz S, Sundström G, Biedemann C, Broström L, Danared H, Mannervik S, Mowat J R and Larsson M 1995 *Phys. Rev. Lett.* **74** 4099
[14] Larson Å, Le Pedellec A, Semaniak J, Strömholm C, Larsson M, Rosen S, Peverall R, Danared H, Djuric N, Dunn G H and Datz S 1998 *Astrophys. J.***505** 459
[15] Thomas R D, Hellberg , Neau A, Rosen S, Larsson M, Vane C R, Bannister M E, Datz S, Petrignani A and van der Zande W J 2005 *Phys. Rev. A* **71** 032711
[16] Rosen S, Derkatch A, Semaniak J, Neau A, Al-Khalili A, Le Padellec A, Vikor L, Thomas R, Danared H, af Ugglas M and Larsson M 2000 *Faraday Discuss.* **115** 295
[17] Zhaunerchyk V, Hellberg F, Ehlerding A, Geppert W D, Larsson M, Vane C R, Bannister M E, Bahati E M, Österdahl F, af Ugglas M and Thomas R D 2005 *Mol. Phys.* **20** 2735
[18] McCall B J, Huneycutt A J, Saykally R J, Djuric N, Dunn G H, Semaniak J, Novotny O, Al-Khalili A, Ehlerding A, Hellberg F, Kalhori S, Neau A, Thomas R D, Paal A, Österdahl F and Larsson M 2004 *Phys. Rev. A* **70** 052716
[19] Neau A, Al Khalili A, Rosen S, Le Padellec A, Derkatch A M, Shi W, Vikor L, Larsson M, Semaniak J, Thomas R, Någård M B, Andersson K, Danared H and af Ugglas M 200 *J. Chem. Phys.* **113** 1762
[20] Tanabe T, Takagi H, Katayama I, Chida K, Watanabe T, Arakaki Y, Haruyama Y, Saito M, Nomura I, Honma T, Noda K and Hosono K 1999 *Phys. Rev. Lett.* **83** 2163
[21] Vikor L, Al-Khalili A, Danared H, Djuric N, Dunn G H, Larsson M, Le Padellec A, Rosen S and af Ugglas M 1999 *Astron. Astrophys.* **344** 1027
[22] Någård M B, Pettersson J B C, Derkatch A M, Al-Khalili A, Neau A, Rosen A, Larsson M, Semaniak J, Danared H, Källberg A, af Ugglas M and Österdahl F 2002 *J. Chem. Phys.* **117** 5264
[23] Friedmann A, Solvia A M, Nizkodorov S A, Bieske E J and Maier J P 1994 *J. Phys. Chem.* **98** 8896
[24] Willitsch S, Innocenti F, Dyke J M and Merkt F 2005 *J. Chem. Phys.* **122** 0243

AUTHOR INDEX